Operator Theory: Advances and Applications
Volume 239

Founded in 1979 by Israel Gohberg

Editors:
Joseph A. Ball (Blacksburg, VA, USA)
Harry Dym (Rehovot, Israel)
Marinus A. Kaashoek (Amsterdam, The Netherlands)
Heinz Langer (Vienna, Austria)
Christiane Tretter (Bern, Switzerland)

Associate Editors:
Vadim Adamyan (Odessa, Ukraine)
Wolfgang Arendt (Ulm, Germany)
Albrecht Böttcher (Chemnitz, Germany)
B. Malcolm Brown (Cardiff, UK)
Raul Curto (Iowa, IA, USA)
Fritz Gesztesy (Columbia, MO, USA)
Pavel Kurasov (Stockholm, Sweden)
Leonid E. Lerer (Haifa, Israel)
Vern Paulsen (Houston, TX, USA)
Mihai Putinar (Santa Barbara, CA, USA)
Leiba Rodman (Williamsburg, VA, USA)
Ilya M. Spitkovsky (Williamsburg, VA, USA)

Honorary and Advisory Editorial Board:
Lewis A. Coburn (Buffalo, NY, USA)
Ciprian Foias (College Station, TX, USA)
J.William Helton (San Diego, CA, USA)
Thomas Kailath (Stanford, CA, USA)
Peter Lancaster (Calgary, Canada)
Peter D. Lax (New York, NY, USA)
Donald Sarason (Berkeley, CA, USA)
Bernd Silbermann (Chemnitz, Germany)
Harold Widom (Santa Cruz, CA, USA)

Subseries
Linear Operators and Linear Systems

Subseries editors:
Daniel Alpay (Beer Sheva, Israel)
Birgit Jacob (Wuppertal, Germany)
André C.M. Ran (Amsterdam, The Netherlands)

Subseries
Advances in Partial Differential Equations

Subseries editors:
Bert-Wolfgang Schulze (Potsdam, Germany)
Michael Demuth (Clausthal, Germany)
Jerome A. Goldstein (Memphis, TN, USA)
Nobuyuki Tose (Yokohama, Japan)
Ingo Witt (Göttingen, Germany)

Robert Denk • Mario Kaip

General Parabolic Mixed Order Systems in L_p and Applications

Robert Denk
FB Mathematik und Statistik
Universität Konstanz
Konstanz, Germany

Mario Kaip
FB Mathematik und Statistik
Universität Konstanz
Konstanz, Germany

ISSN 0255-0156 ISSN 2296-4878 (electronic)
ISBN 978-3-319-01999-4 ISBN 978-3-319-02000-6 (eBook)
DOI 10.1007/978-3-319-02000-6
Springer Cham Heidelberg New York Dordrecht London

Mathematics Subject Classification (2010): 35R35, 35K40, 35Q30, 35Q79, 76T10

Library of Congress Control Number: 2013953241

© Springer International Publishing Switzerland 2013
This work is subject to copyright. All rights are reserved by the Publisher, whether the whole or part of the material is concerned, specifically the rights of translation, reprinting, reuse of illustrations, recitation, broadcasting, reproduction on microfilms or in any other physical way, and transmission or information storage and retrieval, electronic adaptation, computer software, or by similar or dissimilar methodology now known or hereafter developed. Exempted from this legal reservation are brief excerpts in connection with reviews or scholarly analysis or material supplied specifically for the purpose of being entered and executed on a computer system, for exclusive use by the purchaser of the work. Duplication of this publication or parts thereof is permitted only under the provisions of the Copyright Law of the Publisher's location, in its current version, and permission for use must always be obtained from Springer. Permissions for use may be obtained through RightsLink at the Copyright Clearance Center. Violations are liable to prosecution under the respective Copyright Law.
The use of general descriptive names, registered names, trademarks, service marks, etc. in this publication does not imply, even in the absence of a specific statement, that such names are exempt from the relevant protective laws and regulations and therefore free for general use.
While the advice and information in this book are believed to be true and accurate at the date of publication, neither the authors nor the editors nor the publisher can accept any legal responsibility for any errors or omissions that may be made. The publisher makes no warranty, express or implied, with respect to the material contained herein.

Printed on acid-free paper

Springer is part of Springer Science+Business Media (www.birkhauser-science.com)

Dedicated to our families

Birgit, Bernadette, and Lorenz

Christina, Cornelia, and Michael

Contents

Introduction and Outline		**1**
1	**The joint time-space H^∞-calculus**	**11**
1.1	The joint H^∞-calculus for tuples of operators	12
	a) Sectorial and bisectorial operators, $\mathcal{R}$-boundedness	12
	b) Joint H^∞-calculus	17
1.2	Vector-valued Sobolev spaces	27
	a) Interpolation of Banach spaces	27
	b) Retractions and coretractions	34
	c) Definition of Sobolev spaces	36
1.3	The time-space derivative	45
	a) Fourier multipliers	45
	b) Vector-valued space and time derivatives	52
	c) Joint space-time H^∞-calculus	58
2	**The Newton polygon approach for mixed-order systems**	**69**
2.1	Inhomogeneous symbols and the Newton polygon	70
	a) Inhomogeneous symbols and principal parts	71
	b) Newton polygons and order functions	77
2.2	N-parameter-ellipticity and N-parabolicity	91
	a) N-parameter-elliptic symbols and $S_N(L_t \times L_x)$	92
	b) Partition of the co-variable space	94
	c) Equivalent characterization of $S_N(L_t \times L_x)$	98
2.3	H^∞-calculus of N-parabolic mixed-order systems	114
	a) The H^∞-calculus of N-parabolic symbols	115
	b) Mixed-order systems on spaces of mixed scales	123
	c) Remarks on the compatibility condition	132
3	**Triebel-Lizorkin spaces and the L_p-L_q-setting**	**143**
3.1	Vector-valued Triebel-Lizorkin spaces and interpolation	144
3.2	Anisotropic Triebel-Lizorkin spaces and representation by intersections	151

3.3 Auxiliary results on Bessel-valued Triebel-Lizorkin spaces 160
    a)   The joint time-space H^∞-calculus on Bessel-valued Triebel-Lizorkin spaces 161
    b) H^∞-calculus of N-parabolic symbols on Bessel-valued Triebel-Lizorkin spaces 164
3.4 Mixed-order systems on Triebel-Lizorkin spaces 166
3.5 Singular integral operators on L_p-L_q 173
    a)   Singular integral operators 173
    b)   Extension symbols 179

4   Application to parabolic differential equations   187

4.1 The generalized L_p-L_q Stokes problem on $\Omega = \mathbb{R}^n$ 188
    a)   Remarks on homogeneous Sobolev spaces 188
    b)   The generalized Stokes problem 190
4.2 The generalized L_p-L_q thermo-elastic plate equations on $\Omega = \mathbb{R}^n$. 196
4.3 A linear L_p-L_q Cahn-Hilliard-Gurtin problem in $\Omega = \mathbb{R}^n$ 199
4.4 A compressible fluid model of Korteweg type on $\Omega = \mathbb{R}^n$ 202
4.5 A linear three-phase problem on $\Omega = \mathbb{R}^n$ 205
4.6 The spin-coating process 207
4.7 Two-phase Navier-Stokes equations with Boussinesq-Scriven surface and gravity 214
4.8 The L_p-L_q two-phase Stefan problem with Gibbs-Thomson correction 225

List of Figures   229

Bibliography   231

List of symbols   239

Index   247

Introduction and Outline

For more than 50 years, elliptic and parabolic partial differential equations have been treated by an investigation of their symbols. This approach is based on the simple observation that the Fourier transform changes spatial derivatives into multiplication with covariables that can be handled much easier. Uniform estimates on the symbol lead to solvability results in appropriate function spaces. The resulting concept of ellipticity and parabolicity (in the sense of parameter-ellipticity) can be found in classical papers by M.S. Agranovich, M.I. Vishik [AV64], M.S. Agmon [Agm62] and others. Mixed-order systems, also called Douglis-Nirenberg systems, were studied, e.g., by S. Agmon, A. Douglis, and L. Nirenberg [ADN59, ADN64] and by L.R. Volevich [Vol65].

Due to its importance and success in various fields of applications, parabolic and (parameter-)elliptic theory has been, and still is, an active field of research. In the spirit of the classical papers mentioned above, but with modern tools from operator theory and functional analysis, the present book delineates a general approach to non-standard parabolic partial differential equations which are not covered by the classical theory. In particular, we consider problems where the related symbols are not (quasi-)homogeneous in the covariables. In this case the condition of parameter-ellipticity has to be adapted, and the Newton polygon comes into play – a tool which in fact was introduced by I. Newton [New81] and which is widely spread also in algebra. By the Newton polygon approach, many classes of parabolic equations become accessible, e.g., mixed-order systems with general order structure and boundary value problems with dynamic boundary conditions.

The aim of our investigation is to show maximal regularity, in the sense of well-posedness, of the linearized equation in the L_p-Sobolev space setting. We include the case $p \neq 2$ due to its importance for the treatment of nonlinear equations. Besides the Newton polygon idea, our approach is based on functional analytic concepts like the joint H^∞-calculus for tuples of operators, interpolation theory of Banach spaces, and retractions and coretractions. Here we deal with all main scales of L_p-Sobolev spaces: Besov spaces, Bessel potential spaces, and Triebel-Lizorkin spaces, the latter appearing naturally as boundary traces of L_p-L_q-Sobolev spaces. We include the vector-valued case and L_p-L_q-setting for all scales.

The Newton polygon approach developed in the present book gives short and elegant proofs of the well-posedness of linearized parabolic problems. This will be shown in a series of examples and applications, including the generalized Stokes problem, the generalized thermoelastic plate equation, a linear Cahn-Hilliard-Gurtin problem (all in the L_p-L_q-setting), the spin coating process, and the two-phase Stokes equation with Boussinesq-Scriven surface and gravity.

During the last decade, local in time well-posedness results for many nonlinear (in particular, quasi-linear) parabolic equations were obtained by the maximal L_p-regularity approach. We mention the works of D. Bothe, R. Denk, J. Escher, M. Geissert, B. Grec, M. Hieber, J. Prüss, E.V. Radkevich, J. Saal, O. Sawada, G. Simonett, Y. Shibata, and S. Shimizu, cf. [EPS03], [BP07], [GR07], [PSS07], [SS07], [PS09], [PS10], [DGH+11], [SS11b]. As this method based on linearization and maximal regularity for the linearized problem has become sort of standard, we give a short overview of the main steps. This also helps to understand where our results come into play in this process. The maximal regularity approach roughly consists of the following steps:

- For free boundary value problems, the Hanzawa transform can be applied to describe the problem on a time-independent, fixed domain (cf. [Han81], [EPS03], [PSS07], [DGH+11], for instance). This is possible at least for small times where we can write the free boundary (locally) as the graph of an additional unknown function living on the boundary.

  An alternative formulation, which also leads to a fixed domain, uses Lagrangian coordinates, see for example the works of V.A. Solonnikov, Y. Shibata, and S. Shimizu, cf. [Sol84], [Sol03a], [Sol03b], [SS07], [SS11b]. However, in this formulation it is not obvious how to recover the regularity of the free boundary and how to include, e.g., surface tension or additional dynamics on the boundary.

- The next step consists in an analysis of the linearized problem. By localization and perturbation arguments, this is reduced to so-called model problems, i.e., equations in $\mathbb{R}^n$ and boundary value problems in $\mathbb{R}^n_+$ with constant coefficients and without lower-order terms. With the help of Fourier and Laplace transform, the aim is to show well-posedness in suitable L_p-Sobolev spaces.

- Based on the well-posedness of the linearized problem, the nonlinear problem can be solved by a fixed-point argument and the contraction mapping principle. Here the L_p-setting is advantageous as for large p there are better embedding theorems available which help to treat the nonlinearities.

- The local in time well-posedness for the nonlinear problem is the basis for the investigation of global solvability and for long time asymptotics.

Our results deal with the second step, i.e., with maximal regularity for linear model problems. We consider both the L_p- and the L_p-L_q-setting. As handling of the nonlinear terms in the original equations deeply depends on embedding theorems, algebra properties and multiplier results, it seems to be worthwhile to decouple the integrability condition for the time and space variables and to include the case $p \neq q$. As mentioned above, this immediately leads to Triebel-Lizorkin spaces as trace spaces.

The main difference between the equations studied in this book and classical parabolic equations lies in the fact that our applications contain an inherent symbol inhomogeneity. To explain this, let us consider the symbol related (by Fourier transform in space and Laplace transform in time) to the heat equation. This symbol which is given by

$$P(\lambda, \xi) = \lambda + |\xi|^2$$

is quasi-homogeneous in ξ and λ. Therefore, the classical parabolicity condition

$$P(\lambda, \xi) \neq 0, \quad \operatorname{Re} \lambda \geq 0, \ \xi \in \mathbb{R}^n, \ (\lambda, \xi) \neq 0, \tag{1}$$

immediately implies uniform estimates from below. In contrast to this, the symbol given by

$$P(\lambda, \xi) = \lambda + |\xi|^2 \sqrt{\lambda + |\xi|^2},$$

is not quasi-homogeneous. Now condition (1) is no longer sufficient for uniform symbol estimates and solvability results. For such inhomogeneous symbol structures, we introduce the concept of N-parabolicity where "N" stands for "Newton" to indicate the underlying Newton polygon. We will come back to this example in detail in the outline below.

There is a large literature dealing with maximal L_p-regularity for parabolic problems. Here we want to state only a few of these results which are highly related to our approach.

Parabolic problems with boundary dynamics of relaxation type are discussed by R. Denk, J. Prüss, and R. Zacher in [DPZ08]. The authors establish a general L_p-theory for a special class of problems which contain equations in the interior and on the boundary of order one in time. Problems with both free boundary and surface diffusion are not contained in this class. Due to the pressure term and the necessary reductions, linearizations of the two-phase Navier-Stokes equations can also not be handled by the results in [DPZ08]. Our results cover model problems with general boundary dynamics as well as linearization of two-phase Navier-Stokes equations, cf. Section 4.7.

The problems considered in [DV08] exhibit inhomogeneous dynamical boundary conditions as well as additional unknown functions on the boundary. The authors use Newton polygon techniques and N-parabolicity to handle the inherent inhomogeneities of the boundary operators. On the basis of an analysis of the associated mixed-order system on the boundary, the so-called Lopatinskii matrix, the authors derive well-posedness in the Hilbert space L_2. In the Hilbert space setting, all scales of Sobolev spaces coincide, and one can define the space by weight functions which are directly related to symbol estimates. Here we generalize their approach to the general L_p-setting, which is more involved due to the appearance of Besov spaces on the boundary.

In [DHP07] the authors present an L_p-L_q-theory for inhomogeneous boundary conditions, where the boundary operators do not contain time-derivatives and the equation in the interior is of order one in time. A major difficulty of the L_p-L_q-theory developed there is the necessity of vector-valued Triebel-Lizorkin spaces. To our knowledge, boundary value problems with inherent inhomogeneity have not been studied in the L_p-L_q-context so far. For Triebel-Lizorkin spaces and trace results in this direction, we also refer to the works of M.Z. Berkolaiko, and P. Weidemaier in [Ber85], [Ber87a], [Ber87b], [Wei02], [Wei05]. Traces and embedding results involving Triebel-Lizorkin spaces have recently also been analyzed by M. Meyries and M. Veraar in [MV13] where also an application to the Stefan problem can be found. Our approach to vector-valued Triebel-Lizorkin spaces relies on interpolation theory and is based on results by H. Amann, J. Johnson, H.J. Schmeißer, W. Sickel, and H. Triebel in [Ama09], [JS08], [Tri97], [SS05].

In this book, we deal both with scalar symbols and with mixed-order systems, the latter appearing naturally in the context of the so-called reduction to the boundary for boundary value problems. The Newton polygon approach to mixed-order systems was started by S. Gindikin and L.R. Volevich [GV92] and further developed by S. Kozhevnikov [Koz96], R. Denk, R. Mennicken, and L.R. Volevich in, e.g., [DMV98], [DV02a]. A priori-estimates for general mixed-order systems can also be found in papers of R. Denk, M. Dreher, and M. Faierman, cf. [DF10] and [DD11], for example.

The pseudo-differential method to analyze mixed-order systems is used by R. Denk, J. Saal, and J. Seiler in [DSS09] and [DS11]. Both references deal with spaces defined by weight functions, and therefore they cannot handle ground spaces with different scales in time and space. In [DSS08], R. Denk, J. Saal, and J. Seiler use an interpretation of the mixed-order system by a joint H^∞-calculus of the time-derivative and the Laplacian. This approach is highly related to L_p-Fourier multipliers and similar to the approach of the present text. However, we use a joint H^∞-calculus of the time-derivative ∂_t and $\nabla = (\partial_1, \ldots, \partial_n)$ based on results by G. Dore and A. Venni [DV02b, DV05]. In many applications this generalization becomes necessary due to the fact that the systems can include symbols which are not rotation invariant in space.

In the following, we describe the main ideas of the present text which is based on the second author's Ph.D. thesis [Kai12]. As a prototype example which will run like a common thread through this book, we consider the Stefan problem with Gibbs-Thomson correction (see [EPS03], [GR07], [DSS08]). For this free boundary value problem, the model problem in $\mathbb{R}^n_+$ (after Hanzawa transform) is given by

$$\begin{cases} \partial_t u - \Delta u & = 0 & \text{in } \mathbb{R}_+ \times \mathbb{R}^n_+, \\ u + \Delta' h & = g_1 & \text{on } \mathbb{R}_+ \times \mathbb{R}^{n-1}, \\ \partial_t h - \partial_n u & = g_2 & \text{on } \mathbb{R}_+ \times \mathbb{R}^{n-1}, \\ u(t=0) & = 0 & \text{in } \mathbb{R}^n_+, \\ h(t=0) & = 0 & \text{on } \mathbb{R}^{n-1}. \end{cases} \qquad (2)$$

The usual approach in the analysis of such problems is the reduction to the boundary. After employing formally a Laplace transform in t and a partial Fourier transform in $x' = (x_1, \ldots, x_{n-1})$, we obtain an ordinary differential equation for $\hat{u}(\lambda, \xi', \cdot)$ and $\hat{h}(\lambda, \xi', \cdot)$ with fixed (λ, ξ'), which is given by

$$\begin{cases} \omega(\lambda, \xi')^2 \hat{u}(\lambda, \xi', x_n) - \partial_n^2 \hat{u}(\lambda, \xi', x_n) &= 0, \qquad x_n > 0, \\ \hat{u}(\lambda, \xi', 0) - |\xi'|^2 \hat{h}(\lambda, \xi') &= \hat{g}_1(\lambda, \xi'), \\ \lambda \hat{h}(\lambda, \xi') - \partial_n \hat{u}(\lambda, \xi') &= \hat{g}_2(\lambda, \xi'), \end{cases} \qquad (3)$$

where $\omega(\lambda, \xi') := \sqrt{\lambda + |\xi'|^2}$. A stable solution of the first line in (3) is given by

$$\hat{u}(\lambda, \xi', x_n) = \hat{\varphi}(\lambda, \xi') \exp(-\omega(\lambda, \xi') x_n), \qquad x_n > 0$$

with an unknown function $\hat{\varphi}$. The boundary conditions in (3) then yield

$$\begin{pmatrix} 1 & -|\xi'|^2 \\ \omega(\lambda, \xi') & \lambda \end{pmatrix} \begin{pmatrix} \hat{\varphi} \\ \hat{h} \end{pmatrix} = \begin{pmatrix} \hat{g}_1 \\ \hat{g}_2 \end{pmatrix} \qquad (4)$$

for φ and h. One of the key steps in the proof of the local in time well-posedness for the nonlinear Stefan problem (see [EPS03]) lies in the analysis of the operator corresponding to the matrix in (4). Formally, this operator is given as

$$\mathscr{L}(\partial_t, D') := \begin{pmatrix} 1 & \Delta' \\ \sqrt{\partial_t - \Delta'} & \partial_t \end{pmatrix}.$$

Note that $L(\partial_t, D')$ is not a differential operator due to the fact that the entries of the symbol matrix are not polynomials in λ and ξ'. This typically appears when the reduction to the boundary method is used. Moreover, the entries of the matrix $L(\partial_t, D')$ have different orders, and this matrix is a typical example of a (nonstandard) mixed-order system. Therefore, the following questions arise:

- How can we define an operator $\mathscr{L}(\partial_t, D)$ corresponding to a mixed-order symbol $\mathscr{L}(\lambda, \xi)$ with non-polynomial entries?
- How can we find tuples $\mathbb{H}$, $\mathbb{F}$ of L_p- or L_p-L_q-Sobolev spaces which are adapted to the system in the sense that the operator $L(\partial_t, D)$ induces an isomorphism between $\mathbb{H}$ and $\mathbb{F}$ and thus has maximal regularity?
- How can we define parabolicity conditions on $\mathscr{L}(\lambda, \xi)$ which lead to maximal regularity, and can we find equivalent criteria for parabolicity which are easily verified in applications?

The present book will answer these questions for a large class of problems. We describe the main ideas for the development of the theory.

Joint H^∞-calculus for the time-space derivative (Chapter 1). Instead of the pseudo-differential method as used in, e.g., [DS11], we follow another approach based on the joint H^∞-calculus of the tuple $(\partial_t, \partial_1, \ldots, \partial_n)$. This calculus developed by

G. Dore and A. Venni [DV02b, DV05] can be applied whenever the symbol depends holomorphically on λ and z at least in some sectors or bisectors. Therefore, we consider $\mathscr{L}$ as a function of the complex variables λ and $z \in \mathbb{C}^n$. Due to the fact that the time variable runs only on the half-line $(0, \infty)$ whereas the space variables belong to the whole space, there is a difference between the related operators: The time derivative ∂_t enters the H^∞-calculus as a sectorial operator while the space derivatives $\partial_1, \ldots, \partial_n$ induce bisectorial operators.

The existence of a bounded joint H^∞-calculus for the time-space derivative is the main result of Chapter 1 and is stated in Theorem 1.89. Here we consider both Bessel potential spaces and Besov spaces, including vector-valued spaces and L_p-L_q-spaces. Due to the appearance of vector-valued Fourier multipliers, we have to deal with $\mathcal{R}$-boundedness (see Section 1.1). The Sobolev spaces on the half-space $\mathbb{R}^n_+$ are defined with the help of the retraction/coretraction concept (Section 1.2).

The Newton polygon approach and N-parabolicity for mixed-order systems (Chapter 2). For the inverse of the matrix

$$\mathscr{L}(\lambda, z') = \begin{pmatrix} 1 & z_1^2 + \cdots + z_{n-1}^2 \\ \sqrt{\lambda - z_1^2 - \cdots - z_{n-1}^2} & \lambda \end{pmatrix}$$

appearing in the Stefan problem, one has to understand the structure of its determinant

$$\det \mathscr{L}(\lambda, z') = \lambda - (z_1^2 + \cdots + z_{n-1}^2)\sqrt{\lambda - (z_1^2 + \cdots + z_{n-1}^2)}. \tag{5}$$

One can easily observe that $P(z', \lambda) := \det \mathscr{L}(z', \lambda)$ is not a quasi-homogeneous function of λ and z'. More precisely, there exists no relative weight ρ and no degree N such that the equality $P(\eta^\rho \lambda, \eta z) = \eta^N P(\lambda, z)$ holds for all $\eta > 0$. This also implies that there is no well-defined principal symbol of P. The Newton polygon helps to understand the hidden homogeneity structure of the symbol. It is defined by means of the orders in λ and z' appearing in each term in (5). Using

$$\left|\sqrt{\lambda - (z_1^2 + \cdots + z_{n-1}^2)}\right| \approx |\lambda|^{1/2} + |z'|,$$

we obtain the tuples $(0, 1)$, $(2, \frac{1}{2})$, and $(3, 0)$ corresponding to the terms $|\lambda|$, $|\lambda|^{1/2}|z'|^2$, and $|z'|^3$, respectively. The Newton polygon is then defined as the convex hull of these three points and the origin, see Figure 1.

The Newton polygon describes in a geometrical way the inhomogeneous symbol structure and can be used for defining appropriate (non-standard) Sobolev spaces in which the operator induces an isomorphism. For this, we define in Chapter 2 the weight function $W_N = W_N(\lambda, z)$ associated to the Newton polygon N. The classical parabolicity condition $P(\lambda, z) \neq 0$, $\operatorname{Re} \lambda \geq 0$, $z \in (i\mathbb{R})^n$, $(\lambda, z) \neq 0$, is now replaced by an inequality of the form

$$|P(\lambda, z)| \geq C W_N(\lambda, z)$$

Introduction and Outline

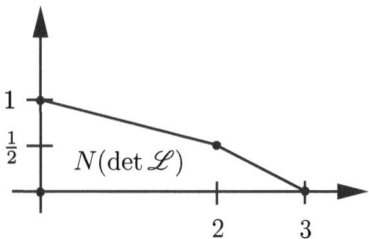

Figure 1: Newton polygon for the Stefan problem with Gibbs-Thomson correction

where λ and z run in suitable open sectors and bisectors, respectively (see Section 2.2). The resulting class of symbols and equations is called N-parabolic. In Section 2.2 c) we give an equivalent condition for N-parabolicity which can easily be verified in applications. Instead of the Newton polygon itself, one can also study the related order function. The concept of order functions, which seems to be new, turns out to be useful when several Newton polygons have to be considered simultaneously.

By the condition of N-parabolicity for the determinant, one can easily generalize the above concept to mixed-order systems. In Section 2.3, we investigate N-parabolic Douglis-Nirenberg systems, Sobolev space tuples related to such systems, and the mapping properties of the operator realizations in L_p-L_q-Sobolev spaces. One of the main theorems of this text states that, given a compatible tuple of Sobolev spaces, the operator $\mathscr{L}(\partial_t, D)$ corresponding to an N-parabolic mixed-order system induces an isomorphism in these spaces (Theorem 2.69).

We remark that, motivated by the applications, we generalized the existing theory in more than one direction: First, the Newton polygons considered here are not necessarily regular, in contrast to the results in [DSS08], [DV08], and [GV92]. Then, we also include spaces like

$$_0B_{pp}^{1-1/(2p)}(\mathbb{R}_+, L_p(\mathbb{R}^n)) \cap L_p(\mathbb{R}_+, B_{pp}^{2-1/p}(\mathbb{R}^n))$$

which appear naturally as trace spaces in boundary value problems. We speak of spaces of mixed scales as L_p belongs to the Bessel potential scale whereas B_{pp} belongs to the Besov scale. In this situation, we need a compatibility condition on the tuples of spaces which is satisfied in many situations and which is discussed in detail in Subsection 2.3 c).

Generalization to Triebel-Lizorkin spaces and L_p-L_q-boundary value problems (Chapter 3). As mentioned above, Triebel-Lizorkin spaces appear inevitably as trace spaces if the boundary value problem is studied in the L_p-L_q-setting with $p \neq q$. Therefore, we prove the analog of the results of Chapter 2 also in the Triebel-Lizorkin setting. The main result on mixed-order systems is formulated in Theorem 3.31.

For the discussion of the compatibility condition in Section 3.4 we use the concept of anisotropic Triebel-Lizorkin spaces with mixed norms as well as the representation of these spaces as an intersection containing vector-valued Triebel-Lizorkin spaces. Our approach to Triebel-Lizorkin spaces is based on interpolation. In the scalar case we have the well-known result

$$F_{pq}^s(\mathbb{R}^n) = \left[H_{p_0}^{s_0}(\mathbb{R}^n), B_{p_1 p_1}^{s_1}(\mathbb{R}^n)\right]_\theta, \quad \theta \in (0,1) \tag{6}$$

with $1/p = (1-\theta)p_0 + \theta p_1$, $1/q = (1-\theta)/2 + \theta p_1$, and $s = (1-\theta)s_0 + \theta s_1$ (cf. [Tri78, Section 2.4.2 (12)]). To the knowledge of the authors, a vector-valued version of (6) has not been discussed so far. In Proposition 3.10 we present an analog representation for Bessel-valued spaces. This paves the way to generalize the results of Section 2.3 to Triebel-Lizorkin spaces by interpolation arguments. Note that there is a coupling between p and q in (6). This coupling also occurs in Chapter 3 and cannot be avoided in the approach via interpolation. In Section 3.5, we include some remarks on singular integral operators and extension operators which are needed in the study of L_p-L_q-boundary value problems.

Applications to general parabolic boundary value problems and mixed-order systems (Chapter 4). Finally, we present a selection of applications which were, in fact, the motivation for the development of the theory. We deal with general parabolic equations and systems where by the word 'general' we understand an inherent inhomogeneous symbol structure as discussed above. In particular, none of these problems are covered by the classical parameter-elliptic and parabolic theory.

The applications we consider can be divided into two groups: The first group consists in partial differential equations on the whole space $\mathbb{R}^n$. Here we directly formulate the equations as a mixed-order system without a reduction to a first-order system. We present the following examples:

- The generalized Stokes system (see [BP07]),
- the generalized thermoelastic plate equations (see, e.g., [MR96] and [DR06]),
- the linear Cahn-Hilliard-Gurtin problem (see [Wil07]),
- a compressible fluid model of Korteweg type (see [Kot08]),
- and a linear three-phase problem which appears in the treatment of a chemical reaction system with electromigration (see [Kot10]).

To our knowledge, most of these examples are studied for the first time in the L_p-L_q-setting.

In the second group one finds boundary value problems on $\mathbb{R}^n_+$ and $\dot{\mathbb{R}}^n$, respectively. This situation is more involved because we have to handle ground spaces which are related to trace spaces. In both cases our approach gives very short and straight proofs of well-posedness. We consider

- a model for the spin-coating process (cf. [DGH[+]11]),

- the two-phase Stokes equations with Boussinesq-Scriven surface and gravity (see [BP10]),
- and the L_p-L_q two-phase Stefan problem with Gibbs-Thomson correction (cf. [EPS03]) as discussed above.

Exemplarily, we formulate the two-phase Stokes equations with Boussinesq-Scriven surface introduced in [BP10]:

$$\begin{cases} \rho\partial_t u - \mu\Delta u + \nabla\pi &= 0 \quad \text{in } \mathbb{R}_+ \times \dot{\mathbb{R}}^{n+1}, \\ \operatorname{div} u &= 0 \quad \text{in } \mathbb{R}_+ \times \dot{\mathbb{R}}^{n+1}, \\ -\mu_s\Delta_x v - \lambda_s\nabla_x\operatorname{div}_x v - [\![\mu\partial_y v]\!] - [\![\mu\nabla_x w]\!] &= g_v \quad \text{on } \mathbb{R}_+ \times \mathbb{R}^n, \\ -2[\![\mu\partial_y w]\!] + [\![\pi]\!] - \sigma\Delta h - Gh &= g_w \quad \text{on } \mathbb{R}_+ \times \mathbb{R}^n, \\ [\![u]\!] &= 0 \quad \text{on } \mathbb{R}_+ \times \mathbb{R}^n, \\ \partial_t h - w + \langle b_0, \nabla \rangle h &= g_h \quad \text{on } \mathbb{R}_+ \times \mathbb{R}^n, \\ u(t=0) &= 0 \quad \text{in } \dot{\mathbb{R}}^{n+1}, \\ h(t=0) &= 0 \quad \text{on } \mathbb{R}^n. \end{cases} \quad (7)$$

In addition to the dynamical boundary condition, (7) also includes a boundary condition of order two. This surface viscosity changes the behavior of the system in the highest order fundamentally. We show in Section 4.7 that the associated mixed-order system on the boundary fits into our theory. The well-posedness in L_p of this problem has not been proved in the literature before. In particular, we give a proof which solves the problem either with or without surface viscosity, i.e., $\lambda_s, \mu_s = 0$ or $\lambda_s \geq 0, \mu_s > 0$, simultaneously.

We emphasize that our method is not at all restricted to these examples, and that the main idea to present these applications is not in all cases to give new results but to provide short and systematical proofs for the well-posedness which serve as prototypes for further applications.

Due to the nature of the problems, some parts of this book are technically involved and require background knowledge from different fields like operator theory, interpolation theory of Banach spaces, and Sobolev space theory. Therefore, we decided to include paragraphs called (somewhat optimistically) 'Motivation'. These texts are emphasized by a frame and should serve as an orientation for the reader and justification of the topics discussed in this section or subsection.

The present book is based on several years of research on the Newton polygon approach which resulted in a series of papers of the first author and in the Ph. D. thesis of the second author. Without the help of many friends and colleagues, this work would not have been possible. We thank M. Hieber, J. Prüss, R. Racke, J. Saal, and J. Seiler for fruitful mathematical discussions and collaborations, H. Amann and H. Triebel for their rich and influential work, and commemorate L.R. Volevich (1934–2007) who introduced the first author into the Newton polygon method and accompanied him for several years in this topic. We also express our gratitude to our friends and colleagues at the University of

Konstanz, T. Moseler, T. Nau, J. Schnur, and O. Weinmann for their substantial support and help. Finally, we are obliged to Springer Basel and to the editors of *Operator Theory: Advances and Applications* for including our book in this series.

Konstanz, September 2013   R. Denk, M. Kaip

Chapter 1

The joint time-space H^∞-calculus

One of the most successful approaches in the theory of partial differential equations is based on the Fourier transform and the symbols of differential operators and their inverses. The Fourier transform maps differentiation into multiplication by the covariable and thus directly leads to a representation of the inverse (i.e., solution) operator. In particular in the case of boundary value problems, non-polynomial symbols like

$$P(\lambda, z) := \lambda - (z_1^2 + \cdots + z_n^2)\sqrt{\lambda - z_1^2 - \cdots - z_n^2} \qquad (1.1)$$

appear. Here the related operator is, at least symbolically, given by $P(\partial_t, \nabla_x) = \partial_t - \Delta\sqrt{\partial_t - \Delta}$. The symbol (1.1) appears in the analysis of the Stefan problem with Gibbs-Thomson correction as the determinant of the Lopatinskii matrix.

The aim of this chapter is to give a precise definition of operators like $P(\partial_t, \nabla_x)$. For this, one can apply the joint H^∞-calculus for sectorial and bisectorial operators as developed, e.g., by G. Dore and A. Venni in [DV05]. Note that P is a holomorphic function of λ and $z = (z_1, \ldots, z_n)$ for

$$|\arg \lambda| + 2 \max_{j=1,\ldots,n} |\arg(z_j) - \tfrac{\pi}{2}| < \pi.$$

As we will see below, the tuple $(\partial_t, \nabla_x) = (\partial_t, \partial_1, \ldots, \partial_n)$ has a bounded joint H^∞-calculus. Here the operators $\partial_j = \partial_{x_j}$ act on the whole real line, have their spectra on the imaginary axis and are bisectorial operators with bisectorial angle 0. In contrast to this, the time derivative ∂_t acts on the half-axis $(0, \infty)$. Therefore, we include zero initial values in the domain of the corresponding operator, the spectrum of which coincides with the complex half-plane $\{\lambda \in \mathbb{C} : \operatorname{Re} \lambda \geq 0\}$. This leads to a sectorial operator with sectorial angle $\tfrac{\pi}{2}$.

The present chapter answers the following questions:

- How can one define (holomorphic) functions of ∂_t and ∇_x (Section 1.1)?

- What are appropriate Sobolev-type spaces (Section 1.2)?
- What results follow for the joint H^∞-calculus for the time-space operator (∂_t, ∇_x) (Section 1.3)?

1.1 The joint H^∞-calculus for tuples of operators

> *Motivation.* The aim of this section is to define functions of operator tuples where the operators are either sectorial or bisectorial and where the function is holomorphic and bounded in corresponding sectors or bisectors. The main result states that the resulting operator is well-defined and bounded. As the proof is based on a Kalton-Weis type theorem, the concept of $\mathcal{R}$-boundedness will play a role. Therefore, we will include a short summary of $\mathcal{R}$-bounded operator families. For the proofs, we will mainly restrict ourselves to giving some references.

Throughout this section, let X and Y be complex Banach spaces.

a) Sectorial and bisectorial operators, $\mathcal{R}$-boundedness

For a linear operator $T\colon X \supseteq D(T) \to X$, the *resolvent set* $\rho(T)$ is given as the set of all complex numbers λ for which $\lambda - T\colon D(T) \to X$ is bijective and $(\lambda-T)^{-1} \in L(X)$, where $L(X)$ stands for the space of all bounded linear operators in X. Note that $\rho(T) \neq \emptyset$ implies that T is closed. By $R(T)$ and $\ker T$ we denote the range and the kernel of T, respectively.

Definition 1.1. For $\theta \in (0, \pi)$ and $\delta \in (0, \frac{\pi}{2})$ we define the *sector* S_θ and the *bisector* Σ_δ by

$$S_\theta := \{r\exp(i\alpha)\colon r \in \mathbb{R}_+,\ \alpha \in (-\theta, \theta)\},$$
$$\Sigma_\delta := \{r\exp(i\alpha)\colon r \in \mathbb{R}\setminus\{0\},\ \alpha \in (\pi/2 - \delta, \pi/2 + \delta)\}.$$

Here, $\mathbb{R}_+ := (0, \infty)$. Let Γ_φ be the curve which is parametrized by $\gamma_\varphi\colon \mathbb{R}\setminus\{0\} \to \mathbb{C}$, $r \mapsto |r|\exp(-i\varphi\,\mathrm{sgn}(r))$ for $\varphi \in (0, \pi)$. The curve Γ_φ is called *admissible* for S_θ if $0 < \theta < \varphi$. An admissible curve for Σ_δ is a curve of the form $\Gamma_\varphi \cup (-\Gamma_\varphi)$ with $\delta + \frac{\pi}{2} < \varphi < \pi$, see Figures 1.1 and 1.2.

Definition 1.2. (i) An operator $T\colon X \supseteq D(T) \to X$ is said to be *sectorial* if

(I) $D(T)$ and $R(T)$ are dense in X,

(II) there exists $\theta \in (0, \pi)$ such that $\rho(T) \supseteq \mathbb{C}\setminus \overline{S}_\theta = -S_{\pi-\theta}$ and

$$\sup_{\lambda \in \mathbb{C}\setminus \overline{S}_\theta} \|\lambda(\lambda - T)^{-1}\|_{L(X)} < \infty. \tag{1.2}$$

1.1. The joint H^∞-calculus for tuples of operators

If T is sectorial, we define the *spectral angle* φ_T as the infimum of all angles $\theta \in (0, \pi)$ such that (1.2) holds.

(ii) An operator $T\colon X \supseteq D(T) \to X$ is said to be *bisectorial* if

(I) $D(T)$ and $R(T)$ are dense in X,

(II) there exists $\delta \in (0, \pi/2)$ such that $\rho(T) \supseteq \mathbb{C} \setminus \overline{\Sigma}_\delta = (-S_{\pi/2-\delta}) \cup S_{\pi/2-\delta}$ and
$$\sup_{\lambda \in \mathbb{C}\setminus\overline{\Sigma}_\delta} \|\lambda(\lambda - T)^{-1}\|_{L(X)} < \infty. \tag{1.3}$$

If T is bisectorial, we define the *spectral angle* $\varphi_T^{(\mathrm{bi})}$ as the infimum of all angles $\delta \in (0, \frac{\pi}{2})$ such that (1.3) holds.

Example 1.3. (i) Let $X = L_2(\mathbb{R}^n)$ and $T := -\Delta$ with domain $D(T) := H^2(\mathbb{R}^n) \subseteq L_2(\mathbb{R}^n)$. Then T is sectorial with spectral angle 0. This can easily be seen from the fact that T is a selfadjoint operator with spectrum $\sigma(T) = [0, \infty)$.

(ii) Let $X = L_2(\mathbb{R})$ and $T := \partial_x$ with domain $D(T) := H^1(\mathbb{R})$. Then T is skew-selfadjoint, i.e., $T^* = -T$, and the spectrum of T coincides with the imaginary axis. From this we obtain that T is bisectorial with spectral angle 0.

(iii) The results from (i) and (ii) hold also in the case when L_2 is replaced by L_p with $1 < p < \infty$. This follows from an application of Michlin's theorem.

Remark 1.4. (i) Note that every bisectorial operator T is also sectorial with $\frac{\pi}{2} - \varphi_T^{(\mathrm{bi})} \leq \varphi_T \leq \frac{\pi}{2} + \varphi_T^{(\mathrm{bi})}$.

(ii) In reflexive Banach spaces, the density of $R(T)$ follows from the injectivity of T. This is an immediate consequence of the following result (see [KW04, Proposition 15.2 c)]): Let T be a linear operator on a reflexive Banach space X with domain $D(T)$ such that there exists $\theta \in (0, \pi)$ for which (1.2) holds. Then we can decompose X into $X = \ker(T) \oplus \overline{R(T)}$.

The following approximation property will be useful later.

Lemma 1.5 ([KW04, Proposition 15.2 a)]). *Let $A\colon X \supseteq D(A) \to X$ be a linear operator with $(-\infty, 0) \subseteq \rho(A)$ satisfying*
$$\sup_{\lambda \in (-\infty, 0)} \|\lambda(\lambda - A)^{-1}\|_{L(X)} < \infty. \tag{1.4}$$

Then we have
$$\begin{aligned}\overline{R(A)} &= \{x \in X \colon \lim_{n\to\infty} -\tfrac{1}{n}(-\tfrac{1}{n} - A)^{-1}x = 0\}, \\ \overline{D(A)} &= \{x \in X \colon \lim_{n\to\infty} -n(-n - A)^{-1}x = x\}.\end{aligned} \tag{1.5}$$

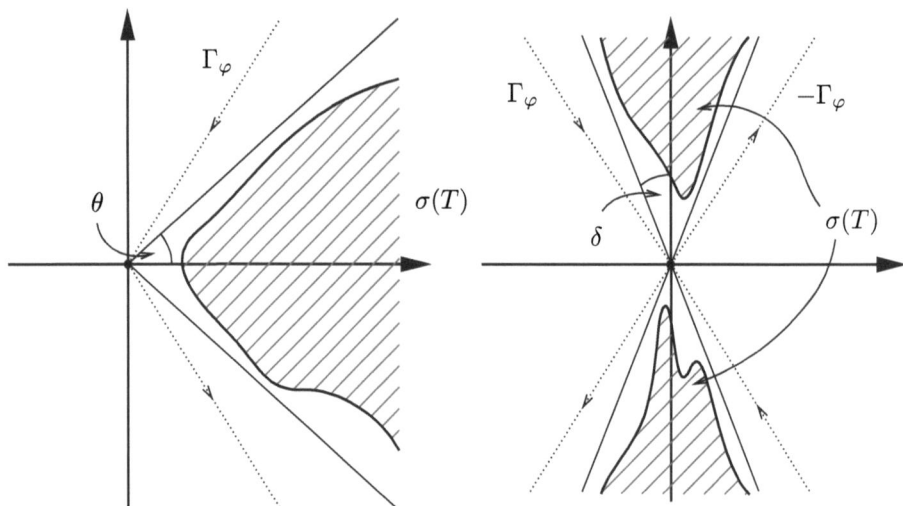

Figure 1.1: Spectrum, S_θ, and admissible curve Γ_φ

Figure 1.2: Spectrum, Σ_δ, and admissable curve $\Gamma_\varphi \cup (-\Gamma_\varphi)$

In the definition of a sectorial operator, the operator family $\{\lambda(\lambda - T)^{-1} : \lambda \in \mathbb{C} \setminus \overline{S}_\theta\}$ is bounded in the operator norm $\|\cdot\|_{L(X)}$. To show maximal regularity in L_p-Sobolev spaces or the existence of a bounded H^∞-calculus, the concept of $\mathcal{R}$-boundedness turns out to be more appropriate. In the following, we restrict ourselves to the main definitions and results. For a more detailed discussion of $\mathcal{R}$-bounded operators, we refer to [DHP03] and [KW04].

In the following definition, $L(X, Y)$ denotes the space of all bounded linear operators from X to Y, and $L_p([0,1], X)$ stands for the standard Bochner space of X-valued L_p-functions.

Definition 1.6 ($\mathcal{R}$-boundedness)**.** Let $\mathcal{T} \subseteq L(X,Y)$ and $p \in [1, \infty)$. Then $\mathcal{T}$ is said to be $\mathcal{R}$-*bounded* if there exists a constant $C > 0$ such that for all $m \in \mathbb{N}$, $(T_k)_{k=1,\ldots,m} \subseteq \mathcal{T}$, and all $(x_k)_{k=1,\ldots,m} \subseteq X$ we have

$$\Big\| \sum_{k=1}^m r_k T_k x_k \Big\|_{L_p([0,1],Y)} \leq C \Big\| \sum_{k=1}^m r_k x_k \Big\|_{L_p([0,1],X)}. \tag{1.6}$$

In this case, $\mathcal{R}_p(\mathcal{T}) := \inf\{C > 0 \colon (1.6) \text{ is satisfied}\}$ is said to be the $\mathcal{R}$-*bound* of $\mathcal{T}$. Here the *Rademacher functions* r_k, $k \in \mathbb{N}$, are given by $r_k \colon [0,1] \to \{-1, 1\}, t \mapsto \operatorname{sign}(\sin(2^k \pi t))$.

Remark 1.7. The proofs of the following results can be found in [DHP03] or [KW04].

1.1. The joint H^∞-calculus for tuples of operators

(i) If $\mathcal{T} \subseteq L(X,Y)$ is $\mathcal{R}$-bounded for one $p \in [1,\infty)$, then (1.6) holds for all $p \in [1,\infty)$ where the $\mathcal{R}$-bounds fulfill $C\mathcal{R}_1(\mathcal{T}) \leq \mathcal{R}_p(\mathcal{T}) \leq C'\mathcal{R}_1(\mathcal{T})$ for some $C, C' > 0$.

(ii) If $\mathcal{T}, \mathcal{S} \subseteq L(X,Y)$ are $\mathcal{R}$-bounded, then $\mathcal{T} + \mathcal{S} := \{T + S \colon T \in \mathcal{T}, S \in \mathcal{S}\}$ is also $\mathcal{R}$-bounded with $\mathcal{R}_p(\mathcal{T} + \mathcal{S}) \leq \mathcal{R}_p(\mathcal{T}) + \mathcal{R}_p(\mathcal{S})$.

(iii) For two given $\mathcal{R}$-bounded families $\mathcal{T}_1 \subseteq L(Z,Y)$ and $\mathcal{T}_2 \subseteq L(X,Z)$ we obtain the $\mathcal{R}$-boundedness of $\mathcal{T}_1\mathcal{T}_2 := \{T_1T_2 \colon T_k \in \mathcal{T}_k, k = 1, 2\} \subseteq L(X,Y)$ with $\mathcal{R}_p(\mathcal{T}_1\mathcal{T}_2) \leq \mathcal{R}_p(\mathcal{T}_1) \cdot \mathcal{R}_p(\mathcal{T}_2)$.

(iv) If $\mathcal{T} \subseteq L(X,Y)$ is $\mathcal{R}$-bounded, then $\mathcal{T}$ is also uniformly bounded. The converse is only true if X and Y are Hilbert spaces.

(v) Let $\mathcal{T} \subseteq L(X,Y)$ be $\mathcal{R}$-bounded. Then the closure in the strong operator topology of the convex hull of $\mathcal{T}$ is also $\mathcal{R}$-bounded with

$$\mathcal{R}_p(\overline{\mathrm{co}(\mathcal{T})}^{\mathrm{strong}}) \leq \mathcal{R}_p(\mathcal{T}).$$

(vi) If $\mathcal{T} \subseteq L(X,Y)$ is $\mathcal{R}$-bounded and $\mathcal{S} \subseteq \mathcal{T}$, then $\mathcal{S}$ is also $\mathcal{R}$-bounded. If $\mathcal{T} \subseteq L(X,Y)$ is finite, then $\mathcal{T}$ is $\mathcal{R}$-bounded.

To show $\mathcal{R}$-boundedness of an operator family, the following results are convenient and, in most cases, can be applied better than directly showing the conditions of the definition.

Theorem 1.8 (**Kahane's contraction principle**, cf. [KW04, Proposition 2.5]). *Let $p \in (1,\infty)$ and $(a_j)_{j \in \mathbb{N}}, (b_j)_{j \in \mathbb{N}} \subseteq \mathbb{C}$ with $|a_j| \leq |b_j|$ for all $j \in \mathbb{N}$. Then we have*

$$\Big\|\sum_{k=1}^n r_k a_k x_k\Big\|_{L_p([0,1],X)} \leq 2 \Big\|\sum_{k=1}^n r_k b_k x_k\Big\|_{L_p([0,1],X)}$$

for all $n \in \mathbb{N}$ and $(x_j)_{j \in \mathbb{N}} \subseteq X$.

Theorem 1.9 (**Square function estimate**, see [DHP03, Remarks 3.2] or [KW04, Remark 2.9]). *Let $(\Omega, \mathcal{A}, \mu)$ be a σ-finite measure space and*

$$\mathcal{T} \subseteq L(L_p(\Omega, \mu, \mathbb{C}))$$

with $1 < p < \infty$. The family $\mathcal{T}$ is $\mathcal{R}$-bounded if and only if there exists $C > 0$ such that

$$\Big\|\Big(\sum_{k=1}^N |T_k f_k|^2\Big)^{1/2}\Big\|_{L_p(\Omega,\mu,\mathbb{C})} \leq C \Big\|\Big(\sum_{k=1}^N |f_k|^2\Big)^{1/2}\Big\|_{L_p(\Omega,\mu,\mathbb{C})} \qquad (1.7)$$

for all $N \in \mathbb{N}$, $(f_k)_k \subseteq L_p(\Omega, \mu, \mathbb{C})$, and $(T_k)_k \subseteq \mathcal{T}$.

For the study of singular integral operators the following result is very helpful because it transfers results on scalar-valued kernels to operator-valued kernels.

Proposition 1.10 ([DHP03, Prop. 4.12], [KW04, Section 5.5]). *Let $1 < p < \infty$, and let I be an arbitrary index set. Let $\mathcal{K} = \{k_i \colon \mathbb{R}_+ \times \mathbb{R}_+ \to L(X) \,|\, i \in I\}$ be a family of kernel functions such that for each $i \in I$ the integral operator*

$$(K_i f)(x_n) := \int_0^\infty k_i(x_n, y_n) f(y_n) dy_n, \quad f \in C^\infty(\mathbb{R}_+, X) \cap L_p(\mathbb{R}_+, X), \quad x_n > 0$$

extends to a bounded operator on $L_p(\mathbb{R}_+, X)$. Suppose that there exists $k_0 \colon \mathbb{R}_+ \times \mathbb{R}_+ \to \mathbb{C}$ such that

(i) *$(K_0 f)(x_n) := \int_0^\infty k_0(x_n, y_n) f(y_n) dy_n$ ($f \in C^\infty(\mathbb{R}_+) \cap L_p(\mathbb{R}_+)$, $x_n > 0$) extends to a bounded operator on $L_p(\mathbb{R}_+)$,*

(ii) *for each $x_n, y_n > 0$, the operator family $\{k_i(x_n, y_n) \colon i \in I\} \subseteq L(X)$ is $\mathcal{R}$-bounded with*

$$\mathcal{R}_p(\{k_i(x_n, y_n) \colon i \in I\}) \leq k_0(x_n, y_n), \quad x_n, y_n > 0.$$

Then $\{K_i \colon i \in I\} \subseteq L_p(\mathbb{R}_+, X)$ is also $\mathcal{R}$-bounded with $\mathcal{R}_p(\{K_i \colon i \in I\}) \leq \|K_0\|_{L(L_p(\mathbb{R}_+))}$.

For the handling of operator-valued L_p-Fourier multipliers, several results demand geometric properties of the underlying Banach space. We introduce two main properties.

Definition 1.11 (**Banach space of class $\mathcal{HT}$**). The *Hilbert transform* Hf of a function $f \in \mathscr{S}(\mathbb{R}, X)$, the Schwartz space of rapidly decreasing X-valued functions, is defined by

$$(Hf)(x) := \tfrac{1}{\pi}\left[PV(\tfrac{1}{x}) * f\right](x) := \tfrac{1}{\pi} \lim_{\varepsilon \searrow 0} \int_{|x-y| \geq \varepsilon} \frac{f(y)}{x-y}\, dy, \quad x \in \mathbb{R}.$$

The Banach space X is called of *class $\mathcal{HT}$* if there exists $p \in (1, \infty)$ such that $Hf \in L_p(\mathbb{R}, X)$ for all $f \in \mathscr{S}(\mathbb{R}, X)$ and

$$\|Hf\|_{L_p(\mathbb{R}, X)} \leq C \|f\|_{L_p(\mathbb{R}, X)}, \quad f \in \mathscr{S}(\mathbb{R}, X).$$

In this case we can extend H to a bounded operator on $L_p(\mathbb{R}, X)$.

Remark 1.12. Banach spaces of class $\mathcal{HT}$ are also called *UMD spaces*. Here UMD stands for the probabilistic property of 'unconditional martingal differences'. A Banach space is of class $\mathcal{HT}$ if and only if it is a UMD space. This equivalence can be found in [Bur81] and [Bou83].

Remark 1.13. The following assertions can be found in [RdF86, Proposition 2] and [Ama95, Theorem 4.5.2], for example.

(i) Banach spaces of class $\mathcal{HT}$ are reflexive.

1.1. The joint H^∞-calculus for tuples of operators

(ii) If X is of class $\mathcal{HT}$ and there exists $T \in L_{\text{Isom}}(X,Y)$, then Y is also of class $\mathcal{HT}$.

(iii) If X is a Banach space of class $\mathcal{HT}$ and (Ω, μ) is a σ-finite measure space, then
$$L_p(\Omega, \mu, X), \quad 1 < p < \infty,$$
is also of class $\mathcal{HT}$.

(iv) If X is of class $\mathcal{HT}$, then we have $\|Hf\|_{L_p(\mathbb{R},X)} \leq C_p \|f\|_{L_p(\mathbb{R},X)}$ for all $f \in \mathscr{S}(\mathbb{R}, X)$ and $p \in (1, \infty)$.

(v) Every Hilbert space is of class $\mathcal{HT}$.

Definition 1.14 (Banach space with property (α)). The Banach space X has *property* (α) if there exists a constant $C > 0$ such that for all $n \in \mathbb{N}$, $(\alpha_{ij})_{i,j=1,\ldots,n} \subseteq \mathbb{C}$, $|\alpha_{ij}| \leq 1$, and all $(x_{ij})_{i,j=1,\ldots,n} \subseteq X$ we have

$$\int_0^1 \int_0^1 \Big\| \sum_{i,j=1}^n r_i(u) r_j(v) \alpha_{ij} x_{ij} \Big\|_X du\, dv$$
$$\leq C \int_0^1 \int_0^1 \Big\| \sum_{i,j=1}^n r_i(u) r_j(v) x_{ij} \Big\|_X du\, dv.$$

Remark 1.15. (i) Let X and Y be Banach spaces. If X has property (α) and there exists $T \in L_{\text{Isom}}(X, Y)$, then Y also has property (α).

(ii) Let X be a Banach space with property (α), and let (Ω, μ) be a σ-finite measure space. Then
$$L_p(\Omega, \mu, X), \quad 1 \leq p < \infty$$
also has property (α), cf. [KW04, Remark 4.10].

(iii) Every Hilbert space has property (α).

b) Joint H^∞-calculus

For the analysis of the time-space derivative, it is essential to consider not a single (sectorial or bisectorial) operator but tuples of operators and the joint H^∞-calculus for such tuples. The concept of H^∞-calculus goes back to A. McIntosh, see for example [McI86] and [CDMY96]. Another comprehensive and more recent work on the H^∞-calculus can be found in [Haa06]. Here we present the basic notation and results by G. Dore and A. Venni [DV05] on the joint H^∞-calculus for sectorial and bisectorial operators.

In the following, X and Y are complex Banach spaces.

Definition 1.16. Let $T_k \colon X \supseteq D(T_k) \to X$, $k = 1, \ldots, N$, be linear operators. The operator tuple $\mathbf{T} := (T_1, \ldots, T_N)$ is called *admissible* if

(i) each of the operators T_k ($k = 1, \ldots, N$) is sectorial or bisectorial,

(ii) for every $j, k = 1, \ldots, N$ the resolvents of T_j and T_k commute.

Note that (ii) means that $(\lambda - T_j)^{-1}(\mu - T_k)^{-1} = (\mu - T_k)^{-1}(\lambda - T_j)^{-1}$ holds for all $\lambda \in \rho(T_j)$ and $\mu \in \rho(T_k)$. For this, it is sufficient that this equality holds for one $\lambda \in \rho(T_j)$ and one $\mu \in \rho(T_k)$ (see [Kat76, Theorem 6.5]).

Let us fix the situation for the remainder of this subsection. Let $\mathbf{T} = (T_1, \ldots, T_N)$ be an admissible operator tuple. If T_k is sectorial, we fix a sector $\Omega_k := S_{\theta_k}$ with $\theta_k > \varphi_{T_k}$ and a curve $\Gamma^{(k)} \subseteq \Omega_k$ of the form $\Gamma^{(k)} = \Gamma_{\varphi_k}$ (see Definition 1.1) with $\varphi_{T_k} < \varphi_k < \theta_k$. Similarly, if T_k is bisectorial, we fix a bisector Σ_{δ_k} with $\delta_k > \varphi_{T_k}^{(bi)}$ and a curve $\Gamma^{(k)} \subseteq \Sigma_k$ of the form $\Gamma^{(k)} = \Gamma_{\varphi_k} \cup (-\Gamma_{\varphi_k})$ with $\varphi_{T_k}^{(bi)} < \varphi_k < \delta_k$. We set $\Omega := \prod_{k=1}^N \Omega_k \subseteq \mathbb{C}^N$ and $\Gamma := \prod_{k=1}^N \Gamma^{(k)}$. Further, let $\mathcal{B}_\mathbf{T}$ be the commutator of $\{(\lambda - T_k)^{-1} \colon \lambda \in \rho(T_k), k \in \{1, \ldots, N\}\} \subseteq L(X)$, that is, the closed subalgebra consisting of all bounded operators that commute with all resolvents $(\lambda - T_k)^{-1}$, $\lambda \in \rho(T_k)$, $k \in \{1, \ldots, N\}$.

Definition 1.17. We define the following spaces of holomorphic functions:

(i) $H(\Omega, Y)$, the vector space of all Y-valued holomorphic functions on Ω,

(ii) $H^\infty(\Omega, Y)$, the Banach space of all Y-valued bounded holomorphic functions on Ω,

(iii) $H_\mathcal{R}^\infty(\Omega, L(X)) := \{f \in H^\infty(\Omega, L(X)) \colon f(\Omega) \subseteq L(X) \text{ is } \mathcal{R}\text{-bounded}\}$,

(iv) $H_0^\infty(\Omega, Y) := \Big\{ f \in H(\Omega, Y) \colon \exists C, s > 0 \, \forall z \in \Omega \colon$
$$\|f(z)\|_Y \leq C \prod_{k=1}^N \big(\min\{|z_k|, |z_k|^{-1}\}\big)^s \Big\},$$

(v) $H_P(\Omega, Y) := \Big\{ f \in H(\Omega, Y) \colon \exists C, s > 0 \, \forall z \in \Omega \colon$
$$\|f(z)\|_Y \leq C \prod_{k=1}^N \big(\max\{|z_k|, |z_k|^{-1}\}\big)^s \Big\}.$$

Note that functions in $H_0^\infty(\Omega, Y)$ show an additional decay at zero and at infinity while functions in $H_P(\Omega, Y)$ are polynomially bounded at zero and infinity.

We will now define $f(\mathbf{T})$ for functions $f \in H_0^\infty(\Omega, Y)$. In many cases, the functions are scalar-valued, i.e., $Y = \mathbb{C}$ (in this case, we will omit Y in the notation). However, if one considers an iterative calculus, Y has to be more general, and therefore we formulate the following definitions and results for closed subalgebras of $\mathcal{B}_\mathbf{T}$. Of course, the scalar-valued setting fits into this generalization via the identification $z \mapsto z \operatorname{id}_Y$.

Definition 1.18 (Operator-valued H^∞-calculus). Let $\mathcal{A} \subseteq \mathcal{B}_\mathbf{T}$ be a closed subalgebra and $f \in H_0^\infty(\Omega, \mathcal{A})$. Then we set

$$f(\mathbf{T}) := \frac{1}{(2\pi i)^N} \int_\Gamma f(z) \prod_{k=1}^N (z_k - T_k)^{-1} dz \in \mathcal{A}.$$

1.1. The joint H^∞-calculus for tuples of operators

Note that the integral converges in the operator norm $\|\cdot\|_{L(X)}$, and $f(\mathbf{T})$ does not depend on Γ by Cauchy's integral theorem.

Lemma 1.19 ([DV05, Lemma 2.7]). *For $n \in \mathbb{N}$ the function*

$$\psi_{n,N} \colon (\mathbb{C} \setminus \{-n, -\tfrac{1}{n}\})^N \to \mathbb{C}, \quad z \mapsto \prod_{k=1}^{N} \frac{n^2 z_k}{(1 + n z_k)(n + z_k)}$$

has the following properties:

  (i) *We have $\psi_{n,N} \in H_0^\infty(\Omega)$.*
  (ii) *For all $f \in H_P(\Omega, Y)$ there exists $m \in \mathbb{N}_0$ such that $\psi_{n,N}^m f \in H_0^\infty(\Omega, Y)$ for all $n \in \mathbb{N}$.*
  (iii) *The operator $\psi_{n,N}(\mathbf{T})$ is injective.*

Definition 1.20 (H_P-calculus). Let $f \in H_P(\Omega, \mathcal{B}_\mathbf{T})$ and $m \in \mathbb{N}_0$ such that $\psi^m f \in H_0^\infty(\Omega, \mathcal{B}_\mathbf{T})$ where $\psi := \psi_{1,N}$. Then we set

$$f(\mathbf{T})x := \psi(\mathbf{T})^{-m}(\psi^m f)(\mathbf{T})x, \quad x \in D(f(\mathbf{T}))$$

with domain $D(f(\mathbf{T})) := \{x \in X \colon (\psi^m f)(\mathbf{T})x \in R(\psi(\mathbf{T})^m)\}$.

Remark 1.21. (i) Every bisectorial operator A is also sectorial. Therefore, we can define $f(A)$ by using the two different curves Γ_s and Γ_bi but both interpretations are equal, i.e., $f(A_\mathrm{s}) = f(A_\mathrm{bi})$ for all $f \in H^\infty(S_\theta, \mathcal{B}_\mathbf{T})$. This can be easily proved by Cauchy's integral theorem.

  (ii) The definition of the domain in Definition 1.20 is not constructive and it is not easy to determine $D(f(\mathbf{T}))$ in a concrete situation.

  (iii) Let $f \in H^\infty(\Omega, \mathcal{B}_\mathbf{T})$ with $f(\mathbf{T}) \in L(X)$. Then we have $\psi_{n,N} f \in H_0^\infty(\Omega, \mathcal{B}_\mathbf{T})$ for all $n \in \mathbb{N}$. For all $x \in X$ we can represent $f(\mathbf{T})x$ by

$$f(\mathbf{T})x = \lim_{n \to \infty} (\psi_{n,N} f)(\mathbf{T})x.$$

  This is the so-called convergence lemma, cf. [DV05, Theorem 4.7], [Haa06, Proposition 5.1.4].

  (iv) Setting $f(\lambda) = \mathrm{id}_X$, $\lambda \in \Omega$, in (iii), we see that for all $x \in X$ we have $\psi_{n,N}(\mathbf{T})x \to x$, $n \to \infty$, in X.

Definition 1.22. Let $\mathcal{A}$ be a closed subalgebra of $\mathcal{B}_\mathbf{T}$. We say that the operator tuple $\mathbf{T} = (T_1, \ldots, T_N)$ admits a *bounded joint $H^\infty(\Omega, \mathcal{A})$-calculus* if there exists $C > 0$ such that $\|f(\mathbf{T})\|_{L(X)} \le C \|f\|_\infty$ for all $f \in H_0^\infty(\Omega, \mathcal{A})$. The tuple $\mathbf{T}$ admits an *$\mathcal{R}$-bounded joint $H^\infty(\Omega, \mathcal{A})$-calculus* if

$$\{f(\mathbf{T}) \colon f \in H_0^\infty(\Omega, \mathcal{A}), \|f\|_\infty \le 1\} \subseteq L(X)$$

is $\mathcal{R}$-bounded. In the case $\mathcal{A} = \mathbb{C}$, we speak of an *($\mathcal{R}$-)bounded joint $H^\infty(\Omega)$-calculus*. In the case of one operator, i.e., for $N = 1$, we speak of an *($\mathcal{R}$-)bounded $H^\infty(\Omega, \mathcal{A})$-calculus*.

Lemma 1.23. *Let $\mathcal{A}$ be a closed subalgebra of $\mathcal{B}_\mathbf{T}$. If $\mathbf{T}$ admits an $\mathcal{R}$-bounded joint $H^\infty(\Omega, \mathcal{A})$-calculus, then*

$$\{f(\mathbf{T}) \colon f \in H^\infty(\Omega, \mathcal{A}), \|f\|_\infty \leq C\} \subseteq L(X)$$

is also $\mathcal{R}$-bounded for all $C > 0$ and we have

$$\mathcal{R}_p(\{f(\mathbf{T}) \colon f \in H^\infty(\Omega, \mathcal{A}), \|f\|_\infty \leq C\})$$
$$\leq C \cdot \mathcal{R}_p(\{g(\mathbf{T}) \colon g \in H_0^\infty(\Omega, \mathcal{A}), \|g\|_\infty \leq 1\}).$$

Proof. We trivially have

$$M := \{(1/C \cdot f)(\mathbf{T}) \colon f \in H_0^\infty(\Omega, \mathcal{A}), \|f\|_\infty \leq C\}$$
$$\subseteq \{g(\mathbf{T}) \colon g \in H_0^\infty(\Omega, \mathcal{A}), \|g\|_\infty \leq 1\}.$$

Hence M is also $\mathcal{R}$-bounded with $\mathcal{R}_p(M) \leq \mathcal{R}_p(\{g(\mathbf{T}) \colon g \in H_0^\infty(\Omega, \mathcal{A}), \|g\|_\infty \leq 1\})$. Remark 1.7 (iii) and (v) and Remark 1.21 (iii) then yield the claim. $\square$

Definition 1.24. Let T be a sectorial operator. Then the H^∞-angle φ_T^∞ is defined as the infimum of the set of all φ such that T admits a bounded $H^\infty(S_\varphi)$-calculus (if this set is non-empty). In the same way, the $\mathcal{R}$-H^∞-angle $\varphi_T^{\mathcal{R},\infty}$ is defined as the infimum over all φ such that T admits an $\mathcal{R}$-bounded $H^\infty(S_\varphi)$-calculus.

Analogously, the angles $\varphi_T^{\infty,(\mathrm{bi})}$ and $\varphi_T^{\mathcal{R},\infty,(\mathrm{bi})}$ are defined for a bisectorial operator T.

Remark 1.25. (i) In general we cannot define an analog to the spectral angle or H^∞-angle in the case of more than one operator (i.e., $N > 1$), because the angles of the sectors or bisectors can be correlated.

(ii) We have $\varphi_T \leq \varphi_T^\infty \leq \varphi_T^{\mathcal{R},\infty}$ (respectively, $\varphi_T^{(\mathrm{bi})} \leq \varphi_T^{\infty,(\mathrm{bi})} \leq \varphi_T^{\mathcal{R},\infty,(\mathrm{bi})}$).

In the following, the notation $A \subseteq B$ for two operators stands for $D(A) \subseteq D(B)$ and $B|_{D(A)} = A$.

Theorem 1.26 ([DV05, Theorem 4.5]). *Let $f, g \in H_P(\Omega, \mathcal{B}_\mathbf{T})$. Then the following assertions hold:*

(i) $f(\mathbf{T})$ *is a closed operator with dense domain.*

(ii) *We have*

$$f(\mathbf{T})g(\mathbf{T}) \subseteq (fg)(\mathbf{T}),$$
$$f(\mathbf{T}) + g(\mathbf{T}) \subseteq (f + g)(\mathbf{T})$$

where the domains are given by

$$D(f(\mathbf{T})g(\mathbf{T})) := \{x \in D(g(\mathbf{T})) \colon g(\mathbf{T})x \in D(f(\mathbf{T}))\},$$
$$D(f(\mathbf{T}) + g(\mathbf{T})) := D(f(\mathbf{T})) \cap D(g(\mathbf{T})).$$

1.1. The joint H^∞-calculus for tuples of operators

(iii) If $g(\mathbf{T}) \in L(X)$, then $(fg)(\mathbf{T}) = f(\mathbf{T})g(\mathbf{T})$ and $(f+g)(\mathbf{T}) = f(\mathbf{T}) + g(\mathbf{T})$.

Theorem 1.27 ([DV05, Theorem 4.9]). *Let $\mathcal{A}$ be a closed subalgebra of $\mathcal{B}_\mathbf{T}$. Then the following statements are equivalent:*

(i) $f(\mathbf{T}) \in L(X)$ *for all* $f \in H^\infty(\Omega, \mathcal{A})$,

(ii) $\mathbf{T}$ *has a bounded joint $H^\infty(\Omega, \mathcal{A})$-calculus.*

The following theorem is a generalization of a result of N.J. Kalton and L. Weis [KW01, Theorem 4.4] to tuples of operators due to G. Dore and A. Venni [DV05]. Kalton-Weis type results are one of the main ingredients for the functional calculus and state that if an operator tuple admits a bounded scalar H^∞-calculus, then we have also a functional calculus for vector-valued functions provided that the range of the functions are not only bounded in operator norm but $\mathcal{R}$-bounded.

Theorem 1.28 ([DV05, Theorem 6.7]). *Let $\widetilde{\Omega}$ be a set of the same type as Ω but with smaller angles. If $\mathbf{T}$ admits a bounded scalar $H^\infty(\widetilde{\Omega})$-calculus, then we have $g(\mathbf{T}) \in L(X)$ for all $g \in H^\infty_\mathcal{R}(\Omega, \mathcal{B}_\mathbf{T})$. In particular, there exists $C = C(\mathbf{T}) > 0$ with*

$$\|g(\mathbf{T})\|_{L(X)} \leq C \cdot \mathcal{R}_2(g(\Omega)).$$

One motivation for the development of a joint H^∞-calculus is given by the possibility of an iteration. The following result shows some compatibility properties.

Lemma 1.29 (Iterative calculus). (i) *For $f \in H^\infty_0(\Omega)$ define $g(z') := f(T_1, z')$ for $z' \in \Omega' := \prod_{k=2}^N \Omega_k$. Then $g \in H^\infty_0(\Omega', \mathcal{B}_\mathbf{T})$ and $f(\mathbf{T}) = g(\mathbf{T}')$ with $\mathbf{T}' := (T_2, \ldots, T_N)$.*

(ii) *Let $J \in \{1, \ldots, N-1\}$, $\Omega' := \prod_{k=1}^J \Omega_k$, and $f \in H_P(\Omega', \mathcal{B}_\mathbf{T})$. For $g: \Omega \to \mathcal{B}_\mathbf{T}$, $z \mapsto f(z_1, \ldots, z_J)$ we have $g \in H_P(\Omega, \mathcal{B}_\mathbf{T})$ and $g(\mathbf{T}) = f(\mathbf{T}')$ with $\mathbf{T}' := (T_1, \ldots, T_J)$.*

Proof. (i) We obviously have $f(\cdot, z') \in H^\infty_0(\Omega_1)$ for all $z' \in \Omega'$ and $g(z') := f(T_1, z') \in \mathcal{B}_\mathbf{T}$ due to Definition 1.18. Then Morera's theorem and Fubini's theorem yield the holomorphy of the function g and even

$$g \in H^\infty_0(\Omega', \mathcal{B}_\mathbf{T}).$$

Hence $g(\mathbf{T}')$ is meaningful. Due to the boundedness of the resolvents we obtain

$$g(\mathbf{T}') = \frac{1}{(2\pi i)^{N-1}} \int_{\Gamma'} g(z') \prod_{k=2}^N (z_k - T_k)^{-1} dz'$$

$$= \frac{1}{(2\pi i)^N} \int_{\Gamma'} \left(\int_{\Gamma_1} f(z)(z_1 - T_1)^{-1} dz_1 \right) \prod_{k=2}^N (z_k - T_k)^{-1} dz'$$

$$= \frac{1}{(2\pi i)^N} \int_{\Gamma'} \int_{\Gamma_1} f(z)(z_1 - T_1)^{-1} \prod_{k=2}^{N} (z_k - T_k)^{-1} dz_1 dz' = f(\mathbf{T})$$

where all integrals converge with respect to $\|\cdot\|_{L(X)}$.

(ii) The easy proof for $J=1$ can be found in [DV05, Theorem 4.12]. It is trivial to extend this to get the result of (ii). $\square$

One disadvantage of the H_P-calculus is the fact that in general no explicit description of the domain of $f(\mathbf{T})$ is available. The following two results show that at least domains can be compared if estimates between the functions are known.

Lemma 1.30. *Let $\mathbf{T}$ admit a bounded $H^\infty(\Omega)$-calculus, and let $f_1, f_2 \in H_P(\Omega)$.*

(i) *If there exists $C > 0$ with $|f_1(z)| \leq C|f_2(z)| \neq 0$ for all $z \in \Omega$, then we have $D(f_2(\mathbf{T})) \subseteq D(f_1(\mathbf{T}))$, $R(f_1(\mathbf{T})) \subseteq R(f_2(\mathbf{T}))$, and*

$$\|f_1(\mathbf{T})x\|_X \leq C'\|f_2(\mathbf{T})x\|_X, \quad x \in D(f_2(\mathbf{T}))$$

for some $C' > 0$.

(ii) *If there exist $C_1, C_2 > 0$ such that $C_1|f_2(z)| \leq |f_1(z)| \leq C_2|f_2(z)|$ and $f_1(z), f_2(z) \neq 0$ for all $z \in \Omega$, then we have $D(f_2(\mathbf{T})) = D(f_1(\mathbf{T}))$, $R(f_1(\mathbf{T})) = R(f_2(\mathbf{T}))$ and*

$$C'_1 \|f_2(\mathbf{T})x\|_X \leq \|f_1(\mathbf{T})x\|_X \leq C'_2 \|f_2(\mathbf{T})x\|_X, \quad x \in D(f_2(\mathbf{T}))$$

for some $C'_1, C'_2 > 0$.

Proof. (i) We have $f_1/f_2 \in H^\infty(\Omega)$ by assumption. Hence we get

$$f_1(\mathbf{T}) = \left(\frac{f_1}{f_2} \cdot f_2\right)(\mathbf{T}) \supseteq \left(\frac{f_1}{f_2}\right)(\mathbf{T}) f_2(\mathbf{T})$$

(cf. Theorem 1.26 (ii)) and $(f_1/f_2)(\mathbf{T}) \in L(X)$. So we obtain the claimed relation between the domains as well as the claimed estimate. On the other hand we have

$$f_1(\mathbf{T}) = \left(f_2 \cdot \frac{f_1}{f_2}\right)(\mathbf{T}) = f_2(\mathbf{T})\left(\frac{f_1}{f_2}\right)(\mathbf{T})$$

(cf. Theorem 1.26 (iii)), which yields the claimed relation between the ranges.

(ii) This follows immediately from (i). $\square$

Lemma 1.31. *Let $g, h \in H_P(\Omega)$ with $g(z), h(z) \neq 0$ for all $z \in \Omega$ and $g^{-1}, h^{-1} \in H^\infty(\Omega)$. If we define $f(z) := g(z) \cdot h(z)$, $z \in \Omega$, we can conclude*

$$f(\mathbf{T}) = g(\mathbf{T})h(\mathbf{T}) = h(\mathbf{T})g(\mathbf{T}),$$

which in particular yields $D(f(\mathbf{T})) = D(g(\mathbf{T})h(\mathbf{T})) = D(h(\mathbf{T})g(\mathbf{T}))$.

1.1. The joint H^∞-calculus for tuples of operators

Proof. It suffices to show $f(\mathbf{T}) = g(\mathbf{T})h(\mathbf{T})$. From Theorem 1.26 we already know

$$f(\mathbf{T}) \supseteq g(\mathbf{T})h(\mathbf{T}), \quad f^{-1}(\mathbf{T}) = g^{-1}(\mathbf{T})h^{-1}(\mathbf{T}) = h^{-1}(\mathbf{T})g^{-1}(\mathbf{T}),$$

which yields $D(g(\mathbf{T})h(\mathbf{T})) \subseteq D(f(\mathbf{T})) = R(f^{-1}(\mathbf{T})) = R(g^{-1}(\mathbf{T})h^{-1}(\mathbf{T})) = R(h^{-1}(\mathbf{T})g^{-1}(\mathbf{T}))$. One can easily show $R(h^{-1}(\mathbf{T})g^{-1}(\mathbf{T})) \subseteq D(g(\mathbf{T})h(\mathbf{T}))$, which finishes the proof. □

At the end of this subsection on the joint H^∞-calculus, we formulate some results on the behaviour with respect to isomorphisms and with respect to shifted operators which will be useful in the applications.

Proposition 1.32 (H^∞**-calculus and isomorphisms**). *Let X, Y be complex Banach spaces such that there exists an isomorphism $\Phi \in L_{\mathrm{Isom}}(Y, X)$. For each $k = 1, \ldots, N$ let*

$$T_k \colon X \supseteq D(T_k) \to X$$

be a linear operator. Define

$$S_k \colon Y \supseteq D(S_k) \to Y, \quad y \mapsto \Phi^{-1} T_k \Phi y$$

where $D(S_k) := \Phi^{-1}(D(T_k))$. Defining $\mathbf{T} := (T_1, \ldots, T_N)$ and $\mathbf{S} := (S_1, \ldots, S_N)$, we get the following assertions:

(i) *If T_k is sectorial, then the operator S_k is also sectorial with $\varphi_{S_k} = \varphi_{T_k}$. In particular, we have $\rho(S_k) = \rho(T_k)$ and $(\lambda - S_k)^{-1} = \Phi^{-1}(\lambda - T_k)^{-1}\Phi$ for all $\lambda \in \rho(T_k)$. The analog results hold in the case of T_k being bisectorial. If $\mathbf{T}$ is admissible, then $\mathbf{S}$ is also admissible.*

(ii) *If $\mathbf{T}$ has a bounded joint $H^\infty(\Omega)$-calculus, then $\mathbf{S}$ also has a bounded joint $H^\infty(\Omega)$-calculus and the representation*

$$f(\mathbf{S}) = \Phi^{-1} f(\mathbf{T}) \Phi$$

holds for all $f \in H_P(\Omega)$.

(iii) *If $\mathbf{T}$ has an $\mathcal{R}$-bounded joint $H^\infty(\Omega)$-calculus, then $\mathbf{S}$ also has an $\mathcal{R}$-bounded joint $H^\infty(\Omega)$-calculus.*

Proof. (i) Without loss of generality let T_k be a sectorial operator. Then we obviously have $\rho(T_k) = \rho(S_k)$ and $(\lambda - S_k)^{-1} = \Phi^{-1}(\lambda - T_k)^{-1}\Phi$ for all $\lambda \in \rho(T_k)$. From this we easily obtain for $\theta \in (0, \pi)$ with $\rho(T_k) = \rho(S_k) \supseteq \mathbb{C} \setminus \overline{S}_\theta$

$$\sup_{\lambda \in \overline{S}_{\theta'}} \|\lambda(\lambda - S_k)^{-1}\|_{L(Y)} \leq C \sup_{\lambda \in \overline{S}_{\theta'}} \|\lambda(\lambda - T_k)^{-1}\|_{L(X)} < \infty$$

for all $\theta' \in (\theta, \pi)$. The density of $D(S_k)$ and $R(S_k)$ is obvious. Therefore, S_k is also sectorial with $\varphi_{T_k} = \varphi_{S_k}$. The claimed admissibility of $\mathbf{S}$ is obvious, too.

(ii) For $f \in H_0^\infty(\Omega)$ we directly derive from (i)

$$f(\mathbf{S})y = \frac{1}{(2\pi i)^N} \int_\Gamma f(z) \prod_{k=1}^N (z_k - S_k)^{-1} y \, dz$$

$$= \frac{1}{(2\pi i)^N} \int_\Gamma f(z) \prod_{k=1}^N \Phi^{-1}(\lambda - T_k)^{-1} \Phi y \, dz$$

$$= \frac{1}{(2\pi i)^N} \Phi^{-1} \left[\int_\Gamma f(z) \prod_{k=1}^N (\lambda - T_k)^{-1} dz \right] \Phi y = \Phi^{-1} f(\mathbf{T}) \Phi y$$

for all $y \in Y$. Hence we have $\|f(\mathbf{S})\|_{L(Y)} \leq C \|f(\mathbf{T})\|_{L(X)} \leq C \|f\|_\infty$ for all $f \in H_0^\infty(\Omega)$. This proves that $\mathbf{S}$ admits a bounded joint $H^\infty(\Omega)$-calculus.

Let $f \in H_P(\Omega)$ and $m \in \mathbb{N}_0$ such that $\psi^m f \in H_0^\infty(\Omega)$. Then we have $\psi(\mathbf{S}) = \Phi^{-1} \psi(\mathbf{T}) \Phi$ and $(\psi^m f)(\mathbf{S}) = \Phi^{-1}(\psi^m f)(\mathbf{T}) \Phi$ due to $\psi \in H_0^\infty(\Omega)$. With this we obtain

$$f(\mathbf{S}) = \psi(\mathbf{S})^{-m}(\psi^m f)(\mathbf{S}) = \Phi^{-1} \psi(\mathbf{T})^{-m} \Phi \Phi^{-1}(\psi^m f)(\mathbf{T}) \Phi$$
$$= \Phi^{-1} f(\mathbf{T}) \Phi.$$

(iii) This easily follows from the representation in (ii) and Remark 1.7 (iii). $\square$

Lemma 1.33 (H^∞-calculus of a shifted operator). *Let $T \colon X \supseteq D(T) \to X$ be a sectorial operator. Then for all $\sigma \geq 0$ the operator $S := \sigma + T$ is also sectorial with $\varphi_S \leq \varphi_T$. In this situation we also have*

$$f(S) = [f(\sigma + \cdot)](T) \tag{1.8}$$

for all $f \in H_P(S_\theta)$ with $\theta > \varphi_T$. If T admits a bounded H^∞-calculus, then S also admits a bounded H^∞-calculus with $\varphi_S^\infty \leq \varphi_T^\infty$. If T even admits an $\mathcal{R}$-bounded H^∞-calculus, then S also admits an $\mathcal{R}$-bounded H^∞-calculus with $\varphi_S^{\mathcal{R},\infty} \leq \varphi_T^{\mathcal{R},\infty}$.

Proof. The sectoriality of S and $\varphi_S \leq \varphi_T$ are obvious. Let $\theta > \varphi_T$ and $g(z) := \sigma + z$ for $z \in S_\theta$. Then we have $g(T) = \sigma + T$ by Theorem 1.26 (iii). Now we directly obtain (1.8) by [KW04, Proposition 15.11], where the authors prove the transformation formula $(f \circ g)(T) = f(g(T))$ for a more general situation. Trivially S also admits a bounded H^∞-calculus since

$$\|f(S)\|_{L(X)} \leq C \|f(\sigma + \cdot)\|_\infty \leq C \|f\|_\infty$$

for all $f \in H_0^\infty(S_\theta)$. The claimed $\mathcal{R}$-boundedness is obvious by $\|f(\sigma + \cdot)\|_\infty \leq \|f\|_\infty$ and Remark 1.7 (vi). $\square$

1.1. The joint H^∞-calculus for tuples of operators

Lemma 1.34 (H^∞-**calculus for a shifted operator tuple**). *Let* $\mathbf{T}$ *be a tuple of operators with a bounded* $H^\infty(\Omega)$-*calculus and* $\mathbf{S} := (\sigma + T_1, T_2, \ldots, T_N)$ *for* $\sigma > 0$. *If* T_1 *is a sectorial operator, then* $\mathbf{S}$ *also admits a bounded* $H^\infty(\Omega)$-*calculus, and*

$$f(\mathbf{S}) = f_\sigma(\mathbf{T}), \quad f_\sigma(z) := f(\sigma + z_1, z_2, \ldots, z_N)$$

for all $f \in H_P(\Omega)$.

Proof. (I) First, we prove the assertion for $f \in H_0^\infty(\Omega)$. For all $z' \in \Omega' := \prod_{k=2}^N \Omega_k$ we define $g(z') := f(\sigma + T_1, z')$. Then Lemma 1.33 and Lemma 1.29 yield $g(z') = f_\sigma(T_1, z')$ and

$$f(\mathbf{S}) = g(\mathbf{T}') = f_\sigma(\mathbf{T})$$

where $\mathbf{T}' := (T_2, \ldots, T_N)$. Obviously, $\|f(\mathbf{S})\|_{L(X)} \leq C\|f_\sigma\|_\infty \leq C\|f\|_\infty$ for all $f \in H_0^\infty(\Omega)$ and thus $\mathbf{S}$ also admits a bounded $H^\infty(\Omega)$-calculus.

(II) Let $f \in H_P(\Omega)$. Then we get

$$f(\mathbf{S}) = \psi(\mathbf{S})^{-m}(\psi^m f)(\mathbf{S}) = \psi_\sigma(\mathbf{T})^{-m}(\psi_\sigma^m f_\sigma)(\mathbf{T})$$
$$\supseteq \psi_\sigma(\mathbf{T})^{-m}\psi_\sigma(\mathbf{T})^m f_\sigma(\mathbf{T}) = f_\sigma(\mathbf{T})$$

due to Theorem 1.26 and the result of part (I). The converse inclusion can be obtained in the same way by

$$f_\sigma(\mathbf{T}) = \psi(\mathbf{T})^{-m}(\psi^m f_\sigma)(\mathbf{T}) = \psi(\mathbf{T})^{-m}\left(\left(\frac{\psi^m}{\psi_\sigma^m}\right) \cdot (\psi_\sigma^m f_\sigma)\right)(\mathbf{T})$$
$$\supseteq \psi(\mathbf{T})^{-m}\left(\frac{\psi^m}{\psi_\sigma^m}\right)(\mathbf{T})(\psi_\sigma^m f_\sigma)(\mathbf{T})$$
$$= \psi(\mathbf{T})^{-m}\left(\frac{\psi^m}{\psi_\sigma^m}\right)(\mathbf{T})(\psi^m f)(\mathbf{S})$$
$$\supseteq \psi(\mathbf{T})^{-m}\psi(\mathbf{T})^m \psi_\sigma^{-m}(\mathbf{T})(\psi^m f)(\mathbf{S})$$
$$= \psi(\mathbf{S})^{-m}(\psi^m f)(\mathbf{S}) = f(\mathbf{S}). \qquad \square$$

In the following, the notation $Y \hookrightarrow X$ means that Y is continuously embedded into X, i.e., there exists a continuous and injective linear mapping from Y to X.

Proposition 1.35 (**Compatibility of the** H^∞-**calculus on spaces of higher regularity**). *Let* X, Y *be complex Banach spaces with* $Y \hookrightarrow X$. *Let*

$$T_k \colon X \supseteq D(T_k) \to X,$$
$$S_k \colon Y \supseteq D(S_k) \to Y, \quad k = 1, \ldots, N$$

be sectorial or bisectorial operators (of the same type) with $D(S_k) \subseteq D(T_k)$ *(with respect to* $Y \hookrightarrow X$) *such that the tuples* $\mathbf{T} := (T_1, \ldots, T_N)$ *and* $\mathbf{S} := (S_1, \ldots, S_N)$

are admissible and $T_k y = S_k y$ for all $y \in D(S_k)$ and $k \in \{1, \ldots, N\}$. Then we obtain
$$f(\mathbf{S}) \subseteq f(\mathbf{T})$$
for all $f \in H_P(\Omega)$ where Ω is admissible for $\mathbf{T}$ and $\mathbf{S}$.

Proof. (I) Let $f \in H_0^\infty(\Omega)$ and $y \in Y$. Then we have

$$\begin{aligned}
f(\mathbf{S})y &= \frac{1}{(2\pi i)^N} \left[\int_\Gamma f(z) \prod_{k=1}^N (z_k - S_k)^{-1} dz \right] y \\
&= \frac{1}{(2\pi i)^N} \int_\Gamma f(z) \prod_{k=1}^N (z_k - S_k)^{-1} y \, dz \\
&= \frac{1}{(2\pi i)^N} \int_\Gamma f(z) \prod_{k=1}^N (z_k - T_k)^{-1} y \, dz \\
&= \frac{1}{(2\pi i)^N} \left[\int_\Gamma f(z) \prod_{k=1}^N (z_k - T_k)^{-1} dz \right] y \\
&= f(\mathbf{T})y
\end{aligned}$$

because of $(z_k - T_k)^{-1}|_Y = (z_k - S_k)^{-1}$ for all $z_k \in \rho(T_k) \cap \rho(S_k)$ and $Y \hookrightarrow X$.

(II) Let $f \in H_P(\Omega)$ and $m \in \mathbb{N}_0$ with $(\psi^m f) \in H_0^\infty(\Omega)$. Using (I) we obtain

$$\begin{aligned}
D(f(\mathbf{S})) &= \{y \in Y : (\psi^m f)(\mathbf{S})y \in R(\psi(\mathbf{S})^m)\} \\
&= \{y \in Y : (\psi^m f)(\mathbf{T})y \in R(\psi(\mathbf{S})^m)\} \\
&\subseteq \{y \in Y : (\psi^m f)(\mathbf{T})y \in R(\psi(\mathbf{T})^m)\} \\
&= Y \cap D(f(\mathbf{T})).
\end{aligned}$$

This yields
$$D(f(\mathbf{S})) \subseteq D(f(\mathbf{T})).$$

It is easy to prove that $\psi(\mathbf{T})^{-m} v = \psi(\mathbf{S})^{-m} v$ for all $v \in R(\psi(\mathbf{S})^m)$. For $y \in D(f(\mathbf{S}))$ we then get

$$f(\mathbf{T})y = \psi(\mathbf{T})^{-m}(\psi^m f)(\mathbf{T})y = \psi(\mathbf{T})^{-m} \underbrace{(\psi^m f)(\mathbf{S})y}_{\in R(\psi(\mathbf{S})^m)} = f(\mathbf{S})y$$

by (I). So we have proved the assertion. $\square$

1.2 Vector-valued Sobolev spaces

Motivation. In this section, we will provide the main spaces for the right-hand sides and the solutions of the general parabolic equations considered in later chapters. In the theory of nonlinear parabolic differential equations, the maximal regularity approach for the linearization turns out to be particularly useful in the setting of L_p-Sobolev spaces. On the other hand, following a standard approach, it is advantageous to consider functions in time and space as vector-valued functions in time (where the values are in an appropriate function class with respect to the space variables). Therefore, we have to consider vector-valued L_p-Sobolev spaces. Here non-integer orders of differentiability appear naturally in the context of boundary value problems and traces with respect to the boundary and with respect to time.

In contrast to the L_2-setting, for $p \ne 2$ there are different types of L_p-Sobolev spaces, the main classes being Besov spaces, Bessel potential spaces, and Triebel-Lizorkin spaces. In the present section we will consider Besov and Bessel potential spaces while Triebel-Lizorkin spaces will be discussed in Chapter 3.

Modern Sobolev space theory is based on two ingredients:

- interpolation theory for Banach spaces,
- and the concept of retractions and coretractions.

Interpolation theory allows us to define and analyze Sobolev spaces of non-integer order, while the concept of retractions is a key method for the step from $\mathbb{R}^n$ to domains. Therefore, this section starts with the discussion of interpolations spaces in Subsection a) and of retractions and coretractions in Subsection b) before we define the Sobolev spaces which will be relevant for our purpose in Subsection c).

a) Interpolation of Banach spaces

We start with the main definitions and results on interpolation spaces, where we refer the reader to the standard references [Tri78], [BL76] for details and proofs.

An interpolation couple $\{X_0, X_1\}$ consists of two complex Banach spaces X_0 and X_1, both linearly and continuously embedded into the same linear complex Hausdorff space. For an interpolation couple, the Banach spaces $X_0 + X_1$ (sum) and $X_0 \cap X_1$ (intersection) are defined in a natural way.

In the following, for interpolation couples $\{X_0, X_1\}$ and $\{Y_0, Y_1\}$ we will write

$$L(\{X_0, X_1\}, \{Y_0, Y_1\}) := \{T\colon X_0 + X_1 \to Y_0 + Y_1 \,|\, T|_{X_k} \in L(X_k, Y_k), k = 0, 1\}.$$

Roughly speaking, an interpolation functor $\mathcal{F}$ maps the interpolation couple $\{X_0, X_1\}$ to an intermediate Banach space $\mathcal{F}(\{X_0, X_1\})$ satisfying

$$X_0 \cap X_1 \hookrightarrow \mathcal{F}(\{X_0, X_1\}) \hookrightarrow X_0 + X_1.$$

A precise definition in the language of categories and (covariant) functors can be found in [Tri78, Subsection 1.2.2].

We remark that in many situations one has $X_1 \hookrightarrow X_0$. In this case, $X_1 \hookrightarrow \mathcal{F}(\{X_0, X_1\}) \hookrightarrow X_0$. A typical example is $X_0 = L_p(\Omega)$ and $X_1 = W_p^k(\Omega)$, the L_p-Sobolev space of order k. Here interpolation leads to Besov or Bessel potential spaces as will be discussed below.

Definition 1.36 (Exact interpolation functor). An interpolation functor $\mathcal{F}$ is called *exact of type* $\theta \in (0,1)$ if for all interpolation couples $\{X_0, X_1\}$, $\{Y_0, Y_1\}$ we have

$$\|T|_{\mathcal{F}(\{X_0, X_1\})}\|_{L(\mathcal{F}(\{X_0, X_1\}), \mathcal{F}(\{Y_0, Y_1\}))} \leq \|T|_{X_0}\|_{L(X_0, Y_0)}^{1-\theta} \|T|_{X_1}\|_{L(X_1, Y_1)}^{\theta}.$$

The most important interpolation functors are the real and the complex interpolation functors. There are several equivalent definitions for them, but we restrict ourselves to one variant. In the following, let $\{X_0, X_1\}$ be an interpolation couple.

Definition 1.37 (Real interpolation functor). For $\theta \in (0,1)$ and $p \in (1, \infty)$ we define

$$K(t, u) := \inf\{\|u_0\|_{X_0} + t\|u_1\|_{X_1} : u_0 + u_1 = u,\ u_i \in X_i\}, \quad t \in \mathbb{R}_+,\ u \in X_0 + X_1.$$

Then the *real interpolation space* $(X_0, X_1)_{\theta, p}$ is defined as the space of all $u \in X_0 + X_1$ satisfying

$$(t \mapsto t^{-\theta} K(t, u)) \in L_p(\mathbb{R}_+, \tfrac{dt}{t}).$$

The norm in $(X_0, X_1)_{\theta, p}$ is defined by

$$\|u\|_{\theta, p} := \left(\int_0^\infty \left(t^{-\theta} K(t, u)\right)^p \tfrac{dt}{t}\right)^{1/p}.$$

Definition 1.38 (Complex interpolation functor). Let $\theta \in (0,1)$, and set $S := (0,1) + i\mathbb{R} \subseteq \mathbb{C}$. Then $F(X_0, X_1)$ is defined as the set of all functions $f \colon \overline{S} \to X_0 + X_1$ satisfying

(i) f is $(X_0 + X_1)$-continuous in $\overline{S}$,

(ii) f is $(X_0 + X_1)$-holomorphic in S,

(iii) $f(k + it) \in X_k$ ($t \in \mathbb{R}$) and $(t \mapsto f(k + it)) \in C(\mathbb{R}, X_k)$ for $k = 0, 1$,

(iv) $\lim_{|t| \to \infty} \|f(k + it)\|_{X_k} = 0$ for $k = 0, 1$.

We endow $F(X_0, X_1)$ with the norm

$$\|f\|_{F(X_0, X_1)} := \max\left\{\sup_{t \in \mathbb{R}} \|f(it)\|_{X_0},\ \sup_{t \in \mathbb{R}} \|f(1 + it)\|_{X_1}\right\}.$$

1.2. Vector-valued Sobolev spaces

Then the *complex interpolation space* $[X_0, X_1]_\theta$ is defined as the set of all $u \in X_0 + X_1$ for which there exists $f \in F(X_0, X_1)$ with $f(\theta) = u$. The norm on $[X_0, X_1]_\theta$ is given by

$$\|u\|_{[X_0,X_1]_\theta} := \inf \{ \|f\|_{F(X_0,X_1)} : f \in F(X_0, X_1),\ f(\theta) = u \}.$$

Theorem 1.39 ([Tri78, Sections 1.3.3, 1.6.2, 1.9.3]). *Let $\theta \in (0,1)$ and $p \in (1, \infty)$. Then the real and the complex interpolation functors $(\cdot, \cdot)_{\theta,p}$ and $[\cdot, \cdot]_\theta$ are both exact of type θ. We have*

$$X_0 \cap X_1 \xhookrightarrow{d} (X_0, X_1)_{\theta,p} \hookrightarrow X_0 + X_1,$$
$$X_0 \cap X_1 \xhookrightarrow{d} [X_0, X_1]_\theta \hookrightarrow X_0 + X_1$$

where $\xhookrightarrow{d}$ stands for dense continuous embedding.

Theorem 1.40 (**Reiteration theorem,** cf. [Tri78, Theorem 1.10.2]). *Let $0 < \theta_0 < \theta_1 < 1$, $\lambda \in (0,1)$, and $p \in (1, \infty)$. Then for $\theta := (1-\lambda)\theta_0 + \lambda \theta_1$ we have*

$$\big((X_0, X_1)_{\theta_0,p}, (X_0, X_1)_{\theta_1,p}\big)_{\lambda,p} = (X_0, X_1)_{\theta,p},$$
$$\big([X_0, X_1]_{\theta_0}, [X_0, X_1]_{\theta_1}\big)_{\lambda,p} = (X_0, X_1)_{\theta,p}$$

with equivalent norms. The same holds if $[X_0, X_1]_{\theta_0}$ is replaced by X_0 (then $\theta_0 := 0$) and if $[X_0, X_1]_{\theta_1}$ is replaced by X_1 (then $\theta_1 := 1$).

Theorem 1.41 (**Interpolation of L_p-spaces,** cf. [Tri78, Theorem 1.18.4]). *Let $\Omega \subseteq \mathbb{R}^n$ be open and $1 < p_0, p_1 < \infty$. Then we have*

$$(L_{p_0}(\Omega, X_0), L_{p_1}(\Omega, X_1))_{\theta,p} = L_p(\Omega, (X_0, X_1)_{\theta,p}),$$
$$[L_{p_0}(\Omega, X_0), L_{p_1}(\Omega, X_1)]_\theta = L_p(\Omega, [X_0, X_1]_\theta)$$

where $1/p = (1-\theta)/p_0 + \theta/p_1$ and $\theta \in (0,1)$.

Theorem 1.42 (**Interpolation of ℓ_p^s-spaces,** cf. [BL76, Theorem 5.6.3]). *We define for a Banach space X, $p \in [1, \infty)$, and $s \in \mathbb{R}$ the weighted ℓ_p-space*

$$\ell_p^s(X) := \Big\{ (x_k)_{k \in \mathbb{N}_0} \subseteq X : \sum_{k=0}^\infty 2^{skp} \|x_k\|_X^p < \infty \Big\},$$
$$\|(x_k)_{k \in \mathbb{N}_0}\|_{\ell_p^s(X)} := \Big(\sum_{k=0}^\infty 2^{skp} \|x_k\|_X^p \Big)^{1/p}.$$

Let $\{X_0, X_1\}$ be an interpolation couple. Then we have

$$[\ell_{p_0}^{s_0}(X_0), \ell_{p_1}^{s_1}(X_1)]_\theta = \ell_p^s([X_0, X_1]_\theta)$$

for all $s_0, s_1 \in \mathbb{R}$, $p_0, p_1 \in (1, \infty)$, $\theta \in (0,1)$ with $1/p = (1-\theta)/p_0 + \theta/p_1$, and $s = (1-\theta)s_0 + \theta s_1$.

Real and complex interpolation is compatible with the geometric properties for Banach spaces as introduced above. More precisely, we have:

Lemma 1.43. *Let $\theta \in (0,1)$ and $p \in (1,\infty)$.*

(i) *If X_0 and X_1 are Banach spaces of class $\mathcal{HT}$, then $(X_0, X_1)_{\theta,p}$ and $[X_0, X_1]_\theta$ are also of class $\mathcal{HT}$.*

(ii) *If X_0 and X_1 are Banach spaces of class $\mathcal{HT}$ with property (α), then the spaces $(X_0, X_1)_{\theta,p}$ and $[X_0, X_1]_\theta$ are also Banach spaces of class $\mathcal{HT}$ with property (α).*

Proof. Part (i) can be found, e.g., in [Ama95, Theorem 4.5.2], part (ii) in [KS12, Theorem 4.5]. □

Theorem 1.44 ($\mathcal{R}$-**boundedness and interpolation,** cf. [KKW06, Prop. 3.7], [KS12, Cor. 3.19])**.** *Let $\{X_0, X_1\}$ and $\{Y_0, Y_1\}$ be interpolation couples of Banach spaces of class $\mathcal{HT}$. For a given family $\mathcal{T} \subseteq L(\{X_0, X_1\}, \{Y_0, Y_1\})$, $p \in (1, \infty)$, and $0 < \theta < 1$ we have:*
If $\mathcal{T}|_{X_k} \subseteq L(X_k, Y_k)$ is $\mathcal{R}$-bounded for $k = 0, 1$, then

$$\mathcal{T}|_{(X_0,X_1)_{\theta,p}} \subseteq L((X_0,X_1)_{\theta,p}, (Y_0,Y_1)_{\theta,p}),$$
$$\mathcal{T}|_{[X_0,X_1]_\theta} \subseteq L([X_0,X_1]_\theta, [Y_0,Y_1]_\theta)$$

are also $\mathcal{R}$-bounded and

$$\mathcal{R}_p\left(\mathcal{T}|_{(X_0,X_1)_{\theta,p}}\right) \leq C \cdot [\mathcal{R}_p(\mathcal{T}|_{X_0})]^{1-\theta} [\mathcal{R}_p(\mathcal{T}|_{X_1})]^\theta,$$
$$\mathcal{R}_p(\mathcal{T}|_{[X_0,X_1]_\theta}) \leq C' \cdot [\mathcal{R}_p(\mathcal{T}|_{X_0})]^{1-\theta} [\mathcal{R}_p(\mathcal{T}|_{X_1})]^\theta,$$

with constants $C, C' > 0$.

Finally, real and complex interpolation is also compatible with the joint H^∞-calculus:

Theorem 1.45 (H^∞-**calculus and interpolation**)**.** *Let $\mathcal{F} \in \{(\cdot,\cdot)_{\theta,p}, [\cdot,\cdot]_\theta\}$, $\theta \in (0,1)$, $p \in (1,\infty)$, and*

$$T_k \colon X_0 \supseteq D(T_k) \to X_0,$$
$$S_k \colon X_1 \supseteq D(S_k) \to X_1,$$

$k = 1, \ldots, N$, *be linear operators satisfying the compatibility conditions $T_k x = S_k x$ for all $x \in D(T_k) \cap D(S_k)$ and*

$$(\lambda - T_k)^{-1} x = (\lambda - S_k)^{-1} x, \quad \lambda \in \rho(T_k) \cap \rho(S_k), \quad x \in X_0 \cap X_1. \tag{1.9}$$

Moreover we define for $k = 1, \ldots, N$ the interpolated operators

$$A_k \colon \mathcal{F}(\{X_0, X_1\}) \supseteq D(A_k) \to \mathcal{F}(\{X_0, X_1\}), \quad D(A_k) := \mathcal{F}(\{D(T_k), D(S_k)\})$$

with $A_k x := T_k x_0 + S_k x_1$ for $x = x_0 + x_1 \in D(A_k) \hookrightarrow D(T_k) + D(S_k)$ with $x_0 \in D(T_k)$, $x_1 \in D(S_k)$. Note that we equip $D(T_k)$ and $D(S_k)$ with the graph norm. Then we have the following results:

1.2. Vector-valued Sobolev spaces

(i) *Let $k = 1, \ldots, N$. If T_k and S_k are both sectorial, then A_k is also sectorial with $\varphi_{A_k} \leq \max\{\varphi_{T_k}, \varphi_{S_k}\}$. The analog result holds for bisectorial operators. If $\mathbf{T} := (T_1, \ldots, T_N)$ and $\mathbf{S} := (S_1, \ldots, S_N)$ are both admissible, then $\mathbf{A} := (A_1, \ldots, A_N)$ is also admissible.*

(ii) *If in the situation of* (i) *both $\mathbf{T}$ and $\mathbf{S}$ admit a bounded joint $H^\infty(\Omega)$-calculus, then the interpolated tuple $\mathbf{A}$ also admits a bounded joint $H^\infty(\Omega)$-calculus, and we have the representation*

$$f(\mathbf{A})x = f(\mathbf{T})x_0 + f(\mathbf{S})x_1, \quad x \in \mathcal{F}(\{X_0, X_1\}), \quad x = x_0 + x_1, \quad x_i \in X_i$$

for all $f \in H^\infty(\Omega)$. For $N = 1$ we even have $\varphi_{A_1}^\infty \leq \max\{\varphi_{T_1}^\infty, \varphi_{S_1}^\infty\}$ (respectively, $\varphi_{A_1}^{\infty,(\mathrm{bi})} \leq \max\{\varphi_{T_1}^{\infty,(\mathrm{bi})}, \varphi_{S_1}^{\infty,(\mathrm{bi})}\}$).

(iii) *Let $\mathbf{T}$ and $\mathbf{S}$ be the same operators as in* (ii) *but with an $\mathcal{R}$-bounded joint $H^\infty(\Omega)$-calculus. If X_0, X_1 are of class $\mathcal{HT}$, then $\mathbf{A}$ also has an $\mathcal{R}$-bounded joint $H^\infty(\Omega)$-calculus. In particular for $N = 1$ we obtain $\varphi_{A_1}^{\mathcal{R},\infty} \leq \max\{\varphi_{T_1}^{\mathcal{R},\infty}, \varphi_{S_1}^{\mathcal{R},\infty}\}$ (respectively, $\varphi_{A_1}^{\mathcal{R},\infty,(\mathrm{bi})} \leq \max\{\varphi_{T_1}^{\mathcal{R},\infty,(\mathrm{bi})}, \varphi_{S_1}^{\mathcal{R},\infty,(\mathrm{bi})}\}$).*

Proof. (i) We have to consider the relation between the resolvents of T_k, S_k and A_k. Without loss of generality we assume that T_k and S_k are sectorial. Let $\lambda \in -S_{\pi-\theta'} \subseteq \rho(T_k) \cap \rho(S_k)$ with $\theta' > \max\{\varphi_{T_k}, \varphi_{S_k}\}$. Then we define the bounded operator

$$R_\lambda^{(k)} \colon X_0 + X_1 \to D(T_k) + D(S_k),$$
$$x_0 + x_1 \mapsto (\lambda - T_k)^{-1}x_0 + (\lambda - S_k)^{-1}x_1.$$

Note that $R_\lambda^{(k)}$ is well defined by (1.9). Hence,

$$(R_\lambda^{(k)})|_{X_0} = (\lambda - T_k)^{-1},$$
$$(R_\lambda^{(k)})|_{X_1} = (\lambda - S_k)^{-1},$$
$$(R_\lambda^{(k)})|_{\mathcal{F}(\{X_0, X_1\})} \in L(\mathcal{F}(\{X_0, X_1\}), D(A_k)).$$

With this we can easily show $\lambda \in \rho(A_k)$ and $(\lambda - A_k)^{-1} = (R_\lambda^{(k)})|_{\mathcal{F}(\{X_0, X_1\})}$, which already yields $-S_{\pi-\theta'} \subseteq \rho(A_k)$. Thus, we derive

$$\|(\lambda - A_k)^{-1}\|_{L(\mathcal{F}(\{X_0, X_1\}))} \leq \|(\lambda - T_k)^{-1}\|_{L(X_0)}^{1-\theta} \|(\lambda - S_k)^{-1}\|_{L(X_1)}^\theta$$

for all $\lambda \in -S_{\pi-\theta'}$ due to Theorem 1.39. We still have to show the density of $R(A_k)$ and $D(A_k)$ in $\mathcal{F}(\{X_0, X_1\})$. Here we use the characterizations

$$\overline{R(A_k)} = \{x \in \mathcal{F}(\{X_0, X_1\}) \colon \lim_{n \to \infty} -\tfrac{1}{n}(-\tfrac{1}{n} - A_k)^{-1}x = 0 \text{ in } \mathcal{F}(\{X_0, X_1\})\},$$
$$\overline{D(A_k)} = \{x \in \mathcal{F}(\{X_0, X_1\}) \colon \lim_{n \to \infty} -n(-n - A_k)^{-1}x = x \text{ in } \mathcal{F}(\{X_0, X_1\})\}$$

given in Lemma 1.5. Let $x \in X_0 \cap X_1$ be arbitrary. Then we get

$$-\tfrac{1}{n}(-\tfrac{1}{n} - A_k)^{-1}x = -\tfrac{1}{n}(-\tfrac{1}{n} - T_k)^{-1}x = -\tfrac{1}{n}(-\tfrac{1}{n} - S_k)^{-1}x,$$
$$-n(-n - A_k)^{-1}x = -n(-n - T_k)^{-1}x = -n(-n - S_k)^{-1}x.$$

Due to $\overline{R(T_k)} = \overline{D(T_k)} = X_0$, $\overline{R(S_k)} = \overline{D(S_k)} = X_1$, and Lemma 1.5 we have

$$\lim_{n\to\infty} -\tfrac{1}{n}(-\tfrac{1}{n} - T_k)^{-1}x = 0 \text{ in } X_0, \quad \lim_{n\to\infty} -\tfrac{1}{n}(-\tfrac{1}{n} - S_k)^{-1}x = 0 \text{ in } X_1,$$
$$\lim_{n\to\infty} -n(-n - T_k)^{-1}x = x \text{ in } X_0, \quad \lim_{n\to\infty} -n(-n - S_k)^{-1}x = x \text{ in } X_1,$$

which yields $-\tfrac{1}{n}(-\tfrac{1}{n} - A_k)^{-1}x \to x$ in $X_0 \cap X_1$ and, by $X_0 \cap X_1 \hookrightarrow \mathcal{F}(\{X_0, X_1\})$, also in $\mathcal{F}(\{X_0, X_1\})$. Thus we have proved $X_0 \cap X_1 \subseteq \overline{R(A_k)}$, $\overline{D(A_k)}$. Therefore, $X_0 \cap X_1 \stackrel{d}{\hookrightarrow} \mathcal{F}(\{X_0, X_1\})$ (cf. Theorem 1.39) yields

$$\overline{R(A_k)} = \overline{D(A_k)} = \mathcal{F}(\{X_0, X_1\}).$$

Hence we have shown that A_k is sectorial with $\varphi_{A_k} \leq \max\{\varphi_{T_k}, \varphi_{S_k}\}$. The claimed admissibility of $\mathbf{A}$ can easily be shown with the representation of the resolvents by $R_\lambda^{(k)}$.

(ii) Let $f \in H_0^\infty(\Omega)$ and define the operator

$$F\colon X_0 + X_1 \to X_0 + X_1,$$
$$x_0 + x_1 \mapsto f(\mathbf{T})x_0 + f(\mathbf{S})x_1,$$

which is well-defined due to (1.9). For $x \in \mathcal{F}(\{X_0, X_1\})$ we obtain

$$f(\mathbf{A})x = \frac{1}{(2\pi i)^N}\Big(\int_\Gamma f(z) \prod_{k=1}^N (z_k - A_k)^{-1} dz\Big)x$$
$$= \frac{1}{(2\pi i)^N} \int_\Gamma f(z) \prod_{k=1}^N (z_k - A_k)^{-1} x\, dz.$$

Here the integral converges in $\mathcal{F}(\{X_0, X_1\})$ and, due to $\mathcal{F}(\{X_0, X_1\}) \hookrightarrow X_0 + X_1$, also in $X_0 + X_1$. In the same way we can show

$$Fx = \frac{1}{(2\pi i)^N} \int_\Gamma f(z) \prod_{k=1}^N R_{z_k}^{(k)} x\, dz, \quad x \in X_0 + X_1$$

(convergence of the integral in $X_0 + X_1$) due to $X_i \hookrightarrow X_0 + X_1$ ($i = 0, 1$). So we derive $F|_{\mathcal{F}(\{X_0, X_1\})} = f(\mathbf{A})$. Additionally, we have $F|_{X_0} = f(\mathbf{T})$ and $F|_{X_1} = f(\mathbf{S})$. Using Theorem 1.39, we deduce

$$\|f(\mathbf{A})\|_{L(\mathcal{F}(\{X_0, X_1\}))} \leq \|f(\mathbf{T})\|_{L(X_0)}^{1-\theta} \|f(\mathbf{S})\|_{L(X_1)}^{\theta} \leq C\|f\|_\infty$$

1.2. Vector-valued Sobolev spaces

for all $f \in H_0^\infty(\Omega)$. So we have proved that $\mathbf{A}$ admits a bounded joint $H^\infty(\Omega)$-calculus and that the claimed representation of $f(\mathbf{A})$ holds.

Now we choose $f \in H^\infty(\Omega)$. According to Remark 1.21 (iii), we have $\mathcal{F}(\{X_0, X_1\}) \hookrightarrow X_0 + X_1$ and $X_i \hookrightarrow X_0 + X_1$, and we get

$$f(\mathbf{A})x = \lim_{n \to \infty} \underbrace{(\psi_{n,N} f)}_{\in H_0^\infty(\Omega)}(\mathbf{A})x = \lim_{n \to \infty} (\psi_{n,N} f)(\mathbf{T})x_0 + \lim_{n \to \infty} (\psi_{n,N} f)(\mathbf{S})x_1$$
$$= f(\mathbf{T})x_0 + f(\mathbf{S})x_1$$

for all $x = x_0 + x_1 \in \mathcal{F}(\{X_0, X_1\})$, $x_i \in X_i$, $i = 0, 1$. Note that here the first limit holds in the topology of $\mathcal{F}(\{X_0, X_1\})$, the second in X_0, and the third in X_1. This yields the asserted representation for $f(\mathbf{A})$ for all functions $f \in H^\infty(\Omega)$.

For $N = 1$ it is obvious that $\varphi_{A_1}^\infty \leq \max\{\varphi_{T_1}^\infty, \varphi_{S_1}^\infty\}$.

(iii) Part (ii) already yields that $\mathbf{A}$ admits a bounded joint $H^\infty(\Omega)$-calculus. By assumption the families

$$\mathcal{T}_1 := \{h(\mathbf{T}) \colon h \in H_0^\infty(\Omega), \|h\|_\infty \leq 1\} \subseteq L(X_0),$$
$$\mathcal{T}_2 := \{h(\mathbf{S}) \colon h \in H_0^\infty(\Omega), \|h\|_\infty \leq 1\} \subseteq L(X_1)$$

are both $\mathcal{R}$-bounded. Thus, the representation of $f(\mathbf{A})$ in part (ii) and Theorem 1.44 yield the $\mathcal{R}$-boundedness of

$$\{h(\mathbf{A}) \colon h \in H_0^\infty(\Omega), \|h\|_\infty \leq 1\} \subseteq L(\mathcal{F}(\{X_0, X_1\})).$$

Therefore $\mathbf{A}$ admits an $\mathcal{R}$-bounded joint $H^\infty(\Omega)$-calculus.

Finally, the assertion on the angle in the case $N = 1$ is obvious. □

Remark 1.46. This theorem about H^∞-calculus and interpolation is essential for our purpose and therefore we want to make some remarks for the special case if X_1 embeds into X_0:

(i) The compatibility conditions for the resolvents in (1.9) are always fulfilled in case of $X_1 \hookrightarrow X_0$. This is due to the fact that we have $(\lambda - S_k)^{-1} = (\lambda - T_k)^{-1}|_{X_1}$ for all $\lambda \in \rho(T_k) \cap \rho(S_k)$ in this case.

(ii) For $X_1 \hookrightarrow X_0$ the representation of $f(\mathbf{A})$ in Theorem 1.45 (ii) is given by

$$f(\mathbf{A}) = f(\mathbf{T})|_{\mathcal{F}(\{X_0, X_1\})}, \quad f \in H^\infty(\Omega).$$

This can be easily seen by $(\lambda - A_k)^{-1} = (\lambda - T_k)^{-1}|_{\mathcal{F}(\{X_0, X_1\})}$ for $\lambda \in \rho(T_k) \cap \rho(S_k)$.

The full strength of Theorem 1.45 will become much clearer in Chapter 3. There we develop a bounded H^∞-calculus of $\nabla_+ = (\partial_t, \nabla)$ on Triebel-Lizorkin spaces. Except for Chapter 3 we always have $X_1 \hookrightarrow X_0$ in every situation.

b) Retractions and coretractions

To define function spaces on domains, the concept of retractions and coretractions can be used. Here the main properties of the function spaces can easily be transferred from the whole space to domains. For our purpose, it is important that the concept is compatible with the real and complex interpolation functors. We will mainly consider the half-space $\mathbb{R}^n_+ = \{x \in \mathbb{R}^n : x_n > 0\}$.

Definition 1.47 (Retraction and coretraction, cf. [Ama95, I.2.3], [Ama09, p. 4]**).** Let X and Y be locally convex spaces. A mapping $r \in L(X, Y)$ is called a *retraction* if there exists $e \in L(Y, X)$ such that $(re)y = y$ for all $y \in Y$. The mapping e is then called a corresponding *coretraction*.

Definition 1.48. Let X be a Banach space and let Y be a locally convex space. For a mapping $\varphi \in L(X, Y)$ we endow the *image space* $\varphi X := \varphi(X)$ with the quotient norm
$$\|u\|_{\varphi X} := \inf\left\{\|f\|_X : f \in \varphi^{-1}(\{u\})\right\}, \quad u \in \varphi X.$$

Remark 1.49. In the situation of Definition 1.48, $(\varphi X, \|\cdot\|_{\varphi X})$ is a Banach space satisfying $\varphi X \hookrightarrow Y$.

Lemma 1.50. *Let X and S be locally convex spaces, and let $r \in L(X, S)$ be a retraction with coretraction $e \in L(S, X)$. Let X_1 and X_2 be Banach spaces with $X_i \subseteq X$. Assume that $r|_{X_i} \in L(X_i, rX_i)$ is a retraction with coretraction $e|_{rX_i} \in L(rX_i, X_i)$, $i = 1, 2$. Then we have*
$$r(X_1 \cap X_2) = (rX_1) \cap (rX_2)$$
with equivalent norms.

Proof. We trivially have $r(X_1 \cap X_2) \subseteq (rX_1) \cap (rX_2)$. For $u \in (rX_1) \cap (rX_2)$ we write $u = (re)u$. As by assumption on e we have $eu \in X_1 \cap X_2$, we see that $r(X_1 \cap X_2) = (rX_1) \cap (rX_2)$ as an equality of sets. Therefore, we only have to show the equivalence of the norms. For $u \in r(X_1 \cap X_2)$ we derive

$$\begin{aligned}
\|u\|_{r(X_1 \cap X_2)} &= \inf\{\|f\|_{X_1 \cap X_2} : f \in X_1 \cap X_2, rf = u\} \\
&= \inf\{\|f\|_{X_1} + \|f\|_{X_2} : f \in X_1 \cap X_2, rf = u\} \\
&\geq \inf\{\|f\|_{X_1} : f \in X_1 \cap X_2, rf = u\} + \inf\{\|g\|_{X_2} : g \in X_1 \cap X_2, rg = u\} \\
&\geq \inf\{\|f\|_{X_1} : f \in X_1, rf = u\} + \inf\{\|g\|_{X_2} : g \in X_2, rg = u\} \\
&= \|u\|_{rX_1} + \|u\|_{rX_2} = \|u\|_{(rX_1) \cap (rX_2)}.
\end{aligned}$$

Using the bounded inverse theorem and Remark 1.49 we conclude the claimed equivalence of norms. □

Lemma 1.51 (Interpolation and retraction). *Let A_0, A_1, Z_0, Z_1 be Banach spaces and $r_k \in L(A_k, Z_k)$, $k = 0, 1$. If r_k is a retraction in $L(A_k, r_k A_k)$ for $k = 0, 1$ with*

1.2. Vector-valued Sobolev spaces

corresponding coretraction $e_k \in L(r_k A_k, A_k)$ *and* $r_0 x = r_1 x$ *for all* $x \in A_0 \cap A_1$, *then we have*

$$r[A_0, A_1]_\theta = [r_0 A_0, r_1 A_1]_\theta, \quad \theta \in (0, 1),$$
$$r(A_0, A_1)_{\theta, p} = (r_0 A_0, r_1 A_1)_{\theta, p}, \quad \theta \in (0, 1), \ 1 < p < \infty$$

with equivalent norms where $r \colon A_0 + A_1 \to r_0 A_0 + r_1 A_1$, $a_0 + a_1 \mapsto r_0 a_0 + r_1 a_1$.

Proof. We only consider the complex interpolation functor as the proof for the real method is essentially the same. We have

$$r \in L([A_0, A_1]_\theta, [r_0 A_0, r_1 A_1]_\theta),$$

which yields

$$r[A_0, A_1]_\theta \hookrightarrow [r_0 A_0, r_1 A_1]_\theta \tag{1.10}$$

according to Remark 1.49. Next, we apply [Tri78, Theorem 1.2.4] with $B_k := r_k A_k$, $k = 0, 1$. This gives

$$\Phi := e|_{[B_0, B_1]_\theta} \in L_{\text{Isom}}\left([B_0, B_1]_\theta, \left(er\left([A_0, A_1]_\theta\right), \|\cdot\|_{[A_0, A_1]_\theta}\right)\right) \tag{1.11}$$

and therefore $e([B_0, B_1]_\theta) = er([A_0, A_1]_\theta)$ as an equality of sets. Due to (1.10) and (1.11) we also obtain $[B_0, B_1]_\theta = r([A_0, A_1]_\theta)$ as an equality of sets. The bounded inverse theorem and (1.10) then yield the assertion. □

Lemma 1.52 (Interpolation and isomorphism). *Let* $A_0, A_1, B_0,$ *and* B_1 *be Banach spaces. If there exist* $r_k \in L_{\text{Isom}}(A_k, B_k)$, $k = 0, 1$, *with* $r_0 x = r_1 x$ *for all* $x \in A_0 \cap A_1$, *then we have* $B_k = r_k A_k$ *with equivalent norms and*

$$r[A_0, A_1]_\theta = [B_0, B_1]_\theta, \quad \theta \in (0, 1),$$
$$r(A_0, A_1)_{\theta, p} = (B_0, B_1)_{\theta, p}, \quad \theta \in (0, 1), \ 1 < p < \infty$$

with equivalent norms where $r \colon A_0 + A_1 \to r_0 A_0 + r_1 A_1$, $a_0 + a_1 \mapsto r_0 a_0 + r_1 a_1$.

Proof. Again, we only consider the complex interpolation functor. Due to $r_k \in L_{\text{Isom}}(A_k, B_k)$ we have

$$\|u\|_{r_k A_k} = \|r_k^{-1} u\|_{A_k} \leq C \|u\|_{B_k}.$$

This yields $r_k A_k = B_k$. Lemma 1.51 then already shows

$$r[A_0, A_1]_\theta = [B_0, B_1]_\theta, \quad \theta \in (0, 1).$$

□

Lemma 1.53. *Let* X, Z *be Banach spaces and* $r \in L(X, Z)$. *If* $Y \overset{\iota}{\hookrightarrow} X$ *for a Banach space* Y, *then we also have*

$$(r \circ \iota) Y \hookrightarrow rX$$

where the injection is given by the identity.

Proof. It is trivial that $(r \circ \iota)Y \subseteq rX$. For $u \in Y$ we get

$$\|(r \circ \iota)u\|_{rX} = \inf\{\|f\|_X \colon f \in r^{-1}(\{(r \circ \iota)u\})\} \leq \|\iota(u)\|_X \leq C\|u\|_Y,$$

which yields $r \circ \iota \in L(Y, rX)$. This directly implies $(r \circ \iota)Y \hookrightarrow rX$ due to Remark 1.49. □

Lemma 1.54. *Let X, Y_0, and Y_1 be Banach spaces with retractions and corresponding coretractions*

$$r_i \colon X \to Y_i, \quad e_i \colon Y_i \to X, \quad i = 0, 1$$

and let $D \subseteq X$ be a dense subspace such that $r_0 f = r_1 f$ for all $f \in D$. Then $Y_0 = Y_1$ with equivalence of norms.

Proof. The retractions r_0 and r_1 are continuous and therefore we obtain $r_0 u = r_1 u$ for all $u \in X$. This yields the equality of the sets Y_0 and Y_1 because the mappings r_0 and r_1 are onto. In particular, we get

$$r_1 e_0 u = r_0 e_0 u = u, \quad u \in Y_0$$

by $r_1 f = r_0 f$ for $f \in X$ and $e_0 u \in X$. With this we obtain

$$\|u\|_{Y_1} = \|r_1 e_0 u\|_{Y_1} \leq C\|e_0 u\|_X \leq C'\|u\|_{Y_0}, \quad u \in Y_0.$$

Using $r_0 e_1 u = u$ for $u \in Y_1$ we can show analogously $\|u\|_{Y_0} \leq C\|u\|_{Y_1}$. So the equivalence of the norms $\|\cdot\|_{Y_0}$ and $\|\cdot\|_{Y_1}$ is proved. □

c) Definition of Sobolev spaces

To define Sobolev spaces, we start with the whole space $\mathbb{R}^n$. In the following, let X be a complex Banach space of class $\mathcal{HT}$ with property (α).

As usual, we denote by $\mathscr{S}(\mathbb{R}^n, X)$ the Schwartz space of smooth rapidly decreasing X-valued functions, equipped with the canonical locally convex topology induced by the family of seminorms

$$p_{\alpha,k}(\varphi) := \sup_{x \in \mathbb{R}^n} (1 + |x|^2)^{k/2} \|\partial^\alpha \varphi(x)\|_X, \quad \varphi \in \mathscr{S}(\mathbb{R}^n, X),$$

with $\alpha \in \mathbb{N}_0^n$, $k \in \mathbb{N}_0$. The space of X-valued tempered distributions is defined by $\mathscr{S}'(\mathbb{R}^n, X) := L(\mathscr{S}(\mathbb{R}^n), X)$. Again this is a locally convex space where the topology is induced by the family of seminorms

$$p_\varphi(f) := \|f(\varphi)\|_X, \quad f \in \mathscr{S}'(\mathbb{R}^n, X)$$

for all $\varphi \in \mathscr{S}(\mathbb{R}^n)$. As usual, we write $\mathscr{S}(\mathbb{R}^n)$ and $\mathscr{S}'(\mathbb{R}^n)$ for $X = \mathbb{C}$. Note that in this case the above topology on $\mathscr{S}'(\mathbb{R}^n)$ coincides with the weak-$*$-topology.

In the same way as in the scalar case, one sees that the Fourier transform

$$(\mathscr{F}f)(\xi) := (2\pi)^{-n/2} \int_{\mathbb{R}^n} f(x) e^{-ix\xi} dx, \quad \xi \in \mathbb{R}^n, \, f \in \mathscr{S}(\mathbb{R}^n, X),$$

1.2. Vector-valued Sobolev spaces

defines an isomorphism $\mathscr{F}\colon \mathscr{S}(\mathbb{R}^n, X) \to \mathscr{S}(\mathbb{R}^n, X)$. Here and in the following, we write $x\xi$ for the standard scalar product of x and ξ in $\mathbb{R}^n$. Therefore, the Fourier transform can be extended by duality to an isomorphism of $\mathscr{S}'(\mathbb{R}^n, X)$, setting $(\mathscr{F}f)(\varphi) := f(\mathscr{F}\varphi)$ for $f \in \mathscr{S}'(\mathbb{R}^n, X)$ and $\varphi \in \mathscr{S}(\mathbb{R}^n)$.

For the definition of Sobolev spaces, we will frequently use the symbol

$$\Lambda_r(z) := \Big(1 - \sum_{k=1}^n z_k^2\Big)^{r/2}, \quad z \in ((\mathbb{C} \setminus \mathbb{R}) \cup \{0\})^n$$

for $r \in \mathbb{R}$. In general, for a scalar-valued symbol $p(z)$ being defined at least on the set $(i\mathbb{R})^n \subseteq \mathbb{C}^n$, we define $\mathrm{op}[p] := \mathscr{F}^{-1} p(i\xi) \mathscr{F}$, i.e.,

$$(\mathrm{op}[p]f)(x) := (2\pi)^{-n/2} \int_{\mathbb{R}^n} e^{ix\xi} p(i\xi)(\mathscr{F}f)(\xi)d\xi, \quad f \in \mathscr{S}(\mathbb{R}^n, X),$$

provided the integral exists for all $f \in \mathscr{S}(\mathbb{R}^n, X)$ (this is the case, for instance, if p is continuous and polynomially bounded on $(i\mathbb{R})^n$). If $\xi \mapsto p(i\xi)$ is smooth and polynomially bounded, $\mathrm{op}[p]$ is well-defined on $\mathscr{S}'(\mathbb{R}^n, X)$ by duality. In particular, we have

$$\mathrm{op}[\Lambda_r] = \mathscr{F}^{-1} \langle \xi \rangle^r \mathscr{F} \quad \text{with } \langle \xi \rangle := \sqrt{1 + |\xi|^2}.$$

We start with Sobolev type spaces on $\mathbb{R}^n$.

Definition 1.55 (Sobolev spaces on $\mathbb{R}^n$). Let $p, q \in (1, \infty)$.

(i) For $k \in \mathbb{N}_0$, the (classical) *Sobolev space* is defined as

$$W_p^k(\mathbb{R}^n, X) := \{f \in L_p(\mathbb{R}^n, X) : \partial^\alpha f \in L_p(\mathbb{R}^n, X) \text{ for all } |\alpha| \leq k\}$$

with norm

$$\|f\|_{W_p^k(\mathbb{R}^n, X)} := \Big(\sum_{|\alpha| \leq k} \|\partial^\alpha f\|_{L_p(\mathbb{R}^n, X)}^p\Big)^{1/p}, \quad f \in W_p^k(\mathbb{R}^n, X).$$

(ii) For $r \in \mathbb{R}$, the *Bessel potential space* is defined as

$$H_p^r(\mathbb{R}^n, X) := \{f \in \mathscr{S}'(\mathbb{R}^n, X) : \mathrm{op}[\Lambda_r]f \in L_p(\mathbb{R}^n, X)\}$$

with norm

$$\|f\|_{H_p^r(\mathbb{R}^n, X)} := \|\mathrm{op}[\Lambda_r]f\|_{L_p(\mathbb{R}^n, X)}, \quad f \in H_p^r(\mathbb{R}^n, X).$$

(iii) Let $\psi \in \mathscr{D}(\mathbb{R}^n)$ with $\psi(x) \in [0,1]$, $x \in \mathbb{R}^n$, $\psi(x) = 1$ if $|x| \leq A$, and $\psi(x) = 0$ if $|x| > B$ for $0 < A < B < \infty$. We define the smooth *dyadic decomposition of unity* by

$$\varphi_0 := \psi, \quad \varphi_1(x) := \varphi_0(x/2) - \varphi_0(x), \quad \varphi_j(x) := \varphi_1(2^{-j+1}x), \quad j \geq 2.$$

For $r \in \mathbb{R}$, the *Besov space* $B_{pq}^r(\mathbb{R}^n, X)$ is defined as the set of all $u \in \mathscr{S}'(\mathbb{R}^n, X)$ with

$$\|u\|_{B_{pq}^r(\mathbb{R}^n,X)} := \left\|\left(2^{rj}\|\mathrm{op}[\varphi_j]u\|_{L_p(\mathbb{R}^n,X)}\right)_{j\in\mathbb{N}_0}\right\|_{\ell_q} < \infty.$$

Theorem 1.56 (Sobolev spaces and interpolation). *Let $p, q \in (1, \infty)$.*

(i) *For $k \in \mathbb{N}_0$ we have $H_p^k(\mathbb{R}^n, X) = W_p^k(\mathbb{R}^n, X)$ with equivalent norms.*

(ii) *For all $r_0, r_1 \in \mathbb{R}$, $r_0 \neq r_1$ and $\theta \in (0, 1)$ we have*

$$\left[H_p^{r_0}(\mathbb{R}^n, X), H_p^{r_1}(\mathbb{R}^n, X)\right]_\theta = H_p^r(\mathbb{R}^n, X)$$

with equivalent norms where $r := (1 - \theta)r_0 + \theta r_1$.

(iii) *For all $r_0, r_1 \in \mathbb{R}$, $r_0 \neq r_1$ and $\theta \in (0, 1)$ we have*

$$\left(H_p^{r_0}(\mathbb{R}^n, X), H_p^{r_1}(\mathbb{R}^n, X)\right)_{\theta, q} = B_{pq}^r(\mathbb{R}^n, X)$$

with equivalent norms where $r := (1 - \theta)r_0 + \theta r_1$.

Proof. See, e.g., [Ama09, Theorem 3.7.1]. □

Remark 1.57. (i) The above statements describe only some properties of the spaces which will be useful for our later purposes. A very detailed and comprehensive work about Banach space valued function spaces can be found in [Ama09]. The author even considers the anisotropic case which is, however, not necessary for our purposes.

(ii) We remark that Theorem 1.56 leads to an alternative but equivalent definition for H_p^r and B_{pq}^r: Starting with the classical spaces $W_p^k(\mathbb{R}^n, X)$ with $k \in \mathbb{N}_0$, the Bessel potential spaces and Besov spaces of order $r > 0$ can be defined by complex and real interpolation, respectively. The spaces with negative order may then be defined by duality.

Lemma 1.58. *Let $p, q \in (1, \infty)$.*

(i) *For every $r \in \mathbb{R}$ we have*

$$\mathscr{S}(\mathbb{R}^n, X) \stackrel{d}{\hookrightarrow} H_p^r(\mathbb{R}^n, X),$$
$$\mathscr{S}(\mathbb{R}^n, X) \stackrel{d}{\hookrightarrow} B_{pq}^r(\mathbb{R}^n, X).$$

(ii) *For every $r, r' \in \mathbb{R}$ we have*

$$\mathrm{op}[\Lambda_r]|_{H_p^{r'+r}(\mathbb{R}^n,X)} \in L_{\mathrm{Isom}}(H_p^{r'+r}(\mathbb{R}^n, X), H_p^{r'}(\mathbb{R}^n, X)),$$
$$\mathrm{op}[\Lambda_r]|_{B_{pq}^{r'+r}(\mathbb{R}^n,X)} \in L_{\mathrm{Isom}}(B_{pq}^{r'+r}(\mathbb{R}^n, X), B_{pq}^{r'}(\mathbb{R}^n, X)).$$

1.2. Vector-valued Sobolev spaces

Proof. (i) See [Ama09, Theorem 2.3.2].

(ii) The first assertion easily follows from the definition of the Bessel potential spaces. The second assertion is then obtained from Theorem 1.56. □

Now we want to define the corresponding Sobolev spaces in the half-space $\mathbb{R}^n_+$. The analog of the Schwartz functions can be defined explicitly:

Definition 1.59. We define the *Schwartz functions* $\mathscr{S}(\mathbb{R}^n_+, X)$ on $\mathbb{R}^n_+$ as the space of all $f \in C^\infty(\mathbb{R}^n_+, X)$ such that

$$\sup_{x \in \mathbb{R}^n_+} (1 + |x|^2)^{k/2} \|\partial^\alpha f(x)\|_X < \infty$$

for all $k \in \mathbb{N}_0$ and $\alpha \in \mathbb{N}_0^n$. Furthermore, we define

$$_0\mathscr{S}(\mathbb{R}^n_+, X) := \{f \in \mathscr{S}(\mathbb{R}^n_+, X) \colon (\partial^\alpha f)|_{\partial \mathbb{R}^n_+} = 0, \alpha \in \mathbb{N}_0^n\}.$$

All spaces are equipped with the canonical locally convex topology.

Lemma 1.60 ([Ama09, Lemma 4.1.1]). *There exists a function $h \in C^\infty(\mathbb{R}_+, \mathbb{R})$ with $h(1/t) = -th(t)$, $t > 0$, and*

$$\int_0^\infty t^s |h(t)|dt < \infty, \quad (-1)^k \int_0^\infty t^k h(t) dt = 1$$

for all $s \in \mathbb{R}$ and $k \in \mathbb{Z}$.

Definition 1.61. (i) The *extension operator* $e^+ \colon \mathscr{S}(\mathbb{R}^n_+, X) \to \mathscr{S}(\mathbb{R}^n, X)$ is defined by

$$[e^+(u)](x) := \begin{cases} u(x), & x \in \overline{\mathbb{R}^n_+}, \\ \int_0^\infty h(t) u(x', -tx_n) dt, & x = (x', x_n) \in \mathbb{R}^n_- \end{cases}$$

and the *pointwise restriction operator* $r^+ \colon \mathscr{S}(\mathbb{R}^n, X) \to \mathscr{S}(\mathbb{R}^n_+, X)$ is defined by $r^+ u := u|_{\mathbb{R}^n_+}$.

(ii) We define the *trivial extension operator* $e_0^+ \colon {}_0\mathscr{S}(\mathbb{R}^n_+, X) \to \mathscr{S}(\mathbb{R}^n, X)$ by

$$[e_0^+(u)](x) := \begin{cases} u(x), & x \in \overline{\mathbb{R}^n_+}, \\ 0, & x \in \mathbb{R}^n_- \end{cases}$$

and the operator $r_0^+ \colon \mathscr{S}(\mathbb{R}^n, X) \to {}_0\mathscr{S}(\mathbb{R}^n_+, X)$, $u \mapsto r^+(1 - e^- r^-)u$. Here r^- and e^- denote the corresponding restriction and extension operators from (i) with respect to $\mathbb{R}^n_-$.

Remark 1.62. The restriction operator $r^+ \in L(\mathscr{S}(\mathbb{R}^n, X), \mathscr{S}(\mathbb{R}^n_+, X))$ is a retraction with coretraction $e^+ \in L(\mathscr{S}(\mathbb{R}^n_+, X), \mathscr{S}(\mathbb{R}^n, X))$. In the same way, $r_0^+ \in L(\mathscr{S}(\mathbb{R}^n, X), {}_0\mathscr{S}(\mathbb{R}^n_+, X))$ is a retraction with coretraction $e_0^+ \in L({}_0\mathscr{S}(\mathbb{R}^n_+, X), \mathscr{S}(\mathbb{R}^n, X))$ (see [Ama09, Lemma 4.1.2, Theorem 4.1.3], [Ama09, (4.1.11), Theorem 4.1.7]).

Next, we define two classes of tempered distributions on half spaces, and by duality we then extend the operators r^+, e^+, r_0^+, and e_0^+ canonically.

Definition 1.63. We define
$$\mathscr{S}'(\mathbb{R}_+^n, X) := L(\mathscr{S}(\mathbb{R}_+^n), X) \quad \text{and} \quad {}_0\mathscr{S}'(\mathbb{R}_+^n, X) := L({}_0\mathscr{S}(\mathbb{R}_+^n), X).$$

Then we lift the operators r^+, e^+, r_0^+, and e_0^+ by

$$\begin{aligned}
e^+ &: {}_0\mathscr{S}'(\mathbb{R}_+^n, X) \to \mathscr{S}'(\mathbb{R}^n, X), & e^+ &:= (r_0^+)', \\
r^+ &: \mathscr{S}'(\mathbb{R}^n, X) \to {}_0\mathscr{S}'(\mathbb{R}_+^n, X), & r^+ &:= (e_0^+)', \\
e_0^+ &: \mathscr{S}'(\mathbb{R}_+^n, X) \to \mathscr{S}'(\mathbb{R}^n, X), & e_0^+ &:= (r^+)', \\
r_0^+ &: \mathscr{S}'(\mathbb{R}^n, X) \to \mathscr{S}'(\mathbb{R}_+^n, X), & r_0^+ &:= (e^+)',
\end{aligned}$$

where $(\cdot)'$ stands for the adjoint operator, e.g., we have $(e^+u)(\varphi) := u(r_0^+\varphi)$ for $\varphi \in \mathscr{S}(\mathbb{R}^n)$ and $u \in {}_0\mathscr{S}'(\mathbb{R}_+^n, X)$.

Remark 1.64. (i) Note that this definition is consistent with Definition 1.61 (i) and (ii) when we keep in mind the embeddings

$$\mathscr{S}(\mathbb{R}^n, X) \hookrightarrow \mathscr{S}'(\mathbb{R}^n, X), \qquad \mathscr{S}(\mathbb{R}_+^n, X) \hookrightarrow {}_0\mathscr{S}'(\mathbb{R}_+^n, X),$$
$${}_0\mathscr{S}(\mathbb{R}_+^n, X) \hookrightarrow \mathscr{S}'(\mathbb{R}_+^n, X).$$

(ii) We have

$$\begin{aligned}
e^+ &\in L({}_0\mathscr{S}'(\mathbb{R}_+^n, X), \mathscr{S}'(\mathbb{R}^n, X)), & r^+ &\in L(\mathscr{S}'(\mathbb{R}^n, X), {}_0\mathscr{S}'(\mathbb{R}_+^n, X)), \\
e_0^+ &\in L(\mathscr{S}'(\mathbb{R}_+^n, X), \mathscr{S}'(\mathbb{R}^n, X)), & r_0^+ &\in L(\mathscr{S}'(\mathbb{R}^n, X), \mathscr{S}'(\mathbb{R}_+^n, X))
\end{aligned}$$

and r_0^+ and r^+ are retractions with corresponding coretractions e_0^+ and e^+, respectively. For this result and further details we refer to [Ama09, Theorem 4.2.2].

The following result shows that e_0^+ and r_0^+ are connected with the support of the distribution.

Lemma 1.65. *For $u \in \mathscr{S}'(\mathbb{R}^n, X)$ the equality*
$$e_0^+ r_0^+ u = u$$
holds if and only if $\operatorname{supp} u \subseteq \overline{\mathbb{R}_+^n}$.

Proof. Analogously to Definition 1.61 and Definition 1.63, we define the operators r^-, r_0^-, e^-, and e_0^- with respect to the half space $\mathbb{R}_-^n$. In [Ama09, p. 81] one can find the decomposition
$$\mathscr{S}'(\mathbb{R}^n, X) = e_0^+ \mathscr{S}'(\mathbb{R}_+^n, X) \oplus e^- {}_0\mathscr{S}'(\mathbb{R}_-^n, X).$$

1.2. Vector-valued Sobolev spaces

Let $u \in \mathscr{S}'(\mathbb{R}^n, X)$. Then there exist $u_1 \in \mathscr{S}'(\mathbb{R}^n_+, X)$ and $u_2 \in {}_0\mathscr{S}'(\mathbb{R}^n_-, X)$ such that $u = e_0^+ u_1 + e^- u_2$. With this decomposition we obtain

$$r_0^+ u = u_1 + r_0^+ e^- u_2, \quad r^- u = r^- e_0^+ u_1 + u_2.$$

It is elementary to show $r_0^+ e^- u_2 = 0$ and $r^- e_0^+ u_1 = 0$ (see [Ama09, Corollary 4.1.9]). Hence we have

$$u = e_0^+ r_0^+ u + e^- r^- u. \tag{1.12}$$

Now assume $\operatorname{supp} u \subseteq \overline{\mathbb{R}^n_+}$. For all $\varphi \in \mathscr{D}(\mathbb{R}^n)$ we have $\operatorname{supp}(e_0^- \varphi) \subseteq \mathbb{R}^n_-$ and therefore $(r^- u)\varphi = u(e_0^- \varphi) = 0$ since $\operatorname{supp} u \cap \operatorname{supp}(e_0^- \varphi) = \emptyset$. From this we derive $e^- r^- u = 0$. Then we obtain $u = e_0^+ r_0^+ u$ from (1.12).

On the other hand, assume $e_0^+ r_0^+ u = u$. Then we have $e^- r^- u = 0$ according to (1.12). So we have $u(e_0^- r_0^- \varphi) = 0$ for all $\varphi \in \mathscr{D}(\mathbb{R}^n_-)$. For all $\varphi \in \mathscr{D}(\mathbb{R}^n_-)$ we derive $e_0^- r_0^- \varphi = \varphi$ and therefore $\operatorname{supp} u \subseteq \overline{\mathbb{R}^n_+}$. $\square$

With the help of the retractions and coretractions above, we can define the Sobolev spaces on $\mathbb{R}^n_+$. For simplicity, we do not consider negative orders of differentiation, so all spaces will be subspaces of $L_p(\mathbb{R}^n_+, X)$. We remark that $H_p^0(\mathbb{R}^n_+, X) = L_p(\mathbb{R}^n_+, X)$ but $B_{pp}^0(\mathbb{R}^n_+, X) \neq L_p(\mathbb{R}^n_+, X)$ in general, and therefore we will not consider $B_{pp}^0(\mathbb{R}^n_+, X)$ in the following.

Definition 1.66 (Sobolev spaces on the half-space). Let $p, q \in (1, \infty)$ and $s > 0$, and let $\mathcal{F} \in \{H_p^s, B_{pq}^s\}$. Then we define

$$\mathcal{F}(\mathbb{R}^n_+, X) := r^+ \mathcal{F}(\mathbb{R}^n, X),$$
$$_0\mathcal{F}(\mathbb{R}^n_+, X) := r_0^+ \mathcal{F}(\mathbb{R}^n, X)$$

with the canonical norms

$$\|u\|_{\mathcal{F}(\mathbb{R}^n_+, X)} := \inf \{\|f\|_{\mathcal{F}(\mathbb{R}^n, X)} : f \in \mathcal{F}(\mathbb{R}^n, X) \text{ with } u = r^+ f\},$$
$$\|u\|_{0\mathcal{F}(\mathbb{R}^n_+, X)} := \inf \{\|f\|_{\mathcal{F}(\mathbb{R}^n, X)} : f \in \mathcal{F}(\mathbb{R}^n, X) \text{ with } u = r_0^+ f\}.$$

We also set $H_p^0(\mathbb{R}^n_+, X) := L_p(\mathbb{R}^n_+, X)$.

Remark 1.67. (i) In the situation of Definition 1.66, we have

$$e^+ \in L(\mathcal{F}(\mathbb{R}^n_+, X), \mathcal{F}(\mathbb{R}^n, X)), \qquad r^+ \in L(\mathcal{F}(\mathbb{R}^n, X), \mathcal{F}(\mathbb{R}^n_+, X)),$$
$$e_0^+ \in L({}_0\mathcal{F}(\mathbb{R}^n_+, X), \mathcal{F}(\mathbb{R}^n, X)), \qquad r_0^+ \in L(\mathcal{F}(\mathbb{R}^n, X), {}_0\mathcal{F}(\mathbb{R}^n_+, X)).$$

Hence r^+ and r_0^+ are retractions on the spaces above with corresponding coretractions e^+ and e_0^+, respectively (see [Ama09, Section 4.4]).

(ii) The results on interpolation of Sobolev spaces carry over to spaces on $\mathbb{R}^n_+$. More precisely, the statements of Theorem 1.56 hold for all $r_0, r_1 \geq 0$ with $H_p^r(\mathbb{R}^n, X)$ being replaced by $H_p^r(\mathbb{R}^n_+, X)$ and $B_{pq}^r(\mathbb{R}^n, X)$ being replaced by $B_{pq}^r(\mathbb{R}^n_+, X)$. In the same way, we may replace $H_p^r(\mathbb{R}^n, X)$ and $B_{pq}^r(\mathbb{R}^n, X)$ by ${}_0 H_p^r(\mathbb{R}^n_+, X)$ and ${}_0 B_{pq}^r(\mathbb{R}^n_+, X)$, respectively.

(iii) The spaces $_0B^s_{pq}(\mathbb{R}_+, X)$ and $_0H^s_p(\mathbb{R}_+, X)$ can also be characterized by vanishing traces, i.e., we have

$$_0H^s_p(\mathbb{R}_+, X) = \begin{cases} \{u \in H^s_p(\mathbb{R}_+, X) : u^{(j)}(0) = 0, \, j = 0, \ldots, k\}, \\ \hspace{3cm} s \in (k + \tfrac{1}{p}, k + 1 + \tfrac{1}{p}), k \in \mathbb{N}_0, \\ H^s_p(\mathbb{R}_+, X), \hspace{1cm} s \in [0, \tfrac{1}{p}), \end{cases}$$

$$_0B^s_{pq}(\mathbb{R}_+, X) = \begin{cases} \{u \in B^s_{pq}(\mathbb{R}_+, X) : u^{(j)}(0) = 0, \, j = 0, \ldots, k\}, \\ \hspace{3cm} s \in (k + \tfrac{1}{p}, k + 1 + \tfrac{1}{p}), k \in \mathbb{N}_0, \\ B^s_{pq}(\mathbb{R}_+, X), \hspace{1cm} s \in (0, \tfrac{1}{p}), \end{cases}$$

cf. [Ama09, Theorem 4.7.1].

(iv) Due to [Ama09, p. 100] we have

$$\mathscr{D}(\mathbb{R}_+, X) \xhookrightarrow{d} {_0H^s_p}(\mathbb{R}_+, X) \text{ for } s \geq 0,$$

$$\mathscr{D}(\mathbb{R}_+, X) \xhookrightarrow{d} {_0B^s_{pq}}(\mathbb{R}_+, X) \text{ for } s > 0.$$

(v) (Sobolev embedding theorem) Let $s > k + \tfrac{1}{p}$ with $k \in \mathbb{N}_0$. Then the embeddings

$$H^s_p(\mathbb{R}_+, X) \hookrightarrow C^k_b([0, \infty), X),$$
$$B^s_{pq}(\mathbb{R}_+, X) \hookrightarrow C^k_b([0, \infty), X)$$

hold. These vector-valued variants of the well-known embedding theorems can be found in [Ama09, Theorem 3.9.1]. Therefore, the traces in (iii) are classical.

For the application of L_p-Fourier multiplier theorems we often need the Bessel potential spaces and Besov spaces to be of class $\mathcal{HT}$ and have property (α). For clarity we state these common facts in the next remark.

Remark 1.68. Let X be a Banach space of class $\mathcal{HT}$, $r \in \mathbb{R}$, and $1 < p, q < \infty$.

(i) The spaces $H^r_p(\mathbb{R}^n, X)$ and $B^r_{pq}(\mathbb{R}^n, X)$ are also of class $\mathcal{HT}$. Using Remark 1.13 (iii), Lemma 1.43 (i), and Lemma 1.58 (ii), this can be easily verified.

(ii) If X additionally has property (α), then $H^r_p(\mathbb{R}^n, X)$ and $B^r_{pq}(\mathbb{R}^n, X)$ also have property (α). This can be seen by Lemma 1.58 (ii), Remark 1.15 (ii) and Lemma 1.43 (ii). For the Besov spaces we can also use a retraction argument instead of the interpolation result in Lemma 1.43 (ii).

1.2. Vector-valued Sobolev spaces

Proposition 1.69. *Let $\{X_0, X_1\}$ be an interpolation couple of Banach spaces of class $\mathcal{HT}$. Then we have*

$$\left[H^s_{p_0}(\mathbb{R}^n_+, X_0), H^s_{p_1}(\mathbb{R}^n_+, X_1)\right]_\theta = H^s_p(\mathbb{R}^n_+, [X_0, X_1]_\theta),$$
$$\left[{}_0H^s_{p_0}(\mathbb{R}^n_+, X_0), {}_0H^s_{p_1}(\mathbb{R}^n_+, X_1)\right]_\theta = {}_0H^s_p(\mathbb{R}^n_+, [X_0, X_1]_\theta),$$
$$(H^s_{p_0}(\mathbb{R}^n_+, X_0), H^s_{p_1}(\mathbb{R}^n_+, X_1))_{\theta,p} = H^s_p(\mathbb{R}^n_+, (X_0, X_1)_{\theta,p}),$$
$$({}_0H^s_{p_0}(\mathbb{R}^n_+, X_0), {}_0H^s_{p_1}(\mathbb{R}^n_+, X_1))_{\theta,p} = {}_0H^s_p(\mathbb{R}^n_+, (X_0, X_1)_{\theta,p})$$

for all $s \geq 0$, $p_0, p_1 \in (1, \infty)$, $\theta \in (0, 1)$, and $1/p = (1-\theta)/p_0 + \theta/p_1$.

Proof. For simplicity we only consider the first case, the other results can be obtained by minor modifications. First, we show $\left[H^s_{p_0}(\mathbb{R}^n, X_0), H^s_{p_1}(\mathbb{R}^n, X_1)\right]_\theta = H^s_p(\mathbb{R}^n, [X_0, X_1]_\theta)$. According to Lemma 1.58 we have the isomorphisms

$$r_0 := (\mathrm{op}_{X_0}[\Lambda_{-s}])|_{L_{p_0}(\mathbb{R}^n, X_0)} \in L_{\mathrm{Isom}}(L_{p_0}(\mathbb{R}^n, X_0), H^s_{p_0}(\mathbb{R}^n, X_0)),$$
$$r_1 := (\mathrm{op}_{X_1}[\Lambda_{-s}])|_{L_{p_1}(\mathbb{R}^n, X_1)} \in L_{\mathrm{Isom}}(L_{p_1}(\mathbb{R}^n, X_1), H^s_{p_1}(\mathbb{R}^n, X_1)).$$

Lemma 1.52 then yields

$$r\left[L_{p_0}(\mathbb{R}^n, X_0), L_{p_1}(\mathbb{R}^n, X_1)\right]_\theta = \left[H^s_{p_0}(\mathbb{R}^n, X_0), H^s_{p_1}(\mathbb{R}^n, X_1)\right]_\theta, \quad \theta \in (0,1). \tag{1.13}$$

The interpolation result of Theorem 1.41 implies

$$[L_{p_0}(\mathbb{R}^n, X_0), L_{p_1}(\mathbb{R}^n, X_1)]_\theta = L_p(\mathbb{R}^n, [X_0, X_1]_\theta). \tag{1.14}$$

Due to

$$rf = r_0 f_0 + r_1 f_1 = \mathscr{F}^{-1}_{X_0} \Lambda_{-s} \mathscr{F}_{X_0} f_0 + \mathscr{F}^{-1}_{X_1} \Lambda_{-s} \mathscr{F}_{X_1} f_1 = \mathscr{F}^{-1}_{X_0+X_1} \Lambda_{-s} \mathscr{F}_{X_0+X_1} f$$

we have

$$r|_{L_p(\mathbb{R}^n, [X_0, X_1]_\theta)} = (\mathrm{op}_{[X_0, X_1]_\theta}[\Lambda_{-s}])|_{L_p(\mathbb{R}^n, [X_0, X_1]_\theta)}$$
$$\in L_{\mathrm{Isom}}(L_p(\mathbb{R}^n, [X_0, X_1]_\theta), H^s_p(\mathbb{R}^n, [X_0, X_1]_\theta)).$$

Using (1.13) and (1.14) we obtain

$$\left[H^s_{p_0}(\mathbb{R}^n, X_0), H^s_{p_1}(\mathbb{R}^n, X_1)\right]_\theta = H^s_p(\mathbb{R}^n, [X_0, X_1]_\theta). \tag{1.15}$$

Applying Lemma 1.51, Remark 1.67 (ii), and (1.15) we then obtain

$$\left[H^s_{p_0}(\mathbb{R}^n_+, X_0), H^s_{p_1}(\mathbb{R}^n_+, X_1)\right]_\theta = H^s_p(\mathbb{R}^n_+, [X_0, X_1]_\theta). \qquad \square$$

Remark 1.70 (Spaces with exponential weights). In many applications, the operators are sectorial or $\mathcal{R}$-sectorial only after a shift, i.e., if one replaces T by $T + \varrho$. Instead of shifting the operator, one can also consider the original operator in spaces with exponential weight. The above properties of Sobolev spaces carry over to their exponentially weighted versions. Here we only give the definition and some remarks on these spaces.

For $\varrho \in \mathbb{R}$ and $u \in L_p(\mathbb{R}_+, X)$, we define $\mathscr{M}_\varrho u$ by

$$(\mathscr{M}_\varrho u)(t) := e^{-\varrho t} u(t), \quad t \in \mathbb{R}_+. \tag{1.16}$$

For $\varrho \geq 0$, the spaces with exponential weight are then defined by

$$\begin{aligned}
{}_0H^s_{p,\varrho}(\mathbb{R}_+, X) &:= \mathscr{M}_{-\varrho}\left({}_0H^s_p(\mathbb{R}_+, X)\right), & s \geq 0, \\
{}_0B^s_{pq,\varrho}(\mathbb{R}_+, X) &:= \mathscr{M}_{-\varrho}\left({}_0B^s_{pq}(\mathbb{R}_+, X)\right), & s > 0,
\end{aligned}$$

with norms

$$\|u\|_{{}_0H^s_{p,\varrho}(\mathbb{R}_+,X)} := \|\mathscr{M}_\varrho u\|_{{}_0H^s_p(\mathbb{R}_+,X)}, \quad u \in {}_0H^s_{p,\varrho}(\mathbb{R}_+, X),\ s \geq 0, \tag{1.17}$$

$$\|f\|_{{}_0B^s_{pq,\varrho}(\mathbb{R}_+,X)} := \|\mathscr{M}_\varrho f\|_{{}_0B^s_{pq}(\mathbb{R}_+,X)}, \quad f \in {}_0B^s_{pq,\varrho}(\mathbb{R}_+, X),\ s > 0. \tag{1.18}$$

It is easily seen that for $\mathcal{F} \in \{H^s_p, B^s_{pq}\}$ we have

$$\mathscr{M}_\varrho|_{\mathcal{F}_\varrho(\mathbb{R}_+,X)} \in L_{\mathrm{Isom}}\left({}_0\mathcal{F}_\varrho(\mathbb{R}_+, X), {}_0\mathcal{F}(\mathbb{R}_+, X)\right)$$

(see [DSS08, Lemma 2.2]), and therefore the interpolation results above also hold for weighted spaces.

In the following, we will frequently use Sobolev spaces in time and space variables which belong to different types, e.g., ${}_0B^s_{pp}(\mathbb{R}_+, H^r_p(\mathbb{R}^n))$. Therefore, we introduce an abbreviation for these spaces.

Definition 1.71 (Spaces of mixed scales). Let $1 < p_0, q_0, p_1, q_1 < \infty$, $s \geq 0$, $\varrho \geq 0$, and $r \in \mathbb{R}$. Let X be a Banach space of class $\mathcal{HT}$ with property (α). For $\mathcal{F} \in \{B_{p_0 q_0}, H_{p_0}\}$ (with $s > 0$ if $\mathcal{F} = B_{p_0 q_0}$) and for $\mathcal{K} \in \{B_{p_1 q_1}, H_{p_1}\}$ we define

$$\begin{aligned}
{}_0\mathcal{F}^s_\varrho(\mathcal{K}^r) &:= {}_0\mathcal{F}^s_\varrho\left(\mathbb{R}_+, \mathcal{K}^r(\mathbb{R}^n, X)\right), \\
{}_0\mathcal{F}^s(\mathcal{K}^r) &:= {}_0\mathcal{F}^s\left(\mathbb{R}_+, \mathcal{K}^r(\mathbb{R}^n, X)\right).
\end{aligned}$$

Note that we always include zero initial values at $t = 0$ and that we do not indicate the dependence on the space X which will be fixed in most situations.

Remark 1.72. In the definition above, for simplicity we always assume that X is a Banach space of class $\mathcal{HT}$ with property (α). This will be the case for all applications. We use the abbreviations ${}_0\mathcal{F}^s(\mathcal{K}^r)$ and ${}_0\mathcal{F}^s_\varrho(\mathcal{K}^r)$ for better readability but one should keep in mind that we have X-valued spaces. In all cases, the proofs of the results below for $\varrho > 0$ will follow from the case $\varrho = 0$ by the isomorphism of Remark 1.70. Therefore, we will formulate most proofs without explicit reference to the case $\varrho > 0$.

1.3 The time-space derivative

Motivation. After introducing function spaces in the previous section, we are now going to show that the time-space derivative $(\partial_t, \nabla_x) = (\partial_t, \partial_{x_1}, \ldots, \partial_{x_n})$ admits a bounded joint H^∞-calculus in these spaces. This means that we may replace the symbols $\lambda, z_1, \ldots, z_n$ by derivatives and obtain a well-defined and (in case of a bounded holomorphic symbol) bounded operator.

The key step to show this is the application of a vector-valued Michlin-type theorem. For this, the notion of a Fourier multiplier will be defined and discussed in Subsection a). The operators ∂_t on one hand and ∂_{x_j}, $j = 1, \ldots, n$, on the other hand, have a different meaning: while ∇_x is considered in the whole space $\mathbb{R}^n$, the time derivative acts on functions defined for $t \in \mathbb{R}_+$, and therefore we include zero initial conditions at $t = 0$ in the domain of the related operator. This implies a change in the spectrum; and in fact, ∂_t is sectorial with angle $\frac{\pi}{2}$ while ∂_{x_j} is bisectorial with angle 0.

The properties of these operators are discussed in Subsection b), while in Subsection c) we will show that (∂_t, ∇_x) has a bounded joint H^∞-calculus in all types of Sobolev spaces.

Throughout this section, let X and Y be complex Banach spaces of class $\mathcal{HT}$ with property (α). We remark that some results below also hold in more general cases.

a) Fourier multipliers

The notion of an L_p-Fourier multiplier is usually connected with operators of the form $\mathscr{F}^{-1} m(\xi) \mathscr{F}$ with $m \in L_\infty(\mathbb{R}^n, L(X, Y))$. As we are considering holomorphic symbols, we will replace ξ by $i\xi$ in our definition.

Definition 1.73 (L_p-**Fourier multiplier**)**.** Let $(i\mathbb{R})^n \subseteq \Omega \subseteq \mathbb{C}^n$, $1 < p < \infty$, and let $m\colon \Omega \to L(X,Y)$ be a function satisfying $(\xi \mapsto m(i\xi)) \in L_\infty(\mathbb{R}^n, L(X,Y))$. Then we define
$$\operatorname{op}[m]\colon \mathscr{S}(\mathbb{R}^n, X) \to L_\infty(\mathbb{R}^n, Y), \ f \mapsto \mathscr{F}^{-1} m(i\xi) \mathscr{F} f.$$
The function m is called an L_p-*Fourier multiplier* if there exists $C_p > 0$ such that

(i) $\operatorname{op}[m] f \in L_p(\mathbb{R}^n, Y)$ for all $f \in \mathscr{S}(\mathbb{R}^n, X)$,

(ii) $\|\operatorname{op}[m] f\|_{L_p(\mathbb{R}^n, Y)} \le C_p \|f\|_{L_p(\mathbb{R}^n, X)}$ for all $f \in \mathscr{S}(\mathbb{R}^n, X)$.

In this case we have a unique continuous extension of $\operatorname{op}[m]$ to an operator in $L(L_p(\mathbb{R}^n, X), L_p(\mathbb{R}^n, Y))$ which will also be denoted by $\operatorname{op}[m]$.

The following theorem is the key result to prove that a given symbol is a Fourier multiplier. In the scalar case $X = Y = \mathbb{C}$, the well-known Michlin theorem (see, e.g., [Tri78, Section 2.2]) gives conditions on m to be a Fourier multiplier. In the vector-valued case, the essential condition is not the norm-boundedness of the symbols and their derivatives but the $\mathcal{R}$-boundedness. The following result due to Weis is the vector-valued analog of Michlin's theorem.

Theorem 1.74 (**Theorem of Weis**, see [KW04, 5.2 b)]). *Let $1 < p < \infty$ and let $\{m_j : j \in I\}$ be a family of functions $m_j : (i\mathbb{R})^n \setminus \{0\} \to L(X,Y)$ satisfying*

$$\xi \mapsto m_j(i\xi) \in C^n(\mathbb{R}^n \setminus \{0\}, L(X,Y)), \quad j \in I.$$

If the set

$$\mathcal{T} := \{\xi^\alpha \partial^\alpha m_j(i\xi) : \xi \in \mathbb{R}^n \setminus \{0\}, \, \alpha \in \{0,1\}^n, \, j \in I\}$$

is $\mathcal{R}$-bounded, then each m_j is a Fourier multiplier, and

$$\{\mathrm{op}[m_j] : j \in I\} \subseteq L(L_p(\mathbb{R}^n, X), L_p(\mathbb{R}^n, Y))$$

is also $\mathcal{R}$-bounded with $\mathcal{R}_p$-bound not greater than $C(p,n,X,Y)\mathcal{R}_p(\mathcal{T})$ for some constant $C(p,n,X,Y) > 0$.

From this we easily obtain:

Proposition 1.75. *Let $m \in H^\infty(\Omega)$ with $\Omega := \prod_{k=1}^n \Sigma_{\delta_k}$. Then $m = m\,\mathrm{id}_X$ is an $L_p(\mathbb{R}^n, X)$-Fourier multiplier for all $p \in (1,\infty)$.*

Proof. According to [DV02b, Lemma 6.29] we have

$$|\xi^\alpha \partial^\alpha m(i\xi)| \leq \alpha! \prod_{k=1}^n (\sin(\delta_k))^{-\alpha_k} \|m\|_\infty, \quad \xi \in \mathbb{R}^n \setminus \{0\}, \, \alpha \in \mathbb{N}_0^n. \quad (1.19)$$

With Kahane's contraction principle (Theorem 1.8) and (1.19) we obtain the $\mathcal{R}$-boundedness of

$$\mathcal{T} := \{\xi^\alpha \partial^\alpha m(i\xi)\,\mathrm{id}_X : \xi \in \mathbb{R}^n \setminus \{0\}, \, \alpha \in \{0,1\}^n\} \subseteq L(X).$$

The assertion is then obtained by Theorem 1.74. $\square$

Lemma 1.76. *Let $1 < p < \infty$, and let $m \in H^\infty(\Omega)$ with $\Omega := \prod_{k=1}^n \Sigma_{\delta_k}$. Then the following assertions hold.*

(i) *The mapping $\mathrm{op}[m] \colon \mathscr{S}(\mathbb{R}^n, X) \to L_\infty(\mathbb{R}^n, X)$ extends to an operator in $L(H_p^r(\mathbb{R}^n, X))$ for every $r \in \mathbb{R}$. Denoting by $\mathrm{op}^{(r,p)}[m]$ and $\mathrm{op}^{(0,p)}[m]$ the extension of $\mathrm{op}[m]$ to $H_p^r(\mathbb{R}^n, X)$ and $L_p(\mathbb{R}^n, X)$, respectively, we get*

$$\mathrm{op}[\Lambda_{-r}]\mathrm{op}^{(0,p)}[m]\mathrm{op}[\Lambda_r]f = \mathrm{op}^{(r,p)}[m]f, \quad f \in H_p^r(\mathbb{R}^n, X). \quad (1.20)$$

(ii) *Let $s, r \in \mathbb{R}$ and $p, q \in (1, \infty)$ with $p = q$ or $s = r$. Then the compatibility*

$$\mathrm{op}^{(r,p)}[m]f = \mathrm{op}^{(s,q)}[m]f, \quad f \in H_p^r(\mathbb{R}^n, X) \cap H_q^s(\mathbb{R}^n, X) \quad (1.21)$$

holds.

1.3. The time-space derivative

Proof. (i) Let $f \in \mathscr{S}(\mathbb{R}^n, X)$. Then we have $\mathrm{op}[m]f \in L_p(\mathbb{R}^n, X)$ according to Definition 1.73 (i) and Proposition 1.75. Due to the holomorphy of $\Lambda_{-|r|}$ we have that $\Lambda_{-|r|}m$ is also a Fourier multiplier. This enables us to prove

$$\mathrm{op}[\Lambda_{-r}]\mathrm{op}[m]\mathrm{op}[\Lambda_r]f = \mathrm{op}[m]f, \quad f \in \mathscr{S}(\mathbb{R}^n, X). \tag{1.22}$$

In particular, we deduce $\mathrm{op}[m]f \in H_p^r(\mathbb{R}^n, X)$ for all $f \in \mathscr{S}(\mathbb{R}^n, X)$. So we obtain

$$\|\mathrm{op}[m]f\|_{H_p^r(\mathbb{R}^n, X)} = \|\mathrm{op}[\Lambda_r]\mathrm{op}[m]f\|_{L_p(\mathbb{R}^n, X)} = \|\mathrm{op}[m]\mathrm{op}[\Lambda_r]f\|_{L_p(\mathbb{R}^n, X)}$$
$$\leq C\|\mathrm{op}[\Lambda_r]f\|_{L_p(\mathbb{R}^n, X)} = C\|f\|_{H_p^r(\mathbb{R}^n, X)}$$

due to $\mathrm{op}[\Lambda_r]f \in \mathscr{S}(\mathbb{R}^n, X)$ and Definition 1.73 (ii). According to Remark 1.67 (iv) we can now extend $\mathrm{op}[m]$ to a bounded operator in $L(H_p^r(\mathbb{R}^n, X))$. The representation in (1.20) then follows from (1.22).

(ii) First, we consider the case $r = s$. Let $f \in H_p^r(\mathbb{R}^n, X) \cap H_q^r(\mathbb{R}^n, X)$ and define the function $h := \mathrm{op}[\Lambda_r]f \in L_p(\mathbb{R}^n, X) \cap L_q(\mathbb{R}^n, X)$. Let $\psi_n \in \mathscr{D}(\mathbb{R}^n, X)$ be a Dirac sequence. With this we get $\varphi_n := \psi_n * h \in \mathscr{D}(\mathbb{R}^n, X)$ and $\varphi_n \to h$ in $L_u(\mathbb{R}^n, X)$ for all $1 < u < \infty$. Hence $\mathrm{op}[\Lambda_{-r}]\varphi_n \to f$ in $H_p^r(\mathbb{R}^n, X) \cap H_q^r(\mathbb{R}^n, X)$ due to Lemma 1.58 (ii). It is easy to see that $\mathrm{op}[\Lambda_{-r}]\varphi_n \in \mathscr{S}(\mathbb{R}^n, X)$ and therefore

$$\mathrm{op}^{(r,p)}[m]f = \lim_{n\to\infty} \mathrm{op}^{(r,p)}[m]\mathrm{op}[\Lambda_{-r}]\varphi_n = \lim_{n\to\infty} \mathrm{op}^{(r,q)}[m]\mathrm{op}[\Lambda_{-r}]\varphi_n$$
$$= \mathrm{op}^{(r,q)}[m]f.$$

Next, we consider the case $p = q$. Without loss of generality, we assume $r < s$, which implies $H_p^s(\mathbb{R}^n, X) \hookrightarrow H_p^r(\mathbb{R}^n, X)$. Let $f \in H_p^s(\mathbb{R}^n, X)$ and $(f_n)_{n \in \mathbb{N}} \subseteq \mathscr{S}(\mathbb{R}^n, X)$ with $f_n \to f$ in $H_p^s(\mathbb{R}^n, X)$ (cf. Remark 1.67 (iv)). Then we obtain $f_n \to f$ in $H_p^r(\mathbb{R}^n, X)$ and therefore

$$\mathrm{op}^{(r,p)}[m]f = \lim_{n\to\infty} \mathrm{op}^{(r,p)}[m]f_n = \lim_{n\to\infty} \mathrm{op}^{(s,p)}[m]f_n = \mathrm{op}^{(s,p)}[m]f. \quad \square$$

Remark 1.77. For $1 < p < \infty$ we set $H_p^{-\infty}(\mathbb{R}^n, X) := \bigcup_{r \in \mathbb{R}} H_p^r(\mathbb{R}^n, X)$. Let $m \in H^\infty(\Omega)$ with $\Omega := \prod_{k=1}^n \Sigma_{\delta_k}$. Then Lemma 1.76 (ii) shows that

$$\mathrm{op}[m] \colon H_p^{-\infty}(\mathbb{R}^n, X) \to H_p^{-\infty}(\mathbb{R}^n, X),$$
$$f \mapsto \mathrm{op}^{(r,p)}[m]f, \quad f \in H_p^r(\mathbb{R}^n, X),$$

is well-defined. By Lemma 1.76 (i) we have

$$\mathrm{op}[m]|_{H_p^r(\mathbb{R}^n, X)} \in L(H_p^r(\mathbb{R}^n, X)), \quad r \in \mathbb{R},$$

and by real interpolation (Theorem 1.56 (iii)) we obtain

$$\mathrm{op}[m]|_{B_{pq}^r(\mathbb{R}^n, X)} \in L(B_{pq}^r(\mathbb{R}^n, X)), \quad r \in \mathbb{R}.$$

The proof of the next result includes ideas from [Tri78, p. 232] and [GGHR06, p. 174]. It shows that the existence of a holomorphic extension of a symbol m to a complex half-space is connected with the fact that the support of a function is preserved by op$[m]$. In the case of functions of time, this can be interpreted as causality of the operator. In the context of pseudodifferential operators, operators with this property are also called Volterra operators, see, e.g., [Kra04], [DS11].

Proposition 1.78 (Holomorphic Fourier multiplier and support). *Let $m \in H^\infty(S_\theta)$ with $\theta > \frac{\pi}{2}$, and let $1 < p, q < \infty$, $\mathcal{F} \in \{B_{pq}, H_p\}$, $s \geq 0$ ($s > 0$ if $\mathcal{F} = B_{pq}$). Then m is a Fourier multiplier on $L_p(\mathbb{R}, X)$ and*

$$\left(r_0^+ \mathrm{op}[m] e_0^+\right)\Big|_{{}_0\mathcal{F}^s(\mathbb{R}_+, X)} \in L({}_0\mathcal{F}^s(\mathbb{R}_+, X)) \tag{1.23}$$

is a well-defined bounded operator. In particular, we have

$$\mathrm{supp}\left(\mathrm{op}[m] e_0^+ f\right) \subseteq [0, \infty)$$

and therefore $e_0^+ r_0^+ \mathrm{op}[m] e_0^+ f = \mathrm{op}[m] e_0^+ f$ for all $f \in {}_0\mathcal{F}^s(\mathbb{R}_+, X)$.

Proof. The assertion in (1.23) is obvious due to Remark 1.67 (i) and Remark 1.77.

Next, we prove the claimed position of the support. Let $\varphi \in \mathscr{D}(\mathbb{R}_+, X)$. Then there exists $r > 0$ such that $\mathrm{supp}\,\varphi \subseteq (0, r)$. For simplicity we do not distinguish between φ and $e_0^+ \varphi$ in the following. It is obvious that $\mathscr{F}\varphi \in \mathscr{S}(\mathbb{R}, X)$ and therefore $m(i\cdot)(\mathscr{F}\varphi) \in L_1(\mathbb{R}, X)$. Hence, we can use the explicit formula

$$\left[\mathscr{F}^{-1}(m(i\cdot)(\mathscr{F}\varphi))\right](t) = (2\pi)^{-1/2} \int_{-\infty}^{\infty} m(i\lambda)(\mathscr{F}\varphi)(\lambda) e^{it\lambda} d\lambda$$

for the inverse Fourier transform for all $t \in \mathbb{R}$. The function $\mathscr{F}\varphi$ can be canonically extended to $\mathbb{C}$ by

$$(\mathscr{F}\varphi)(z) := (2\pi)^{-1/2} \int_{-\infty}^{\infty} \varphi(t) e^{-itz} dt, \quad z \in \mathbb{C}.$$

The complex function $\mathscr{F}\varphi$ is an entire function, and the vector-valued Paley-Wiener Theorem (see [Jef04, Proposition 2.1], for instance) yields the estimate

$$\|(\mathscr{F}\varphi)(z)\|_X \leq C(1 + |z|)^{-1} \exp(r|\mathrm{Im}\,z|), \quad z \in \mathbb{C}. \tag{1.24}$$

Using Cauchy's integral theorem we will show

$$\left[\mathscr{F}^{-1}(m(i\cdot)(\mathscr{F}\varphi))\right](t) = (2\pi)^{-1/2} \int_{-\infty}^{\infty} m(i\lambda + \gamma)(\mathscr{F}\varphi)(\lambda - i\gamma) e^{it(\lambda - i\gamma)} d\lambda \tag{1.25}$$

for all $\gamma > 1$ and $t \in \mathbb{R}$. In order to do this we define the closed curve

$$\Gamma(\varepsilon, R) := (-R, -\varepsilon) \cup \{\varepsilon e^{i(\theta + \pi)} : \theta \in [0, \pi]\} \cup (\varepsilon, R)$$
$$\cup \{R - i\eta : \eta \in (0, \gamma)\}$$
$$\cup \{-\eta - i\gamma : \eta \in [-R, R]\}$$
$$\cup \{-R - i(\gamma - \eta) : \eta \in (0, \gamma)\} \subseteq \mathbb{C}$$

1.3. The time-space derivative

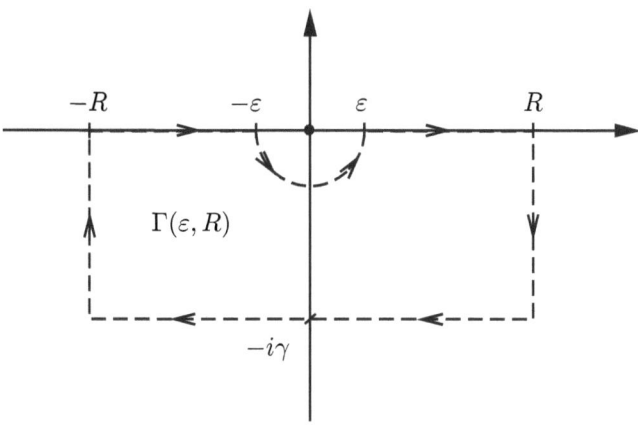

Figure 1.3: Path of integration $\Gamma(\varepsilon, R)$

for $0 < \varepsilon < 1$ and large $1 < R < \infty$ (cf. Figure 1.3).

Due to the holomorphy of all involved functions and Cauchy's integral theorem we obtain

$$\int_{\Gamma(\varepsilon,R)} m(iz)(\mathscr{F}\varphi)(z)e^{itz}dz = 0. \tag{1.26}$$

(I) For the parametrization $\gamma_\varepsilon(\theta) := \varepsilon e^{i(\theta+\pi)}$, $\theta \in [0, \pi]$, we have

$$\left\| \int_{\{\varepsilon e^{i(\theta+\pi)}:\, \theta \in [0,\pi]\}} m(iz)(\mathscr{F}\varphi)(z)e^{itz}dz \right\|_X$$

$$= \left\| \int_0^\pi m(i\gamma_\varepsilon(\theta))(\mathscr{F}\varphi)(\gamma_\varepsilon(\theta))e^{it\gamma_\varepsilon(\theta)}\gamma_\varepsilon'(\theta)d\theta \right\|_X$$

$$\leq \varepsilon \int_0^\pi |m(i\gamma_\varepsilon(\theta))| \cdot \|(\mathscr{F}\varphi)(\gamma_\varepsilon(\theta))\|_X |e^{it\gamma_\varepsilon(\theta)}|d\theta$$

$$\leq \varepsilon \|m\|_\infty C(t)$$

due to $m \in H^\infty(S_\theta)$ and the boundedness of

$$(\theta, \varepsilon) \mapsto \|(\mathscr{F}\varphi)(\gamma_\varepsilon(\theta))\|_X |e^{it\gamma_\varepsilon(\theta)}|.$$

We get

$$\lim_{\varepsilon \to 0} \int_{\{\varepsilon e^{i(\theta+\pi)}:\, \theta \in [0,\pi]\}} m(iz)(\mathscr{F}\varphi)(z)e^{itz}dz = 0.$$

(II) Concerning the parametrization $\gamma_R(\eta) := R - i\eta$, $\eta \in [0, \gamma]$, we have

$$\left\| \int_{\{R-i\eta:\, \eta \in (0,\gamma)\}} m(iz)(\mathscr{F}\varphi)(z)e^{itz}dz \right\|_X$$

$$= \left\| \int_0^\gamma m(i\gamma_R(\eta))(\mathscr{F}\varphi)(\gamma_R(\eta))e^{it\gamma_R(\eta)}d\eta \right\|_X$$

$$\leq \|m\|_\infty \int_0^\gamma \|(\mathscr{F}\varphi)(\gamma_R(\eta))\|_X e^{t\eta} d\eta$$

$$\leq C\|m\|_\infty (1+R)^{-1} \int_0^\gamma e^{(r+t)\eta} d\eta,$$

where we have used (1.24). With this we obtain

$$\lim_{R\to\infty} \int_{\{R-i\eta:\ \eta\in(0,\gamma)\}} m(iz)(\mathscr{F}\varphi)(z)e^{itz}dz = 0.$$

The same arguments also yield

$$\lim_{R\to\infty} \int_{\{-R-i(\gamma-\eta):\ \eta\in(0,\gamma)\}} m(iz)(\mathscr{F}\varphi)(z)e^{itz}dz = 0.$$

With (I), (II) and (1.26) we obtain

$$0 = \lim_{R\to\infty} \lim_{\varepsilon\to 0} \int_{\Gamma(\varepsilon,R)} m(iz)(\mathscr{F}\varphi)(z)e^{itz}dz$$

$$= \int_{-\infty}^\infty m(i\lambda)(\mathscr{F}\varphi)(\lambda)e^{it\lambda}d\lambda - \int_{-\infty}^\infty m(i\lambda+\gamma)(\mathscr{F}\varphi)(\lambda-i\gamma)e^{it(\lambda-i\gamma)}d\lambda,$$

which proves (1.25) for all $\gamma > 0$. For $t < 0$ and all $\gamma > 1$ we obtain

$$\left\|[\mathscr{F}^{-1}(m(i\cdot)(\mathscr{F}\varphi))](t)\right\|_X \leq (2\pi)^{-1/2}\|m\|_\infty e^{t\gamma} \int_{-\infty}^\infty \|(\mathscr{F}\varphi)(\lambda-i\gamma)\|_X d\lambda.$$

Using integration by parts and $\gamma > 1$ it is easy to see that

$$\|(\mathscr{F}\varphi)(\lambda-i\gamma)\|_X = (2\pi)^{-1/2} \left\| \int_0^\infty \varphi(x)\partial_x^2(e^{-ix(\lambda-i\gamma)})dx \right\|_X \cdot |\lambda-i\gamma|^{-2}$$

$$\leq (2\pi)^{-1/2} \left\| \int_0^\infty \varphi''(x)e^{-ix(\lambda-i\gamma)}dx \right\|_X \cdot (\lambda^2+1)^{-1}$$

$$\leq (2\pi)^{-1/2}\|\varphi''\|_\infty \int_0^\infty |e^{-ix(\lambda-i\gamma)}|dx \cdot (\lambda^2+1)^{-1}$$

$$\leq (2\pi)^{-1/2}\|\varphi''\|_\infty \cdot (\lambda^2+1)^{-1}$$

and therefore we derive

$$\left\|[\mathscr{F}^{-1}(m(i\cdot)(\mathscr{F}\varphi))](t)\right\|_X \leq \frac{1}{2}\|\varphi''\|_\infty \|m\|_\infty \lim_{\gamma\to\infty} e^{t\gamma} = 0, \quad t < 0,$$

which yields

$$\operatorname{supp}(\operatorname{op}[m]\varphi) \subseteq [0,\infty).$$

1.3. The time-space derivative

Let $f \in {}_0\mathcal{F}^s(\mathbb{R}_+, X)$. By Remark 1.67 (iv) there exists $(\varphi_k)_{k\in\mathbb{N}} \subseteq \mathscr{D}(\mathbb{R}_+, X)$ with
$$f = \lim_{k\to\infty} \varphi_k \text{ in } {}_0\mathcal{F}^s(\mathbb{R}_+, X).$$
Lemma 1.65, Remark 1.67 (i), and Remark 1.77 (ii) yield
$$\text{op}[m]e_0^+ f = \lim_{k\to\infty} \text{op}[m]e_0^+ \varphi_k = \lim_{k\to\infty} e_0^+ r_0^+ \text{op}[m]e_0^+ \varphi_k = e_0^+ r_0^+ \text{op}[m]e_0^+ f$$
where the limits hold in $\mathcal{F}^s(\mathbb{R}, X)$. We finally derive $\text{supp}(\text{op}[m]f) \subseteq [0, \infty)$ with Lemma 1.65. $\square$

The Fourier multipliers considered above worked in the whole space $\mathbb{R}^n$. With respect to the time variable, however, we will consider the half-line $\mathbb{R}_+$, including vanishing trace condition at $t = 0$. We can easily define Fourier multipliers in $\mathbb{R}_+$ by using the extension and restriction operators e_0^+ and r_0^+. Note that we want to stay in the spaces with zero trace at $t = 0$. For this, the property of the support discussed in Proposition 1.78 will be essential.

For $s \in \mathbb{R}$ and $\lambda \in \mathbb{C} \setminus (-\infty, 0)$, we define the symbol
$$\Psi_s(\lambda) := (1 + \lambda)^s.$$
Symbolically, we want to replace λ by ∂_t. But we have to take into account the initial condition at $t = 0$. Therefore, we define
$$\text{op}_+[\Psi_s] := r_0^+ \text{op}[\Psi_s]e_0^+ = r_0^+ \mathscr{F}^{-1}\Psi_s(i\cdot)\mathscr{F}e_0^+$$
on $L_p(\mathbb{R}_+, X)$.

Proposition 1.79 (Properties of $\text{op}_+[\Psi_s]$). *For $1 < p, q < \infty$ and $s \geq 0$ we have*
$$\text{op}_+[\Psi_s]\big|_{{}_0 H_p^{s'+s}(\mathbb{R}_+, X)} \in L_{\text{Isom}}({}_0 H_p^{s'+s}(\mathbb{R}_+, X), {}_0 H_p^{s'}(\mathbb{R}_+, X)), \quad s' \geq 0,$$
$$\text{op}_+[\Psi_s]\big|_{{}_0 B_{p,q}^{s'+s}(\mathbb{R}_+, X)} \in L_{\text{Isom}}({}_0 B_{p,q}^{s'+s}(\mathbb{R}_+, X), {}_0 B_{p,q}^{s'}(\mathbb{R}_+, X)), \quad s' > 0.$$

Proof. Similar to the argumentation for Λ_r we obtain
$$\text{op}[\Psi_r]\big|_{H_p^{r'+r}(\mathbb{R}, X)} \in L_{\text{Isom}}(H_p^{r'+r}(\mathbb{R}, X), H_p^{r'}(\mathbb{R}, X)), \quad r', r \in \mathbb{R}. \tag{1.27}$$
From (1.27) and Remark 1.67 (i) we get
$$\text{op}_+[\Psi_s]\big|_{{}_0 H_p^{s'+s}(\mathbb{R}_+, X)} \in L({}_0 H_p^{s'+s}(\mathbb{R}_+, X), {}_0 H_p^{s'}(\mathbb{R}_+, X)),$$
$$\text{op}_+[\Psi_{-s}]\big|_{{}_0 H_p^{s'+s}(\mathbb{R}_+, X)} \in L({}_0 H_p^{s'}(\mathbb{R}_+, X), {}_0 H_p^{s'+s}(\mathbb{R}_+, X))$$
for $s, s' \geq 0$. For all $f \in {}_0 H_p^{s'}(\mathbb{R}_+, X)$ we then conclude
$$\text{op}_+[\Psi_s]\text{op}_+[\Psi_{-s}]f = r_0^+ \text{op}[\Psi_s]e_0^+ r_0^+ \text{op}[\Psi_{-s}]e_0^+ f = r_0^+ \text{op}[\Psi_s]\text{op}[\Psi_{-s}]e_0^+ f$$
$$= r_0^+ e_0^+ f = f$$

due to Proposition 1.78. For all $g \in {}_0H_p^{s'+s}(\mathbb{R}_+, X)$ we get

$$\mathrm{op}_+[\Psi_{-s}]\mathrm{op}_+[\Psi_s]g = r_0^+ \mathrm{op}[\Psi_{-s}]e_0^+ r_0^+ \mathrm{op}[\Psi_s]e_0^+ g.$$

In the same way as in the proof of Proposition 1.78, we get $\mathrm{supp}(\mathrm{op}[\Psi_s]e_0^+g) \subseteq [0, \infty)$. The polynomial boundedness of the symbol Ψ_s can be handled by stronger Paley-Wiener estimates of the form

$$\|(\mathcal{F}\varphi)(z)\|_X \leq C(1+|z|)^{\lceil s \rceil + 1} \exp(r|\operatorname{Im} z|), \quad z \in \mathbb{C}.$$

Hence, we also obtain $\mathrm{op}_+[\Psi_{-s}]\mathrm{op}_+[\Psi_s]g = g$. This yields the assertion for the Bessel potential spaces. The remaining assertion follows by interpolation. □

Remark 1.80. The above results hold analogously in exponentially weighted spaces. For $\varrho \geq 0$, we define

$$\mathrm{op}_+^{(\varrho)}[\Psi_s] := \mathscr{M}_\varrho^{-1} \mathrm{op}_+[\Psi_s(\cdot + \varrho)] \mathscr{M}_\varrho$$

in $L_{p,\varrho}(\mathbb{R}_+, X)$. Then the statements in Proposition 1.79 hold for $\mathrm{op}_+^{(\varrho)}[\Psi_s]$ if we replace all spaces by their exponentially weighted versions.

b) Vector-valued space and time derivatives

Now we are going to study realizations of the space derivatives $\partial_j = \partial_{x_j}$, $j = 1, \ldots, n$, in $\mathbb{R}^n$ and of the time derivative ∂_t in $\mathbb{R}_+$. We will show that the corresponding operators have good mapping properties in all scales of Sobolev spaces considered above. To simplify the notation, we will fix $r \in \mathbb{R}$, $p, q \in (1, \infty)$, and $\mathcal{K} \in \{B_{pq}, H_p\}$. As before, X is assumed to be a Banach space of class $\mathcal{HT}$ with property (α). The $\mathcal{K}^r(\mathbb{R}^n, X)$-realization of ∂_j is defined as the operator $\mathcal{D}_j$ given by

$$\mathcal{D}_j \colon \mathcal{K}^r(\mathbb{R}^n, X) \supseteq D(\mathcal{D}_j) \to \mathcal{K}^r(\mathbb{R}^n, X), \tag{1.28}$$
$$u \mapsto \partial_j u$$

where $D(\mathcal{D}_j) := \{u \in \mathcal{K}^r(\mathbb{R}^n, X) \colon \partial_j u \in \mathcal{K}^r(\mathbb{R}^n, X)\}$.

We are interested in a bounded joint H^∞-calculus for the tuple $\mathcal{D} := (\mathcal{D}_1, \ldots, \mathcal{D}_n)$. Note that for better readability, we do not indicate the ground space, in this case $\mathcal{K}^r(\mathbb{R}^n, X)$, in the notation $\mathcal{D}$. Sometimes, we will also use the more precise notation $\mathcal{D}^{\mathcal{N}}$ for the $\mathcal{N}$-realization of $\nabla_x = (\partial_1, \ldots, \partial_n)$.

The following result is a slight generalization of [DV02b, Theorem 7.1, Remark 7.5].

Theorem 1.81 (**Joint $\mathcal{R}$-bounded H^∞-calculus for $\mathcal{D}$**)**.** Let $\delta_j \in (0, \frac{\pi}{2})$ for $j = 1, \ldots, n$ and $\Omega_x := \prod_{j=1}^n \Sigma_{\delta_j}$.

1.3. The time-space derivative

(i) *The vector-valued partial derivative operator $\mathcal{D}_j$ is bisectorial on $\mathcal{K}^r(\mathbb{R}^n, X)$ with $\varphi_{\mathcal{D}_j}^{(bi)} = 0$ for all $j = 1, \ldots, n$. The resolvents are given by*

$$(\lambda - \mathcal{D}_j)^{-1} = \operatorname{op}[R_{\lambda,j}]|_{\mathcal{K}^r(\mathbb{R}^n, X)}, \quad \lambda \in \mathbb{C} \setminus (i\mathbb{R}) \tag{1.29}$$

with $R_{\lambda,j}(z) := (\lambda - z_j)^{-1}$, $z \in i\mathbb{R}^n$. In particular, the tuple $\mathcal{D}$ is admissible.

(ii) *The operator tuple $\mathcal{D}$ has an $\mathcal{R}$-bounded joint $H^\infty(\Omega_x)$-calculus. Moreover, we obtain the representation*

$$f(\mathcal{D}) = \operatorname{op}[f]|_{\mathcal{K}^r(\mathbb{R}^n, X)} \in L(\mathcal{K}^r(\mathbb{R}^n, X)) \tag{1.30}$$

for all $f \in H^\infty(\Omega_x)$.

Proof. (i) (I) Let $j \in \{1, \ldots, n\}$ be arbitrary. The case $r = 0$ and $\mathcal{K} = H_p$ (i.e., the $L_p(\mathbb{R}^n, X)$-realization) can be found in detail in [DV02b, Theorem 7.1, Remark 7.5]. The asserted representation of the resolvent can be obtained easily by standard arguments. The resolvents commute due to the representation as Fourier multipliers (cf. [DV02b, Theorem 7.2]). Therefore, the tuple $\mathcal{D}$ is admissible in the sense of Definition 1.16.

(II) Now let $\mathcal{K} = H_p$. We will make use of the operator $\Phi := \operatorname{op}[\Lambda_r] \in L(H_p^r(\mathbb{R}^n, X), L_p(\mathbb{R}^n, X))$. In the following, we will write $\mathcal{D}_{j,r}$ for the $H_p^r(\mathbb{R}^n, X)$-realization of ∂_j. In order to apply Proposition 1.32 (i) we have to show $D(\mathcal{D}_{j,r}) = \Phi^{-1}(D(\mathcal{D}_{j,0}))$ and $\mathcal{D}_{j,r}u = \Phi^{-1}\mathcal{D}_{j,0}\Phi u$ for all $u \in H_p^r(\mathbb{R}^n, X)$.
For $u \in D(\mathcal{D}_{j,r}) \subseteq H_p^r(\mathbb{R}^n, X)$ we have $\Phi u \in L_p(\mathbb{R}^n, X)$ and $\partial_j \Phi u = \Phi \partial_j u \in L_p(\mathbb{R}^n, X)$ according to Lemma 1.58 (ii). This implies $u = \Phi^{-1}\Phi u \in \Phi^{-1}(D(\mathcal{D}_{j,0}))$.
Let $f \in D(\mathcal{D}_{j,0}) \subseteq L_p(\mathbb{R}^n, X)$. We have $\Phi^{-1}f \in H_p^r(\mathbb{R}^n, X)$ and therefore $\partial_j \Phi^{-1}f = \Phi^{-1}\partial_j f \in H_p^r(\mathbb{R}^n, X)$. This yields $\Phi^{-1}f \in D(\mathcal{D}_{j,r})$.
Hence we obtain $D(\mathcal{D}_{j,r}) = \Phi^{-1}(D(\mathcal{D}_{j,0}))$. It is obvious that $\mathcal{D}_{j,r}u = \Phi^{-1}\mathcal{D}_{j,0}\Phi u$ for all $u \in H_p^r(\mathbb{R}^n, X)$, and therefore Proposition 1.32 (i) and (I) yield the assertion for this case. The representation of the resolvent follows from Proposition 1.32 (i) in combination with Remark 1.77 (i).

(III) Now consider $\mathcal{K} = B_{pq}$, i.e., let $\mathcal{D}_j$ be the $B_{pq}^r(\mathbb{R}^n, X)$-realization of ∂_j. Theorem 1.56 (iii) yields that Besov spaces can be obtained as real interpolation spaces of Bessel potential spaces. The bisectoriality on Besov spaces $B_{pq}^r(\mathbb{R}^n, X)$ can then be derived by Theorem 1.45 (i) (respectively, Remark 1.46) and (II) as soon as

$$(D(\mathcal{D}_{j,-m}), D(\mathcal{D}_{j,m}))_{\theta,q} = D(\mathcal{D}_j)$$

has been shown, with $m := \min\{k \in \mathbb{N} : |r| < k\}$ and $\theta := (r+m)/(2m)$. In this case we also have

$$B_{pq}^r(\mathbb{R}^n, X) = (H_p^{-m}(\mathbb{R}^n, X), H_p^m(\mathbb{R}^n, X))_{\theta,q}.$$

According to $1 \in \rho(\mathcal{D}_{j,-m}) \cap \rho(\mathcal{D}_{j,m})$ and the representation of the resolvents in (II) we get

$$(1 - \mathcal{D}_{j,-m})^{-1} \in L(H_p^{-m}(\mathbb{R}^n, X), D(\mathcal{D}_{j,-m})),$$
$$(1 - \mathcal{D}_{j,-m})^{-1}|_{H_p^m(\mathbb{R}^n,X)} = (1 - \mathcal{D}_{j,m})^{-1} \in L(H_p^m(\mathbb{R}^n, X), D(\mathcal{D}_{j,m})),$$

which yield

$$A := (1 - \mathcal{D}_{j,-m})^{-1}|_{B_{pq}^r(\mathbb{R}^n, X)}$$
$$\in L(B_{pq}^r(\mathbb{R}^n, X), (D(\mathcal{D}_{j,-m}), D(\mathcal{D}_{j,m}))_{\theta,q}).$$

We have $\partial_j \in L(D(\mathcal{D}_{j,m}), H_p^m(\mathbb{R}^n, X)) \cap L(D(\mathcal{D}_{j,-m}), H_p^{-m}(\mathbb{R}^n, X))$ and therefore

$$\partial_j \in L((D(\mathcal{D}_{j,-m}), D(\mathcal{D}_{j,m}))_{\theta,q}, B_{pq}^r(\mathbb{R}^n, X)).$$

The definition of $D(\mathcal{D}_j)$ then implies $(D(\mathcal{D}_{j,-m}), D(\mathcal{D}_{j,m}))_{\theta,q} \subseteq D(\mathcal{D}_j)$. Furthermore, we derive

$$A(1 - \mathcal{D}_j)u = (1 - \mathcal{D}_{j,-m})^{-1}(1 - \mathcal{D}_j)u$$
$$= (1 - \mathcal{D}_{j,-m})^{-1}(1 - \mathcal{D}_{j,-m})u$$
$$= u$$

for all $u \in D(\mathcal{D}_j) \subseteq D(\mathcal{D}_{j,-m})$. So we get

$$D(\mathcal{D}_j) \subseteq R(A) \subseteq (D(\mathcal{D}_{j,-m}), D(\mathcal{D}_{j,m}))_{\theta,q}.$$

Altogether we have shown

$$D(\mathcal{D}_j) = (D(\mathcal{D}_{j,-m}), D(\mathcal{D}_{j,m}))_{\theta,q}.$$

Remark 1.46 (ii) then yields the asserted representation of the resolvents.

(ii) (I) If $r = 0$ and $\mathcal{K} = H_p$ (i.e., in the case of the $L_p(\mathbb{R}^n, X)$-realization), the existence of a bounded H^∞-calculus is a direct consequence of [DV02b, Theorem 7.3, Remark 7.5]. The representation of $f(\mathcal{D})$ ($f \in H_0^\infty(\Omega_x)$) by Fourier-multipliers can be found there, too. Due to the representation of $f(\mathcal{D})$ as a Fourier multiplier we can prove the $\mathcal{R}$-boundedness of the family

$$\{f(\mathcal{D}) : f \in H_0^\infty(\Omega_x), \|f\|_\infty \leq 1\} \subseteq L_p(\mathbb{R}^n, X)$$

by Theorem 1.74, (1.19), and Theorem 1.8.

Let $f \in H^\infty(\Omega_x)$. Then Theorem 1.27 implies $f(\mathcal{D}) \in L(L_p(\mathbb{R}^n, X))$. Using Remark 1.21 (iii), Lemma 1.19 (i), and Proposition 1.75 we obtain

$$f(\mathcal{D})u = \lim_{j \to \infty} (\psi_{j,n} f)(\mathcal{D})u = \lim_{j \to \infty} \text{op}[\psi_{j,n} f]u$$
$$= \lim_{j \to \infty} \text{op}[\psi_{j,n}]\text{op}[f]u = \lim_{j \to \infty} \psi_{j,n}(\mathcal{D})\text{op}[f]u = \text{op}[f]u$$

1.3. The time-space derivative

for all $u \in L_p(\mathbb{R}^n, X)$. So we have proved the claimed representation (1.30) for all $f \in H^\infty(\Omega_x)$.

(II) Proposition 1.32 (iii) with $\Phi := \Lambda_k$ and the same arguments as in part (i) show that $\mathcal{D}$ admits an $\mathcal{R}$-bounded $H^\infty(\Omega_x)$-calculus on $H_p^r(\mathbb{R}^n, X)$ for all $r \in \mathbb{R}$.

(III) Repeating the arguments from part (i), Theorem 1.45 (iii), Remark 1.46, and (II) we can show that $\mathcal{D}$ admits an $\mathcal{R}$-bounded $H^\infty(\Omega_x)$-calculus on $B_{pq}^r(\mathbb{R}^n, X)$ for all $r \in \mathbb{R}$, and

$$f(\mathcal{D}) = [f(\mathcal{D}_{1,-m}, \ldots, \mathcal{D}_{n,-m})]|_{B_{pq}^r(\mathbb{R}^n, X)} = \mathrm{op}[f]|_{B_{pq}^r(\mathbb{R}^n, X)}$$

for all $f \in H^\infty(\Omega_x)$. □

Similarly to the space derivatives ∂_j, $j = 1, \ldots, n$, considered above, we will now analyze the realization of the time derivative ∂_t. For this, we fix $s \geq 0$, $p, q \in (1, \infty)$ and $\mathcal{F} \in \{B_{pq}, H_p\}$ where we assume $s > 0$ if $\mathcal{F} = B_{pq}$. Then the ${}_0\mathcal{F}^s(\mathbb{R}_+, X)$-realization of ∂_t is defined as the operator $\mathcal{D}_t$ given by

$$\mathcal{D}_t \colon {}_0\mathcal{F}^s(\mathbb{R}_+, X) \supseteq D(\mathcal{D}_t) \to {}_0\mathcal{F}^s(\mathbb{R}_+, X), \tag{1.31}$$
$$u \mapsto \partial_t u$$

where $D(\mathcal{D}_t) := {}_0\mathcal{F}^{s+1}(\mathbb{R}_+, X)$. Again we also use the more precise notation $\mathcal{D}_t^\mathcal{N}$ for the $\mathcal{N}$-realization of ∂_t. Note that for $u \in D(\mathcal{D}_t)$ we have $\partial_t u \in {}_0\mathcal{F}^s(\mathbb{R}_+, X)$ by Proposition 1.78.

We will also consider the time-derivative operator on the whole space defined by

$$\widetilde{\mathcal{D}_t} \colon \mathcal{F}^s(\mathbb{R}, X) \supseteq D(\widetilde{\mathcal{D}_t}) \to \mathcal{F}^s(\mathbb{R}, X), \quad u \mapsto \partial_t u$$

where $D(\widetilde{\mathcal{D}_t}) := \mathcal{F}^{s+1}(\mathbb{R}, X)$.

In the following we are interested in the spectral properties of $\mathcal{D}_t$.

Lemma 1.82. *For all $f \in {}_0\mathcal{F}^s(\mathbb{R}_+, X)$ we have $e_0^+ \mathcal{D}_t f = \widetilde{\mathcal{D}_t} e_0^+ f$.*

Proof. Due to Remark 1.67 (i) we have

$$A := e_0^+ \mathcal{D}_t - \widetilde{\mathcal{D}_t} e_0^+ \in L({}_0\mathcal{F}^{s+1}(\mathbb{R}_+, X), \mathcal{F}^s(\mathbb{R}, X))$$

and $A\varphi = 0$ for all $\varphi \in \mathscr{D}(\mathbb{R}_+)$. The assertion then follows from the density of $\mathscr{D}(\mathbb{R}_+)$ in ${}_0\mathcal{F}^{s+1}(\mathbb{R}_+, X)$, cf. Remark 1.67 (iv). □

In the following theorem, we use the notation σ_r, σ_p, and σ_c for the residual spectrum, the point spectrum, and the continuous spectrum of an operator, respectively.

Theorem 1.83 (Spectrum of the time derivative). *For the $_0\mathcal{F}^s(\mathbb{R}_+, X)$-realization $\mathcal{D}_t$ of ∂_t we have*

$$\rho(\mathcal{D}_t) = \{z \in \mathbb{C} \colon \operatorname{Re}(z) < 0\},$$
$$\sigma_r(\mathcal{D}_t) = \{z \in \mathbb{C} \colon \operatorname{Re}(z) > 0\},$$
$$\sigma_p(\mathcal{D}_t) = \emptyset,$$
$$\sigma_c(\mathcal{D}_t) = i\mathbb{R}.$$

In particular, $\mathcal{D}_t$ is a sectorial operator with $\varphi_{\mathcal{D}_t} = \frac{\pi}{2}$ and

$$(\lambda - \mathcal{D}_t)^{-1} u = r_0^+ (\lambda - \widetilde{\mathcal{D}_t})^{-1} e_0^+ u, \quad u \in {_0\mathcal{F}^s(\mathbb{R}_+, X)},$$

for all λ with $\operatorname{Re}(\lambda) < 0$.

Proof. Let $u \in D(\mathcal{D}_t)$ and $\lambda \in \mathbb{C}$ with $(\lambda - \mathcal{D}_t)u = 0$. According to Sobolev's embedding theorem (cf. Remark 1.67 (v)) we have $_0\mathcal{F}^{s+1}(\mathbb{R}_+, X) \hookrightarrow {_0\mathcal{F}^1(\mathbb{R}_+, X)} \hookrightarrow C_b([0, \infty), X)$. Due to the characterization in Remark 1.67 (iii) and (v) we have the classical vanishing trace $u(0) = 0$. In fact, we even have $u \in C_b^1([0, \infty), X)$ since $\partial_t u = \lambda u \in C_b([0, \infty), X)$. For arbitrary $x' \in X'$ (the dual space of X) we define the complex-valued function $\widetilde{u} := x' \circ u \in C_b^1([0, \infty))$. The function $\widetilde{u}$ solves the scalar ordinary differential equation

$$\lambda \widetilde{u} - \partial_t \widetilde{u} = 0, \quad \widetilde{u}(0) = 0.$$

With the classical theory of ordinary differential equation we trivially obtain $\widetilde{u} = 0$ and therefore also $u = 0$. Hence $(\lambda - \mathcal{D}_t)$ is one-to-one for all $\lambda \in \mathbb{C}$, which yields $\sigma_p(\mathcal{D}_t) = \emptyset$.

Let $\operatorname{Re}\lambda < 0$. Then we can use the results about the space-derivative given above. From (1.29) we obtain

$$(\lambda - \widetilde{\mathcal{D}_t})^{-1} = \operatorname{op}[R_\lambda]|_{\mathcal{F}^s(\mathbb{R}_+, X)}, \quad R_\lambda(z) := (\lambda - z)^{-1}, \quad z \in i\mathbb{R}.$$

The function R_λ has a holomorphic extension to a bisector, i.e., $R_\lambda \in H^\infty(\Sigma_\delta)$ for small $\delta > 0$. Therefore we can apply Proposition 1.78 for $\operatorname{op}[R_\lambda]$. It can be easily seen by Lemma 1.82 and Remark 1.67 (i) that

$$(\lambda - \mathcal{D}_t) r_0^+ (\lambda - \widetilde{\mathcal{D}_t})^{-1} e_0^+ f$$
$$= r_0^+ e_0^+ (\lambda - \mathcal{D}_t) r_0^+ (\lambda - \widetilde{\mathcal{D}_t})^{-1} e_0^+ f$$
$$= r_0^+ (\lambda - \widetilde{\mathcal{D}_t}) e_0^+ r_0^+ (\lambda - \widetilde{\mathcal{D}_t})^{-1} e_0^+ f$$
$$= r_0^+ (\lambda - \widetilde{\mathcal{D}_t})(\lambda - \widetilde{\mathcal{D}_t})^{-1} e_0^+ f = f$$

for all $f \in {_0\mathcal{F}^s(\mathbb{R}_+, X)}$. In the same way we obtain

$$r_0^+ (\lambda - \widetilde{\mathcal{D}_t})^{-1} e_0^+ (\lambda - \mathcal{D}_t) f = f, \quad f \in {_0\mathcal{F}^{s+1}(\mathbb{R}_+, X)}$$

1.3. The time-space derivative

by Lemma 1.82. Altogether we obtain $\{z \in \mathbb{C} \colon \operatorname{Re}(z) < 0\} \subseteq \rho(\mathcal{D}_t)$ and

$$(\lambda - \mathcal{D}_t)^{-1} = r_0^+ (\lambda - \widetilde{\mathcal{D}_t})^{-1} e_0^+$$

for all λ with $\operatorname{Re}\lambda < 0$. With this representation of the resolvents and Remark 1.67 (i) we derive, for every $\theta' \in (\pi/2, \pi)$,

$$\sup_{\lambda \in \mathbb{C} \setminus \overline{S}_{\theta'}} \|\lambda(\lambda - \mathcal{D}_t)^{-1}\|_{L(_0\mathcal{F}^s(\mathbb{R}_+, X))} \\ \leq C \sup_{\lambda \in \mathbb{C} \setminus \overline{S}_{\theta'}} \|\lambda(\lambda - \widetilde{\mathcal{D}_t})^{-1}\|_{L(\mathcal{F}^s(\mathbb{R}, X))} < \infty \tag{1.32}$$

from the bisectoriality of $\widetilde{\mathcal{D}_t}$.

For λ with $\operatorname{Re}(\lambda) > 0$ we have to determine the range of $\lambda - \mathcal{D}_t$. For this, we consider the adjoint operator $(\lambda - \mathcal{D}_t)'$. Due to the embedding $_0\mathcal{F}^s(\mathbb{R}_+, X) \hookrightarrow L_p(\mathbb{R}_+, X)$ and Hölder's inequality we can define a functional in $(_0\mathcal{F}^s(\mathbb{R}_+, X))'$ by

$$F_\lambda \colon {}_0\mathcal{F}^s(\mathbb{R}_+, X) \to \mathbb{C},$$
$$f \mapsto \int_0^\infty \langle f(t), e^{-\lambda t} x' \rangle_{X \times X'} dt$$

for $x' \in X'$. We easily obtain $\int_0^\infty e^{-\lambda t} \varphi'(t) dt = \int_0^\infty \lambda e^{-\lambda t} \varphi(t) dt$ for all $\varphi \in \mathscr{D}(\mathbb{R}_+, X)$. The density of the test functions and $_0\mathcal{F}^{s+1}(\mathbb{R}_+, X) \hookrightarrow {}_0H_p^1(\mathbb{R}_+, X)$ show that this is even true for all $\varphi \in D(\mathcal{D}_t) = {}_0\mathcal{F}^{s+1}(\mathbb{R}_+, X)$. Thus, we can conclude

$$F_\lambda((\lambda - \mathcal{D}_t)f) = x' \left(\int_0^\infty \lambda e^{-\lambda t} f(t) dt - \int_0^\infty e^{-\lambda t} f'(t) dt \right) = 0$$

for all $f \in D(\mathcal{D}_t)$. Hence we obtain $F_\lambda \in D((\lambda - \mathcal{D}_t)')$ and

$$0 \neq F_\lambda \in \ker((\lambda - \mathcal{D}_t)') = R(\lambda - \mathcal{D}_t)^\perp. \tag{1.33}$$

Here

$$R(\lambda - \mathcal{D}_t)^\perp := \{ F \in \left({}_0\mathcal{F}^s(\mathbb{R}_+, X) \right)' : \mathcal{F}|_{R(\lambda - \mathcal{D}_t)} = 0 \}$$

denotes the annihilator of $R(\lambda - \mathcal{D}_t)$ and the equality in (1.33) follows from [Kat76, p. 168]. Consequently, $R(\lambda - \mathcal{D}_t)$ cannot be dense in $_0\mathcal{F}^s(\mathbb{R}_+, X)$ and we conclude $\{z \in \mathbb{C} \colon \operatorname{Re}(z) > 0\} \subseteq \sigma_r(\mathcal{D}_t)$. As the spectrum is closed, this implies $i\mathbb{R} \subseteq \sigma(\mathcal{D}_t)$.

Due to $\sigma_p(\mathcal{D}_t) = \emptyset$ the operator $\mu i - \mathcal{D}_t$ is injective for all $\mu \in \mathbb{R}$. According to (1.32) and $|\lambda| \leq C_{\theta'} |\lambda + i\mu|$, $\lambda \in \mathbb{C} \setminus \overline{S}_{\theta'}$, the operator $\mathcal{D}_t - i\mu$ satisfies the resolvent estimates in Lemma 1.4. Therefore we obtain the density of $R(\mathcal{D}_t - i\mu)$ in $_0\mathcal{F}^s(\mathbb{R}_+, X)$ by reflexivity of X (cf. Remark 1.13 (i)). Hence we have $i\mathbb{R} \subseteq \sigma_c(\mathcal{D}_t)$.

Altogether we have $\{z \in \mathbb{C} \colon \operatorname{Re}(z) < 0\} \subseteq \rho(\mathcal{D}_t)$, $\sigma_p(\mathcal{D}_t) = \emptyset$, $\{z \in \mathbb{C} \colon \operatorname{Re}(z) > 0\} \subseteq \sigma_r(\mathcal{D}_t)$, and $i\mathbb{R} \subseteq \sigma_c(\mathcal{D}_t)$. Thus we have obtained the full characterization of the spectrum. □

Thanks to our previous results we easily obtain the existence of an $\mathcal{R}$-bounded H^∞-calculus of $\mathcal{D}_t$. This was already stated in [DSS08, Proposition 2.7] but our approach is slightly different.

Theorem 1.84. *The vector-valued time-derivative operator $\mathcal{D}_t$ has an $\mathcal{R}$-bounded H^∞-calculus in ${}_0\mathcal{F}^s(\mathbb{R}_+, X)$ with $\varphi_{\mathcal{D}_t}^{\mathcal{R},\infty} = \frac{\pi}{2}$. For all $f \in H^\infty(S_\theta)$ with $\theta \in (\frac{\pi}{2}, \pi)$ we can represent $f(\mathcal{D}_t)$ by*

$$f(\mathcal{D}_t) = \left(r_0^+ f(\widetilde{\mathcal{D}_t}) e_0^+\right)\big|_{{}_0\mathcal{F}^s(\mathbb{R}_+, X)} = \left(r_0^+ \mathrm{op}[f] e_0^+\right)\big|_{{}_0\mathcal{F}^s(\mathbb{R}_+, X)}. \tag{1.34}$$

Proof. Due to Theorem 1.83 the resolvent of $\mathcal{D}_t$ can be written as

$$(\lambda - \mathcal{D}_t)^{-1} u = r_0^+ (\lambda - \widetilde{\mathcal{D}_t})^{-1} e_0^+ u, \quad u \in {}_0\mathcal{F}^s(\mathbb{R}_+, X), \quad \mathrm{Re}\,\lambda < 0.$$

Using Remark 1.21 (iii) and Definition 1.18 we infer

$$f(\mathcal{D}_t) u = r_0^+ f(\widetilde{\mathcal{D}_t}) e_0^+ u, \quad u \in {}_0\mathcal{F}^s(\mathbb{R}_+, X)$$

for all $f \in H_0^\infty(S_\theta) \subseteq H^\infty(\Sigma_{\theta-\pi/2})$ and $\theta \in (\frac{\pi}{2}, \pi)$. So we derive the representation (1.34) for all $f \in H_0^\infty(S_\theta)$ by (1.30). As in part (I) of the proof of Theorem 1.81 (ii) we can also verify this representation for all $f \in H^\infty(S_\theta)$.

According to Theorem 1.81 (ii) we have the $\mathcal{R}$-boundedness of

$$\{f(\widetilde{\mathcal{D}_t}) : f \in H_0^\infty(S_\theta), \|f\|_\infty \leq 1\} \subseteq L(\mathcal{F}^s(\mathbb{R}, X)).$$

The boundedness of all involved operators and Remark 1.7 (iii) then yield the $\mathcal{R}$-boundedness of

$$\{f(\mathcal{D}_t) : f \in H_0^\infty(S_\theta), \|f\|_\infty \leq 1\} \subseteq L({}_0\mathcal{F}^s(\mathbb{R}_+, X))$$

for all $\theta \in (\frac{\pi}{2}, \pi)$. Thus, we even get $\varphi_{\mathcal{D}_t}^{\mathcal{R},\infty} = \frac{\pi}{2}$. □

c) Joint space-time H^∞-calculus

Now we are going to show that the tuple (∂_t, ∇_x) admits a bounded joint H^∞-calculus. We will consider the operators in the space ${}_0\mathcal{F}^s(\mathbb{R}_+, \mathcal{K}^r(\mathbb{R}^n, X))$ where $\mathcal{F}, \mathcal{K} \in \{B_{pq}, H_p\}$ as before. As $\mathcal{K}^r(\mathbb{R}^n, X)$ is a Banach space of class $\mathcal{HT}$ with property (α), too, the results from Subsection b) can be applied directly to $\mathcal{D}_t$. For the space derivatives $\mathcal{D} = (\mathcal{D}_1, \ldots, \mathcal{D}_n)$, we know the existence of an $\mathcal{R}$-bounded joint H^∞-calculus in the space $\mathcal{K}^r(\mathbb{R}^n, X)$. We now have to consider the natural extension to ${}_0\mathcal{F}^s(\mathbb{R}_+, \mathcal{K}^r(\mathbb{R}^n, X))$, i.e., now the space derivatives act on functions of time and space. Note that we identify a function $u = u(t, x)$ with the vector-valued function $t \mapsto u(t, \cdot)$, and therefore the natural extension of $\mathcal{D}_j$ can be written as $t \mapsto (\mathcal{D}_j \circ u)(t, \cdot)$.

In the following, we fix $p_0, p_1, q_0, q_1 \in (1, \infty)$, $\mathcal{F} \in \{B_{p_0 q_0}, H_{p_0}\}$, $\mathcal{K} \in \{B_{p_1 q_1}, H_{p_1}\}$, $s \geq 0$ ($s > 0$ if $\mathcal{F} = B_{p_0 q_0}$), and $r \in \mathbb{R}$.

We start with a definition of the natural extension of an operator.

1.3. The time-space derivative

Definition and Lemma 1.85 (Natural extension). *Let X and Y be Banach spaces of class $\mathcal{HT}$. For a densely defined closed linear operator $A\colon X \supseteq D(A) \to Y$ the natural extension of A to $_0\mathcal{F}^s(\mathbb{R}_+, X)$ given by*

$$A^+ \colon {_0\mathcal{F}^s}(\mathbb{R}_+, X) \supseteq D(A^+) \to {_0\mathcal{F}^s}(\mathbb{R}_+, Y),$$
$$u \mapsto A \circ u$$

is a well-defined operator with dense domain $D(A^+) := {_0\mathcal{F}^s}(\mathbb{R}_+, D(A))$.

Proof. First, we consider the case $s \in \mathbb{N}_0$ and $\mathcal{F} = H_{p_0}$. We have $\partial_t^k(A \circ u) = A \circ (\partial_t^k u)$ for all $k \leq s$ and all $u \in D(A^+) = {_0H_{p_0}^s}(\mathbb{R}_+, D(A))$. Thus, we obtain

$$\partial_t^k u \in {_0H_{p_0}^{s-k}}(\mathbb{R}_+, D(A)), \quad k \leq s$$

for $u \in D(A^+)$. We have $\lim_{t \to 0} \partial_t^k u(t) = 0$ in $D(A)$ for all $k \leq s-1$, which leads to

$$\lim_{t \to 0} \partial_t^k(A \circ u)(t) = \lim_{t \to 0} A(\partial_t^k u(t)) = 0, \quad k \leq s-1.$$

So we have proved $A \circ u \in {_0H_{p_0}^s}(\mathbb{R}_+, Y)$ for $u \in {_0H_{p_0}^s}(\mathbb{R}_+, D(A))$ and

$$A^+ \in L({_0H_{p_0}^s}(\mathbb{R}_+, D(A)), {_0H_{p_0}^s}(\mathbb{R}_+, Y)) \tag{1.35}$$

with $\|A^+\|_{L({_0H_{p_0}^s}(\mathbb{R}_+, D(A)), {_0H_{p_0}^s}(\mathbb{R}_+, Y))} \leq 1$. Note that $D(A)$ is equipped with the graph norm.

Next, we consider the case $s \geq 0$ and $\mathcal{F}$ as in Definition 1.71. We directly get

$$A^+ \in L({_0\mathcal{F}^s}(\mathbb{R}_+, D(A)), {_0\mathcal{F}^s}(\mathbb{R}_+, Y)) \tag{1.36}$$

by an interpolation argument, (1.35), and Theorem 1.56. The real and complex interpolation methods are exact (cf. Theorem 1.39) and therefore we also have

$$\|A^+\|_{L({_0\mathcal{F}^s}(\mathbb{R}_+, D(A)), {_0\mathcal{F}^s}(\mathbb{R}_+, Y))} \leq 1. \tag{1.37}$$

Altogether we conclude that A^+ is a well-defined operator.

We have to prove the density of the domain. For this we need the concept of tensor products. For a Banach space Z we define

$$\mathscr{D}(\mathbb{R}_+) \otimes Z := \operatorname{span}\{\varphi \cdot z \colon \varphi \in \mathscr{D}(\mathbb{R}_+),\, z \in Z\} \subseteq \mathscr{D}(\mathbb{R}_+, Z)$$

endowed with the subspace topology of the locally convex space $\mathscr{D}(\mathbb{R}_+, Z)$. One can easily show

$$\mathscr{D}(\mathbb{R}_+) \otimes D(A) \stackrel{d}{\hookrightarrow} \mathscr{D}(\mathbb{R}_+) \otimes X, \tag{1.38}$$

and in [Ama03, Theorem 1.3.6 (i)] one can find the dense embedding

$$\mathscr{D}(\mathbb{R}_+) \otimes X \stackrel{d}{\hookrightarrow} \mathscr{D}(\mathbb{R}_+, X). \tag{1.39}$$

Altogether we obtain $\mathscr{D}(\mathbb{R}_+) \otimes D(A) \overset{d}{\hookrightarrow} {}_0\mathcal{F}^s(\mathbb{R}_+, X)$ due to (1.38), (1.39), and Remark 1.67 (iv). So we have proved

$$\mathscr{D}(\mathbb{R}_+) \otimes D(A) \overset{d}{\hookrightarrow} {}_0\mathcal{F}^s(\mathbb{R}_+, X),$$

which yields the density of ${}_0\mathcal{F}^s(\mathbb{R}_+, D(A))$ in ${}_0\mathcal{F}^s(\mathbb{R}_+, X)$ because of $\mathscr{D}(\mathbb{R}_+) \otimes D(A) \subseteq {}_0\mathcal{F}^s(\mathbb{R}_+, D(A))$. □

Lemma 1.86. (i) *In the same situation as in Definition and Lemma 1.85 we have*

$$A^+ \in L({}_0\mathcal{F}^s(\mathbb{R}_+, X), {}_0\mathcal{F}^s(\mathbb{R}_+, Y))$$

if $A \in L(X, Y)$. In particular, we have

$$A^+ \in L_{\mathrm{Isom}}({}_0\mathcal{F}^s(\mathbb{R}_+, X), {}_0\mathcal{F}^s(\mathbb{R}_+, Y))$$

if and only if $A \in L_{\mathrm{Isom}}(X, Y)$. In both cases we also have

$$\|A^+\|_{L({}_0\mathcal{F}^s(\mathbb{R}_+, X), {}_0\mathcal{F}^s(\mathbb{R}_+, Y))} \leq \|A\|_{L(X,Y)}.$$

(ii) *Let $\mathcal{T} \subseteq L(X, Y)$ be $\mathcal{R}$-bounded. Then the family of extended operators*

$$\mathcal{T}^+ := \{T^+ : T \in \mathcal{T}\} \subseteq L({}_0\mathcal{F}^s(\mathbb{R}_+, X), {}_0\mathcal{F}^s(\mathbb{R}_+, Y))$$

is also $\mathcal{R}$-bounded.

Proof. (i) The boundedness of A^+ follows from (1.37) and the equivalence of the norms $\|\cdot\|_{D(A)}$ and $\|\cdot\|_X$. The estimate can be obtained by the same arguments as in the proof of Definition and Lemma 1.85.

If A is bijective, it is obvious that $(A^{-1})^+$ is the inverse of A^+ and therefore A^+ is also bijective. Let A^+ be bijective and let $x \in X$ with $Ax = 0$. For $\varphi \in \mathscr{D}(\mathbb{R}_+) \setminus \{0\}$ we then derive $A^+(\varphi \cdot x) = 0$, which yields $\varphi \cdot x = 0$. Thus we get $x = 0$. Let $y \in Y$ and $\varphi \in \mathscr{D}(\mathbb{R}_+) \setminus \{0\}$. Then there exists $f \in {}_0\mathcal{F}^s(\mathbb{R}_+, X)$ with $A^+f = \varphi \cdot y$. For $t_0 \in \mathbb{R}_+$ with $\varphi(t_0) \neq 0$ we then derive $A(f(t_0)/\varphi(t_0)) = y$. Hence A is also bijective.

(ii) First, we consider the case $s \in \mathbb{N}_0$ and $\mathcal{F} = H_{p_0}$. Let $m \in \mathbb{N}$, $(T_k)_{k=1,\ldots,m} \subseteq$

1.3. The time-space derivative

$\mathcal{T}$, and $(f_k)_{k=1,\ldots,m} \subseteq {}_0H^s_{p_0}(\mathbb{R}_+, X)$. Then we have

$$\Big\| \sum_{k=1}^m r_k T_k^+ f_k \Big\|_{L_{p_0}([0,1], {}_0H^s_{p_0}(\mathbb{R}_+,Y))}$$

$$= \Big(\int_0^1 \sum_{j=1}^s \Big\| \sum_{k=1}^m r_k(\omega) T_k^+ \Big(\Big(\frac{d}{dt}\Big)^j f_k\Big) \Big\|_{L_{p_0}(\mathbb{R}_+,Y)}^{p_0} d\omega \Big)^{1/p_0}$$

$$= \Big(\sum_{j=1}^s \int_{\mathbb{R}_+} \int_0^1 \Big\| \sum_{k=1}^m r_k(\omega) T_k \Big[\Big(\Big(\frac{d}{dt}\Big)^j f_k\Big)(t)\Big] \Big\|_Y^{p_0} d\omega\, dt \Big)^{1/p_0}$$

$$\leq \mathcal{R}_{p_0}(\mathcal{T}) \Big(\sum_{j=1}^s \int_{\mathbb{R}_+} \int_0^1 \Big\| \sum_{k=1}^m r_k(\omega) \Big(\Big(\frac{d}{dt}\Big)^j f_k\Big)(t) \Big\|_X^{p_0} d\omega\, dt \Big)^{1/p_0}$$

$$= \Big\| \sum_{k=1}^m r_k f_k \Big\|_{L_{p_0}([0,1], {}_0H^s_{p_0}(\mathbb{R}_+,X))}$$

by Fubini's theorem and the $\mathcal{R}$-boundedness of $\mathcal{T}$. An interpolation argument based on Theorem 1.44 and Remark 1.67 (ii) then yields the assertion for the whole fractional Bessel potential and Besov scale. $\square$

The following result shows that the natural extension is compatible with the H^∞-calculus.

Lemma 1.87. *Let X be a Banach space of class $\mathcal{HT}$ and let $\mathbf{T} = (T_1, \ldots, T_N)$ be an admissible tuple on X with $T_k \colon X \supseteq D(T_k) \to X$, $k = 1, \ldots, N$, and let $\Omega \subseteq \mathbb{C}^n$ be chosen in accordance to $\mathbf{T}$ as in the text after Definition 1.16. The following permanence properties for the natural extension $\mathbf{T}^+ := (T_1^+, \ldots, T_N^+)$ on ${}_0\mathcal{F}^s(\mathbb{R}_+, X)$ hold.*

(i) *If T_k is sectorial, then T_k^+ is also sectorial with $\varphi_{T_k^+} = \varphi_{T_k}$ and*

$$(\lambda - T_k^+)^{-1} = \big[(\lambda - T_k)^{-1}\big]^+$$

for all $\lambda \in \rho(T_k)$. The analog result holds for bisectorial operators.

(ii) *The natural extension $\mathbf{T}^+$ is an admissible tuple on ${}_0\mathcal{F}^s(\mathbb{R}_+, X)$.*

(iii) *We have*
$$f(\mathbf{T}^+) = f(\mathbf{T})^+$$
for all $f \in H_0^\infty(\Omega)$.

(iv) *If $\mathbf{T}$ admits a bounded $H^\infty(\Omega)$-calculus, then $\mathbf{T}^+$ also admits a bounded $H^\infty(\Omega)$-calculus with*
$$f(\mathbf{T}^+) = f(\mathbf{T})^+$$
for all $f \in H^\infty(\Omega)$.

(v) *If* **T** *admits an $\mathcal{R}$-bounded $H^\infty(\Omega)$-calculus, then* **T**$^+$ *also admits an $\mathcal{R}$-bounded $H^\infty(\Omega)$-calculus.*

Proof. (i) Let T_k be sectorial or bisectorial. Then T_k^+ is also a densely defined operator by Lemma and Definition 1.85. Lemma 1.86 directly yields $\rho(T_k) = \rho(T_k^+)$ and $(\lambda - T_k^+)^{-1} = \left[(\lambda - T_k)^{-1}\right]^+$ for $\lambda \in \rho(T_k)$. Now Lemma 1.86 yields the resolvent estimate for T_k^+ and thus $\varphi_{T_k^+} = \varphi_{T_k}$ (respectively, $\varphi_{T_k^+}^{(\text{bi})} = \varphi_{T_k}^{(\text{bi})}$).

The operator T_k is injective according to Remark 1.4 (ii) and therefore it is obvious that T_k^+ is also injective. Hence $R(T_k^+)$ is dense in ${}_0\mathcal{F}^s(\mathbb{R}_+, X)$ due to Remark 1.4 (ii) again.

(ii) All T_k^+ are sectorial or bisectorial so we only have to prove that the resolvents commute. Using (i) we obtain

$$(\lambda - T_k^+)^{-1}(\mu - T_j^+)^{-1} = [(\lambda - T_k)^{-1}]^+[(\mu - T_j)^{-1}]^+$$
$$= [(\lambda - T_k)^{-1}(\mu - T_j)^{-1}]^+ = [(\mu - T_j)^{-1}(\lambda - T_k)^{-1}]^+$$
$$= (\mu - T_j^+)^{-1}(\lambda - T_k^+)^{-1}$$

for all $\lambda \in \rho(T_k)$, $\mu \in \rho(T_j)$ and $j, k = 1, \ldots, N$.

(iii) Let $f \in H_0^\infty(\Omega)$. By approximating the integrand by step functions, it is easy to see that the following equalities hold (see also [Kai12, Lemma 2.44] for details):

$$(f(\mathbf{T}^+)u)(t) = (2\pi i)^{-N}\left(\left[\int_\Gamma f(z)\prod_{k=1}^N [(z_k - T_k)^{-1}]^+ dz\right]u\right)(t)$$
$$= (2\pi i)^{-N}\left(\int_\Gamma f(z)\prod_{k=1}^N [(z_k - T_k)^{-1}]^+ u\, dz\right)(t)$$
$$= (2\pi i)^{-N}\int_\Gamma f(z)\prod_{k=1}^N (z_k - T_k)^{-1} u(t)\, dz$$
$$= (2\pi i)^{-N}\int_\Gamma f(z)\prod_{k=1}^N (z_k - T_k)^{-1} dz\, u(t) = (f(\mathbf{T})^+ u)(t)$$

for $u \in {}_0\mathcal{F}^s(\mathbb{R}_+, X)$ and almost all $t \in \mathbb{R}_+$. Hence, we obtain $f(\mathbf{T}^+)u = f(\mathbf{T})^+ u$.

(iv) Lemma 1.86 and (iii) directly yield that $\mathbf{T}^+$ admits a bounded $H^\infty(\Omega)$-calculus. Remark 1.21 (iii) and Theorem 1.27 imply $f(\mathbf{T}^+) = f(\mathbf{T})^+$ for all $f \in H^\infty(\Omega)$.

(v) This easily follows from (iii) and Lemma 1.86 (ii). □

1.3. The time-space derivative

For the remainder of this subsection, let X be a Banach space of class $\mathcal{HT}$ with property (α).

Corollary 1.88. *Let $\mathcal{D}_t$ be the ${}_0\mathcal{F}^s(\mathbb{R}_+, \mathcal{K}^r(\mathbb{R}^n, X))$-realization of ∂_t and $\mathcal{D}^+$ be the natural extension of the $\mathcal{K}^r(\mathbb{R}^n, X)$-realization $\mathcal{D}$ to ${}_0\mathcal{F}^s(\mathbb{R}_+, \mathcal{K}^r(\mathbb{R}^n, X))$. Then we have the following results:*

(i) *The sectorial operator $\mathcal{D}_t$ admits an $\mathcal{R}$-bounded $H^\infty(S_\theta)$-calculus for all $\theta \in (\pi/2, \pi)$.*

(ii) *The operator tuple $\mathcal{D}^+$ admits a bounded joint $H^\infty(\Omega_x)$-calculus where $\Omega_x := \prod_{j=1}^n \Sigma_{\delta_j}$ and $\delta_j \in (0, \pi/2)$, $j = 1, \ldots, n$.*

(iii) *The tuple $(\mathcal{D}_t, \mathcal{D}^+)$ is admissible.*

Proof. In parts these results have already been stated in Theorem 1.84 and Theorem 1.81 (ii). Using Lemma 1.87, the assertions in (i) and (ii) follow. The resolvents commute due to the representation as Fourier multipliers, thus the admissibility follows. □

For simplicity of notation, we will from now on again write $\mathcal{D}$ instead of $\mathcal{D}^+$, so all operators will be defined on ${}_0\mathcal{F}^s(\mathbb{R}_+, \mathcal{K}^r(\mathbb{R}^n, X))$. The resulting tuple will be denoted by

$$\mathcal{D}_+ := (\mathcal{D}_t, \mathcal{D}) = (\mathcal{D}_t, \mathcal{D}_1, \ldots, \mathcal{D}_n).$$

This is the ${}_0\mathcal{F}^s(\mathbb{R}_+, \mathcal{K}^r(\mathbb{R}^n, X))$-realization of

$$\nabla_+ := (\partial_t, \nabla_x).$$

Again we will also use the more precise notation $\mathcal{D}_+^\mathcal{N}$ for the $\mathcal{N}$-realization of ∇_+.

One of the main results of this chapter states that $\mathcal{D}_+$ admits a bounded joint H^∞-calculus:

Theorem 1.89 (Joint time-space H^∞-calculus). *The tuple $\mathcal{D}_+$ admits a bounded joint $H^\infty(\Omega)$-calculus on ${}_0\mathcal{F}^s(\mathbb{R}_+, \mathcal{K}^r(\mathbb{R}^n, X))$ where $\Omega := S_\theta \times \prod_{k=1}^n \Sigma_{\delta_k}$ with $\theta \in (\pi/2, \pi)$, $\delta_k \in (0, \pi/2)$, $k = 1, \ldots, n$.*

Proof. Let $\Omega_t := S_\theta$, $\Omega_x := \prod_{j=1}^n \Sigma_{\delta_j}$, and let $h \in H_0^\infty(\Omega)$. Then we define

$$g\colon \Omega_x \to L\big({}_0\mathcal{F}^s(\mathbb{R}_+, \mathcal{K}^r(\mathbb{R}^n, X))\big),$$
$$z \mapsto \|h\|_\infty^{-1} \cdot h(\mathcal{D}_t, z),$$

which is meaningful because of $h(\cdot, z) \in H_0^\infty(\Omega_t)$, $z \in \Omega_x$. Due to Lemma 1.29 we have $g \in H_0^\infty(\Omega_x, \mathcal{B}_{\mathcal{D}_+})$ and

$$\|h\|_\infty^{-1} \cdot \|h(\cdot, z)\|_\infty \leq 1, \quad z \in \Omega_x.$$

Hence, Corollary 1.88 (i) yields the $\mathcal{R}$-boundedness of $g(\Omega_x)$ and therefore $g \in H_\mathcal{R}^\infty(\Omega_x, \mathcal{B}_{\mathcal{D}_+})$. Note that $\mathcal{R}_2(g(\Omega_x))$ can be estimated from above by a constant which is independent of h.

From Corollary 1.88 (ii) we get that $\mathcal{D}$ has a bounded joint $H^\infty(\Omega'_x)$-calculus for $\Omega'_x := \prod_{j=1}^n \Sigma_{\delta'_j}$ with $\delta'_j \in (0, \delta_j)$, $j = 1, \ldots, n$. Therefore, we can conclude with Theorem 1.28 that

$$\|g(\mathcal{D})\|_{L(_0\mathcal{F}^s(\mathbb{R}_+, \mathcal{K}^r(\mathbb{R}^n, X)))} \leq C_1 \cdot \mathcal{R}_2(g(\Omega_x)) \leq C_2 \tag{1.40}$$

with $C_1 = C_1(\mathcal{D}) > 0$, $C_2 = C_2(\mathcal{D}) > 0$. Due to Lemma 1.29 we obtain

$$g(\mathcal{D}) = \|h\|_\infty^{-1} \cdot h(\mathcal{D}_+).$$

So we get

$$\|h(\mathcal{D}_+)\|_{L(_0\mathcal{F}^s(\mathbb{R}_+, \mathcal{K}^r(\mathbb{R}^n, X))} \leq C_2 \cdot \|h\|_\infty, \quad h \in H_0^\infty(\Omega)$$

by (1.40). □

Remark 1.90. Let $p_0, p_1 \in (1, \infty)$. Similarly to Proposition 1.35, for $f \in H_P(\Omega)$ we can define the maximal realization

$$f(\mathcal{D}_+) \colon \bigcup_{s \in \mathbb{R}} D(f(\mathcal{D}_+^{(s)})) \to \bigcup_{s \in \mathbb{R}} L_{p_0}(\mathbb{R}_+, H_{p_1}^s(\mathbb{R}^n, X))$$

where $f(\mathcal{D}_+^{(s)})$ is defined by the joint H^∞-calculus of the $L_{p_0}(\mathbb{R}_+, H_{p_1}^s(\mathbb{R}^n, X))$-realization of ∇_+. In fact, for $u \in D(f(\mathcal{D}_+^{(s)}))$ the assignment

$$f(\mathcal{D}_+)u := f(\mathcal{D}_+^{(s)})u$$

is well-defined since

$$f(\mathcal{D}_+^{(s)})u = f(\mathcal{D}_+^{(t)})u$$

holds for all $s \leq t$ and $u \in D(f(\mathcal{D}_+^{(t)}))$.

Remark 1.91. It is easily seen that the analog of Theorem 1.89 also holds in the more general situation where we include a shift in the operator and exponential weights in the spaces. To include a shift in the operator (see Lemma 1.34), we replace $\mathcal{D}_t$ by $\sigma + \mathcal{D}_t$ with $\sigma \geq 0$. Finally, we can consider exponentially weighted spaces as in Remark 1.70. With the help of the isomorphism $\mathscr{M}_\varrho$ defined in (1.16), we can easily extend the results of Theorem 1.89 to the spaces $_0\mathcal{F}^s_\varrho(\mathbb{R}_+, \mathcal{K}^r(\mathbb{R}^n, X))$.

These straightforward generalizations yield that for every $\varrho \geq 0$ and $\sigma \geq 0$, the $_0\mathcal{F}^s_\varrho(\mathbb{R}_+, \mathcal{K}^r(\mathbb{R}^n, X))$-realization $\mathcal{D}^{(\varrho)}_{+,\sigma}$ of $(\sigma + \partial_t, \nabla_x)$ admits a bounded joint H^∞-calculus in $_0\mathcal{F}^s_\varrho(\mathbb{R}_+, \mathcal{K}^r(\mathbb{R}^n, X))$. For $f \in H_P(\Omega)$, we also have

$$f(\mathcal{D}^{(\varrho)}_{+,\sigma}) = f_\sigma(\mathcal{D}^{(\varrho)}_{+,0})$$

with f_σ being defined in Lemma 1.34. Similarly, the equality

$$f(\mathcal{D}^{(\varrho)}_{+,0})u = \mathscr{M}_\varrho^{-1} f(\mathcal{D}^{(0)}_{+,\varrho}) \mathscr{M}_\varrho u$$

holds for all $u \in {_0\mathcal{F}^s(\mathbb{R}_+, \mathcal{K}^r(\mathbb{R}^n, X))}$. For details, we refer to [Kai12].

1.3. The time-space derivative

Having established the joint H^∞-calculus for $\mathcal{D}_+ = (\mathcal{D}_t, \mathcal{D}_1, \ldots, \mathcal{D}_n)$, we can now replace $f(\lambda, z)$ by $f(\mathcal{D}_t, \mathcal{D}_1, \ldots, \mathcal{D}_n)$ for suitable functions f to obtain well-defined operators in all scales of Sobolev spaces considered above. In particular, the symbols $\Lambda_r(z) = \left(1 - \sum_{k=1}^n z_k^2\right)^{1/2}$ and $\Psi_s(\lambda) = (1+\lambda)^s$ belong to $H_P(\Omega)$. The following proposition states that the resulting operators are even isomorphisms in the corresponding spaces. This fact which is based on the representation as Fourier multipliers in Theorem 1.81 and Theorem 1.84 will be a basis for the Newton polygon approach in Chapter 2.

Proposition 1.92. *For arbitrary $r' \in \mathbb{R}$, $s' > 0$, and $s, r \geq 0$, the symbols*

$$\Lambda_r(z) := \left(1 - \sum_{k=1}^n z_k^2\right)^{r/2}, \quad z \in \Sigma_\delta^n, \quad \delta > 0,$$

$$\Psi_s(\lambda) := (1+\lambda)^s, \quad \lambda \in S_\theta, \quad \theta > \pi/2$$

give rise to isomorphisms

$$\Lambda_r(\mathcal{D}_+) = \mathrm{op}[\Lambda_r]\big|_{_0\mathcal{F}^{s'}(\mathbb{R}_+, \mathcal{K}^{r'+r}(\mathbb{R}^n, X))}$$
$$\in L_{\mathrm{Isom}}\left({}_0\mathcal{F}^{s'}(\mathbb{R}_+, \mathcal{K}^{r'+r}(\mathbb{R}^n, X)), {}_0\mathcal{F}^{s'}(\mathbb{R}_+, \mathcal{K}^{r'}(\mathbb{R}^n, X))\right),$$

$$\Psi_s(\mathcal{D}_+) = \mathrm{op}_+[\Psi_s]\big|_{_0\mathcal{F}^{s+s'}(\mathbb{R}_+, \mathcal{K}^{r'}(\mathbb{R}^n, X))}$$
$$\in L_{\mathrm{Isom}}\left({}_0\mathcal{F}^{s+s'}(\mathbb{R}_+, \mathcal{K}^{r'}(\mathbb{R}^n, X)), {}_0\mathcal{F}^{s'}(\mathbb{R}_+, \mathcal{K}^{r'}(\mathbb{R}^n, X))\right)$$

where the domains are given by

$$D(\Lambda_r(\mathcal{D}_+)) = {}_0\mathcal{F}^{s'}(\mathbb{R}_+, \mathcal{K}^{r'+r}(\mathbb{R}^n, X)),$$
$$D(\Psi_s(\mathcal{D}_+)) = {}_0\mathcal{F}^{s+s'}(\mathbb{R}_+, \mathcal{K}^{r'}(\mathbb{R}^n, X)).$$

We also have

$$(\Lambda_r(\mathcal{D}_+))^{-1} = (\Lambda_r^{-1})(\mathcal{D}_+), \quad (\Psi_s(\mathcal{D}))^{-1} = (\Psi_s^{-1})(\mathcal{D}).$$

Proof. (i) We will use the abbreviation ${}_0\mathcal{F}^{s'}(\mathcal{K}^{r'}) := {}_0\mathcal{F}^{s'}(\mathbb{R}_+, \mathcal{K}^{r'}(\mathbb{R}^n, X))$ from Definition 1.71. It is obvious that $\Lambda_r^{-1} \in H^\infty(\Sigma_\delta^n)$ and $\Lambda_r \in H_P(\Sigma_\delta^n)$. From Lemma 1.29, Lemma 1.87 (iv), and (1.30) we see that $\Lambda_r^{-1}(\mathcal{D}_+)$ is the natural extension of $\Lambda_r^{-1}(\mathcal{D})$ and therefore

$$\Lambda_r^{-1}(\mathcal{D}_+) = \mathrm{op}[\Lambda_r^{-1}]\big|_{_0\mathcal{F}^{s'}(\mathcal{K}^{r'})} \in L\left({}_0\mathcal{F}^{s'}(\mathcal{K}^{r'})\right).$$

We even get

$$(\Lambda_r^{-1})(\mathcal{D}_+) \in L_{\mathrm{Isom}}\left({}_0\mathcal{F}^{s'}(\mathcal{K}^{r'}), {}_0\mathcal{F}^{s'}(\mathcal{K}^{r+r'})\right)$$

by Lemma 1.58 and Lemma 1.86. Theorem 1.26 implies that $\Lambda_r(\mathcal{D}_+) \in L({}_0\mathcal{F}^{s'}(\mathcal{K}^{r'}))$ is invertible with $[\Lambda_r(\mathcal{D}_+)]^{-1} = (\Lambda_r^{-1})(\mathcal{D}_+)$ and therefore

$$D(\Lambda_r(\mathcal{D}_+)) = R(\Lambda_r^{-1}(\mathcal{D}_+)) = {}_0\mathcal{F}^{s'}(\mathcal{K}^{r+r'})$$

and $\Lambda_r(\mathcal{D}_+) = \mathrm{op}[\Lambda_r]\big|_{_0\mathcal{F}^{s'}(\mathcal{K}^{r+r'})} \in L_{\mathrm{Isom}}\left({}_0\mathcal{F}^{s'}(\mathcal{K}^{r+r'}), {}_0\mathcal{F}^{s'}(\mathcal{K}^{r'})\right).$

(ii) We have $\Psi_s \in H_P(S_\theta)$ and $\Psi_s^{-1} \in H^\infty(S_\theta)$. In the same way as in part (i) we get
$$\Psi_s^{-1}(\mathcal{D}_+) = r_0^+ \mathrm{op}[\Psi_s^{-1}]e_0^+\big|_{{}_0\mathcal{F}^{s'}(\mathcal{K}^{r'})} = \mathrm{op}_+[\Psi_s^{-1}] = (\mathrm{op}_+[\Psi_s])^{-1} \in L\big({}_0\mathcal{F}^{s'}(\mathcal{K}^{r'})\big)$$
by Lemma 1.29 and Theorem 1.84. From Proposition 1.79 we get
$$\Psi_s^{-1}(\mathcal{D}_+) \in L_{\mathrm{Isom}}\big({}_0\mathcal{F}^{s'}(\mathcal{K}^{r'}), {}_0\mathcal{F}^{s+s'}(\mathcal{K}^{r'})\big).$$
We conclude $[\Psi_s(\mathcal{D}_+)]^{-1} = \Psi_s^{-1}(\mathcal{D}_+)$ and
$$D(\Psi_s(\mathcal{D}_+)) = R(\Psi_s^{-1}(\mathcal{D}_+)) = {}_0\mathcal{F}^{s+s'}(\mathcal{K}^{r'})$$
as in part (i). $\square$

Example 1.93 (Laplace operator). Let $r \in \mathbb{R}$ and $s > 0$. Consider the symbol $\lambda(z) := \sum_{j=1}^n z_j^2$ for $z \in \Sigma_\delta^n$. Then $\lambda \in H_P(\Sigma_\delta^n)$, and the ${}_0\mathcal{F}^s(\mathbb{R}_+, \mathcal{K}^r(\mathbb{R}^n, X))$-realization of $\lambda(\mathcal{D}_+)$ coincides with the realization of the Laplacian Δ in this space which is given as
$$\Delta \colon {}_0\mathcal{F}^s(\mathbb{R}_+, \mathcal{K}^r(\mathbb{R}^n, X)) \supseteq D(\Delta) \to {}_0\mathcal{F}^s(\mathbb{R}_+, \mathcal{K}^r(\mathbb{R}^n, X)),$$
$$u \mapsto \Delta_x u$$
where $D(\Delta) := {}_0\mathcal{F}^s(\mathbb{R}_+, \mathcal{K}^{r+2}(\mathbb{R}^n, X))$.

Proof. Note that we have $\lambda(z) = 1 - \Lambda_2(z)$ for $z \in \Omega_x$. Theorem 1.26 (iii) directly yields
$$\lambda(\mathcal{D}_+) = 1 - \Lambda_2(\mathcal{D}_+)$$
and therefore $D(\lambda(\mathcal{D}_+)) = D(\Lambda_2(\mathcal{D}_+)) = \mathcal{F}^s(\mathcal{K}^{r+2})$ due to Proposition 1.92. In particular, we see that
$$-\lambda(\mathcal{D}_+)u = \Lambda_2(\mathcal{D}_+)u - u = \mathrm{op}[\Lambda_2] \circ u - u = (1 - \Delta) \circ u - u$$
$$= -\Delta \circ u = -\Delta_x u$$
for all $u \in \mathcal{F}^s(\mathcal{K}^{r+2})$. $\square$

Example 1.94 (Hilbert transform). The function
$$f \colon \Sigma_\delta \to \mathbb{C}, \quad z \mapsto \frac{z}{\sqrt{-z^2}}$$
is an element of $H^\infty(\Sigma_\delta)$. Note that we have $f(z) = i \cdot \mathrm{sign}\,(\mathrm{Im}\,z)$ for $z \in \Sigma_\delta$ due to the definition of the square root $\sqrt{\zeta} = \sqrt{|\zeta|}\exp(i(\arg \zeta)/2)$ with $\arg \zeta \in (-\pi, \pi]$. For the $L_p(\mathbb{R}, X)$-realization of $f(\mathcal{D})$ we obtain
$$f(\mathcal{D}) = \mathrm{op}[f]|_{L_p(\mathbb{R}, X)} = \mathscr{F}^{-1} m \mathscr{F} \in L(L_p(\mathbb{R}, X)),$$
$$m(\xi) := i\xi/|\xi| = i \cdot \mathrm{sign}\,\xi, \quad \xi \in \mathbb{R} \setminus \{0\}$$

1.3. The time-space derivative

due to the representation (1.30). It is shown in [Ama95, Section 4.3] that $\mathscr{F}^{-1}m\mathscr{F}$ coincides with the Hilbert transform which was defined in Definition 1.11. This implies that $f(\mathcal{D})$ is the Hilbert transform on $L_p(\mathbb{R}, X)$. Note that we generally assumed that X is of class $\mathcal{HT}$.

Remark 1.95. Analog representations hold for Riesz transforms and the Helmholtz projection in the whole space. For the definition and further results on Riesz transforms we refer to Definition 4.36 and [Ste70, Ch. III].

Chapter 2

The Newton polygon approach for mixed-order systems

Solvability results in elliptic and parabolic theory are in many cases based on estimates of the symbols of the related operators. In classical elliptic and parabolic theory, these symbols are (quasi-)homogeneous. In this case, the condition that the symbol does not vanish automatically implies uniform estimates on its inverse. In several classes of applications, however, the symbol does not have a quasi-homogeneous structure. During the 1990's, the Newton polygon approach was developed to treat parabolic equations with inhomogeneous symbol structure, see the book of S. Gindikin and L.R. Volevich [GV92], the papers [DMV98] and [DV08] and the references therein. The idea of the Newton polygon approach is to replace the condition of non-vanishing symbols by an inequality where the right-hand side is given by the structure of the equation. The resulting class of equations is called N-parabolic where the letter "N" stands for "Newton". In fact, it was I. Newton in 1669 who used the idea of polyhedra to understand the homogeneities of polynomials of several variables, see [New81]. Since then, the Newton polygon has been widely used in algebra and analysis.

Whereas the approach by S. Gindikin and L.R. Volevich to N-parabolic equations was L_2-based, for the investigation of nonlinear problems also L_p-results are important. During the last years, several papers on the L_p-theory of N-parabolic problems and the application to free boundary value problems appeared, see, e.g., [DSS08]. In the present chapter, we will generalize the class of treatable equations and symbols in comparison to known results. Moreover, we will include so-called non-regular Newton polygons which will play a role in the treatment of the linear L_p-L_q-Cahn-Hilliard-Gurtin Problem on $\mathbb{R}^n$ in Section 4.3, for example.

2.1 Inhomogeneous symbols and the Newton polygon

Motivation. The simplest example of an elliptic operator is the Laplacian Δ in $\mathbb{R}^n$ with the symbol $P(z) = z_1^2 + \cdots + z_n^2$, i.e., here we have $\Delta = \operatorname{op}[P] = -\mathscr{F}^{-1}|\xi|^2\mathscr{F}$ in the sense of Definition 1.73. Note that, as we are dealing with a holomorphic calculus, we defined the symbol in a way that ∂_j has the symbol z_j instead of $i\xi_j$ which is also common in elliptic literature.

The symbol $P(z) = z_1^2 + \cdots + z_n^2$ is homogeneous of degree 2 in the sense that $P(\eta z) = \eta^2 P(z)$ for $\eta > 0$. For homogeneous symbols of degree N, the classical ellipticity condition $P(z) \neq 0$ ($z \neq 0$) implies an inequality of the form $|P(z)| \geq C|z|^N$. In the same way, the parabolic operator $\partial_t - \Delta$ related to the heat equation has the symbol $P(\lambda, z) = \lambda - z_1^2 - \cdots - z_n^2$ which is quasi-homogeneous of degree 2 in the sense that

$$P(\eta^2\lambda, \eta z) = \eta^2 P(\lambda, z), \quad \eta > 0,\ \lambda \in \mathbb{C},\ z \in \mathbb{C}^n.$$

Again, the standard parabolicity (parameter-ellipticity) condition

$$P(\lambda, z) \neq 0, \quad (\lambda, z) \in (S_\theta \times \mathbb{R}^n) \setminus \{(0,0)\}$$

leads to uniform estimates from below.

In contrast to this, the symbol

$$P(\lambda, z) = \lambda - (z_1^2 + \cdots + z_n^2)\sqrt{\lambda - z_1^2 - \cdots - z_n^2}$$

which appears in the analysis of the Stefan problem, is not quasi-homogeneous. More precisely, there exists no relative weight ρ and no degree N such that the equality $P(\eta^\rho \lambda, \eta z) = \eta^N P(\lambda, z)$ holds. This also implies that there is no well-defined principal symbol of P. In the Newton polygon approach, we will formulate a condition of parabolicity (in the sense of parameter-ellipticity) for general symbols with inhomogeneous structure. It is also possible to define a family of principal symbols which will give us a simple characterization of N-parabolicity in the subsequent section.

In the present section, we start with the discussion of quasi-homogeneous symbols and principal parts of operators in Subsection a). This allows us to define the symbol class for N-parabolic operators. The definition of the Newton polygon and related notions are formulated and analyzed in Subsection b), where also the important concept of order functions is introduced. The notion of order functions will play a role in later estimates on the symbols and their inverses.

2.1. Inhomogeneous symbols and the Newton polygon

a) Inhomogeneous symbols and principal parts

We start with a definition of quasi-homogeneous functions.

Definition 2.1. (i) A *cone* in $\mathbb{C}^n$ is a subset $L \subseteq \mathbb{C}^n$ such that $\eta z \in L$ for all $\eta \geq 0$ and $z \in L$. If $L \subseteq \mathbb{C}^n$ is a cone, then a function $\psi \colon L \setminus \{0\} \to \mathbb{C}$ is called *homogeneous of degree* $N \in \mathbb{R}$ if

$$\psi(\eta z) = \eta^N \psi(z), \quad \eta > 0, \ z \in L \setminus \{0\}.$$

We write $S^{(N)}(L)$ for the set of all $\psi \in C(L, \mathbb{C})$ which are homogeneous of degree N and for which we have $\psi(z) \neq 0$ for all $z \in L$.

(ii) Let $\rho > 0$ and $N \in \mathbb{R}$, and let $L_t \subseteq \mathbb{C}$ and $L_x \subseteq \mathbb{C}^n$ be closed cones. Then a function $\psi \colon (L_t \times L_x) \setminus \{(0,0)\} \to \mathbb{C}$ is called *ρ-homogeneous of degree* N if

$$\psi(\eta^\rho \lambda, \eta z) = \eta^N \psi(\lambda, z), \quad \eta > 0, \ (\lambda, z) \in (L_t \times L_x) \setminus \{(0,0)\}.$$

Functions which are ρ-homogeneous are also called *quasi-homogeneous*, and ρ is called the *weight* of λ relative to z. We define $S^{(\rho,N)}(L_t \times L_x)$ as the set of all functions $\psi \in C(L_t \times L_x, \mathbb{C})$ which are ρ-homogeneous of degree N and which satisfy $\psi(\lambda, z) \neq 0$ for all $(\lambda, z) \in (L_t \times L_x) \setminus \{(0,0)\}$.

In the following, let $\rho > 0$, $N \geq 0$, and $L_t \subseteq \mathbb{C}$ and $L_x \subseteq \mathbb{C}^n$ be closed cones.

Remark 2.2. If $L \subseteq \mathbb{C}^n$ is a closed cone and $\psi \in S^{(0)}(L)$, then ψ is a constant. This follows immediately from the fact that ψ is, by homogeneity of degree 0, constant along all rays ηz, $\eta \in (0, \infty)$ with $z \in L$ and by the continuity in 0.
Similarly, any $\psi \in S^{(\rho,0)}(L_t \times L_x)$ is constant. If $\psi \in S^{(\rho,N)}(L_t \times L_x)$ with $N > 0$, then $\psi(0) = 0$ follows again from homogeneity and continuity.

Example 2.3. (i) Let

$$\omega(\lambda, z) := \sqrt{\lambda - z_1^2 - \cdots - z_n^2} = \sqrt{\lambda + |z|_-^2}, \quad |z|_- := \sqrt{-\sum_{k=1}^n z_k^2}$$

where $\lambda \in L_t := \overline{S}_\theta$, $\theta \in (0, \pi)$ and $z \in L_x := \overline{\Sigma}_\delta^n$, $\delta \in (0, \pi/2)$ with $\theta + 2\delta < \pi$. Then we have $\omega \in S^{(2,1)}(L_t \times L_x)$. Note that we define the complex square root $\sqrt{z}$ as $\sqrt{|z|} \exp(\frac{1}{2} i \arg(z))$ with $\arg z \in (-\pi, \pi]$.

(ii) Let

$$\psi(z) := \frac{z_k}{|z|_-}, \quad z \in L_x \setminus \{0\}, \quad L_x := \overline{\Sigma}_\delta^n, \ \delta > 0.$$

Then ψ is homogeneous of degree 0 but $\psi \notin S^{(0)}(L_x)$ as ψ cannot be extended continuously to L_x.

(iii) Let $\psi(\lambda, z) := |z|^N + |\lambda|^{N/\rho}$ with $N > 0$. Then $\psi \in S^{(\rho,N)}(L_t \times L_x)$. This function is a canonical ρ-homogeneous function and will play an important role for symbol estimates below.

Remark 2.4. Let $\psi\colon L_t \times L_x \setminus \{0\} \to \mathbb{C}$ be continuous and ρ-homogeneous of degree N. Due to the compactness of the set

$$\{(\lambda, z) \in L_t \times L_x \colon |z|^N + |\lambda|^{N/\rho} = 1\},$$

there exists $C > 0$ with

$$|\psi(\lambda, z)| \leq C\big(|z|^N + |\lambda|^{N/\rho}\big), \quad (\lambda, z) \in L_t \times L_x.$$

If $\psi \in S^{(\rho,N)}(L_t \times L_x)$, then

$$C_1\big(|z|^N + |\lambda|^{N/\rho}\big) \leq |\psi(\lambda, z)| \leq C_2\big(|z|^N + |\lambda|^{N/\rho}\big), \quad (\lambda, z) \in L_t \times L_x,$$

holds with constants $C_1, C_2 > 0$. The analog results hold for homogeneous functions $\psi\colon L_t \setminus \{0\} \to \mathbb{C}$ and $\psi\colon L_x \setminus \{0\} \to \mathbb{C}$.

As we treat inhomogeneous symbols, we consider relative weights $\gamma > 0$ of λ with respect to z. For every weight, we obtain a γ-principal part and a γ-order of the symbol. For instance, for the symbol $\omega(\lambda, z) = \sqrt{\lambda - z_1^2 - \cdots - z_n^2} = \sqrt{\lambda + |z|_-^2}$ and $\gamma \in (0, 2)$, the order of the term $|z|_-^2$ is 2, and therefore this term dominates the term λ which has, by definition, order γ. Therefore, for $\gamma < 2$ the principal part of $\omega(\lambda, z)$ equals $|z|_-$. For $\gamma = 2$, the principal part of ω coincides with ω, while for $\gamma > 2$ the γ-principal part of ω is given as $\sqrt{\lambda}$.

We start with the formal definition:

Definition 2.5 (γ-order and γ-principal part of ρ-homogeneous symbols). Let ψ be a symbol in $S^{(\rho,N)}(L_t \times L_x)$. Then we define for all $\gamma > 0$ the γ-order by

$$d_\gamma(\psi) := \max\left\{N, \frac{N}{\rho}\gamma\right\}$$

and the γ-principal part by

$$\pi_\gamma \psi \colon (L_t \times L_x) \to \mathbb{C},$$
$$(\lambda, z) \mapsto \lim_{\eta \to \infty} \eta^{-d_\gamma(\psi)} \psi(\eta^\gamma \lambda, \eta z).$$

In the same way we define the ∞-order

$$d_\infty(\psi) := \frac{N}{\rho}$$

and the ∞-principal part

$$\pi_\infty \psi \colon (L_t \times L_x) \to \mathbb{C},$$
$$(\lambda, z) \mapsto \lim_{\eta \to \infty} \eta^{-d_\infty(\psi)} \psi(\eta\lambda, z).$$

2.1. Inhomogeneous symbols and the Newton polygon

Example 2.6. Let $\theta \in (0, \pi)$ and $\delta \in (0, \frac{\pi}{2})$ with $\theta + 2\delta < \pi$. As in Example 2.3 (i) we consider the symbol

$$\omega(\lambda, z) := \sqrt{\lambda + |z|^2_-}, \quad (\lambda, z) \in \overline{S}_\theta \times \overline{\Sigma}^n_\delta.$$

Then we have

$$d_\gamma(\omega) = \max\{1, \gamma/2\}, \quad \gamma > 0, \qquad d_\infty(\omega) = 1/2,$$

$$\pi_\gamma \omega(\lambda, z) = \begin{cases} |z|_-, & \gamma < 2, \\ \sqrt{\lambda + |z|^2_-}, & \gamma = 2, \\ \sqrt{\lambda}, & \gamma > 2, \end{cases} \quad (\lambda, z) \in \overline{S}_\theta \times \overline{\Sigma}^n_\delta,$$

and $\pi_\infty \omega(\lambda, z) = \sqrt{\lambda}$ for $(\lambda, z) \in \overline{S}_\theta \times \overline{\Sigma}^n_\delta$. So the formal definition of the γ-principal part coincides with the intuitive discussion above.

We remark that the γ-principal part can be computed explicitly:

Lemma 2.7. *For $\psi \in S^{(\rho, N)}(L_t \times L_x)$ and $\gamma \in (0, \infty]$ we have*

$$\pi_\gamma \psi(\lambda, z) = \begin{cases} \psi(0, z), & \gamma < \rho, \\ \psi(\lambda, z), & \gamma = \rho, \\ \psi(\lambda, 0), & \gamma > \rho, \end{cases} \quad (\lambda, z) \in (L_t \times L_x).$$

Proof. The assertion is obvious for $(\lambda, z) = (0, 0)$, so we only have to consider $(\lambda, z) \in (L_t \times L_x) \setminus \{(0, 0)\}$. We have to distinguish three different cases:

(I) $\gamma > \rho$: Without loss of generality we assume $\gamma < \infty$. Let $(\lambda, z) \in (L_t \times L_x) \setminus \{(0, 0)\}$, $\eta > 0$, and $\alpha := (\eta^\rho |z|^\rho + \eta^\gamma |\lambda|)^{1/\rho}$. Then we get

$$\psi(\eta^\gamma \lambda, \eta z) = \alpha^N \psi(\alpha^{-\rho} \eta^\gamma \lambda, \alpha^{-1} \eta z)$$

and

$$\alpha^{-1} \eta z = (|z|^\rho + \eta^{\gamma-\rho} |\lambda|)^{-1/\rho} z \to 0, \quad \eta \to \infty,$$

$$\alpha^{-\rho} \eta^\gamma \lambda = (\eta^{\rho-\gamma} |z|^\rho + |\lambda|)^{-1} \lambda \to m(\lambda) := \begin{cases} \frac{\lambda}{|\lambda|}, & \lambda \neq 0 \\ 0, & \lambda = 0 \end{cases}, \quad \eta \to \infty.$$

Hence we obtain

$$\eta^{-\gamma N/\rho} \psi(\eta^\gamma \lambda, \eta z) = (\eta^{\rho-\gamma} |z|^\rho + |\lambda|)^{N/\rho} \psi(\alpha^{-\rho} \eta^\gamma \lambda, \alpha^{-1} \eta z)$$
$$\to |\lambda|^{N/\rho} \psi(m(\lambda), 0) = \psi(\lambda, 0)$$

for $\eta \to \infty$. This yields $\pi_\gamma \psi(\lambda, z) = \psi(\lambda, 0)$.

(II) $\gamma < \rho$: Analogously, with $\eta > 0$, $(\lambda, z) \in (L_t \times L_x) \setminus \{(0,0)\}$, and $\alpha := (\eta^\rho |z|^\rho + \eta^\gamma |\lambda|)^{1/\rho}$ we get

$$\eta^{-N} \psi(\eta^\gamma \lambda, \eta z) = (|z|^\rho + \eta^{\gamma-\rho}|\lambda|)^{N/\rho} \psi(\alpha^{-\rho} \eta^\gamma \lambda, \alpha^{-1} \eta z) \to |z|^N \psi(0, m(z))$$

for $\eta \to \infty$. This yields $\pi_\gamma \psi(\lambda, z) = \psi(0, z)$.

(III) $\gamma = \rho$: For $\eta > 0$ and $(\lambda, z) \in (L_t \times L_x) \setminus \{(0,0)\}$ we obviously have $\eta^{-N} \psi(\eta^\rho \lambda, \eta z) = \psi(\lambda, z)$. □

The inhomogeneous symbols which we consider consist of a sum of terms where each term is a product of quasi-homogenous functions in λ, z, and (λ, z). A typical example is again the symbol from the Stefan problem

$$P(\lambda, z) = \lambda - (z_1^2 + \cdots + z_n^2)\sqrt{\lambda - z_1^2 - \cdots - z_n^2}.$$

This is the sum of two terms where the first one is a function of λ which belongs to $S^{(1)}(L_t)$ while the second term is the product of a function of z belonging to $S^{(2)}(L_x)$ and a function of (λ, z) belonging to $S^{(2,1)}(L_t \times L_x)$.

According to Definition 1.17 (i) and (v), $H(\mathring{L}_t \times \mathring{L}_x)$ and $H_P(\mathring{L}_t \times \mathring{L}_x)$ stand for the space of all holomorphic functions in $\mathring{L}_t \times \mathring{L}_x$ and the space of all holomorphic and polynomially bounded (at zero and infinity) functions in $\mathring{L}_t \times \mathring{L}_x$, respectively.

Definition 2.8 (Symbol class $\widetilde{S}(L_t \times L_x)$). For $\rho > 0$ we define $\widetilde{S}(L_t \times L_x)$ as the set of all functions

$$P: L_t \times L_x \to \mathbb{C} \qquad (2.1)$$
$$(\lambda, z) \mapsto \sum_{\ell \in I} \tau_\ell(\lambda, z) \varphi_\ell(\lambda) \psi_\ell(z)$$

where $I = I_P$ is an arbitrary finite set and

$$\tau_\ell \in S^{(\rho, N_\ell)}(L_t \times L_x) \cap H(\mathring{L}_t \times \mathring{L}_x), \quad N_\ell \geq 0,$$
$$\varphi_\ell \in S^{(M_\ell)}(L_t) \cap H(\mathring{L}_t), \quad M_\ell \geq 0,$$
$$\psi_\ell \in S^{(L_\ell)}(L_x) \cap H(\mathring{L}_x), \quad L_\ell \geq 0$$

for all $\ell \in I$. Note that we do not include ρ in the notation as ρ will be fixed in all applications.

It is easy to see that the symbol class $\widetilde{S}(L_t \times L_x)$ is included in the space of polynomially bounded holomorphic functions $H_P(\mathring{L}_t \times \mathring{L}_x)$.

Example 2.9. (i) The symbol class $\widetilde{S}(L_t \times L_x)$ includes all complex polynomials in λ and z.

2.1. Inhomogeneous symbols and the Newton polygon

(ii) Let $|z|_- = (-\sum_{k=1}^n z_k^2)^{1/2}$, $\omega = \omega(\lambda, z) := \sqrt{\lambda + |z|_-^2}$, and

$$P(\lambda, z) := \lambda^3 + \lambda^2 |z|_- \omega + 5\lambda^2 |z|_-^2 + 4\lambda |z|_-^4 + \sigma\lambda |z|_-^3 + \sigma |z|_-^4 \omega + \sigma |z|_-^5$$

(cf. the spin coating model in Section 4.6 below). Then we have $P \in \widetilde{S}(\overline{S}_\theta \times \overline{\Sigma}_\delta^n)$ for $\theta \in (0, \pi)$, $\delta \in (0, \pi/2)$ with $\theta + 2\delta < \pi$.

(iii) Let $\omega_i = \omega_i(\lambda, z) := \sqrt{\rho_i \lambda + \mu_i |z|_-^2}$ ($i = 1, 2$) where $\rho_i, \mu_i > 0$, and $(\lambda, z) \in \overline{S}_\theta \times \overline{\Sigma}_\delta^n$ with θ, δ as in (ii). Then we obtain $P \in \widetilde{S}(\overline{S}_\theta \times \overline{\Sigma}_\delta^n)$ where

$$P(\lambda, z) := \lambda^4 + \omega_1 \omega_2 \lambda |z|_- + \omega_1 \lambda + 1$$

for $(\lambda, z) \in \overline{S}_\theta \times \overline{\Sigma}_\delta^n$. Typically, symbols of this form occur in the treatment of two-phase problems since the phases have different viscosities and densities, cf. Section 4.7.

We want to mention that the representation of the symbol in (2.1) is not unique. Consider the two representations of the symbol $P(\lambda, z) = (z^2 - \lambda) - z^2 + \lambda + 1 = 1$, for example. Obviously, $(\lambda, z) \mapsto z^2 - \lambda$ is 2-homogeneous of order 2 (i.e., $\rho = 2$, $N = 2$) and $z^2 - \lambda \neq 0$ for appropriate sets L_t and L_x. These ambivalent representations of the symbol P yield that the next definitions depend on the representation of P.

Definition 2.10 (γ-order and γ-principal part of symbols in $\widetilde{S}(L_t \times L_x)$). For all $\gamma > 0$ we define the γ-*order* of the symbol (2.1) by

$$d_\gamma(P) := \max_{\ell \in I_P} (d_\gamma(\tau_\ell) + \gamma M_\ell + L_\ell),$$

and for $(\lambda, z) \in L_t \times L_x$ the γ-*principal part* is defined by

$$\pi_\gamma P(\lambda, z) := \lim_{\eta \to \infty} \eta^{-d_\gamma(P)} P(\eta^\gamma \lambda, \eta z) = \sum_{\ell \in I_\gamma} [\pi_\gamma \tau_\ell](\lambda, z) \varphi_\ell(\lambda) \psi_\ell(z)$$

with $I_\gamma := I_\gamma(P) := \{\ell \in I_P : d_\gamma(\tau_\ell) + \gamma M_\ell + L_\ell = d_\gamma(P)\}$. In the same way we define the ∞-*order* by

$$d_\infty(P) := \max_{\ell \in I_P} (M_\ell + N_\ell/\rho),$$

and for $(\lambda, z) \in L_t \times L_x$ the ∞-*principal part* is defined by

$$\pi_\infty P(\lambda, z) := \lim_{\eta \to \infty} \eta^{-d_\infty(P)} P(\eta\lambda, z) = \sum_{\ell \in I_\infty} \tau_\ell(\lambda, 0) \varphi_\ell(\lambda) \psi_\ell(z)$$

with $I_\infty := \{\ell \in I : M_\ell + N_\ell/\rho = d_\infty(P)\}$.

In order to avoid a dependence on the representation we introduce the concept of regular representations of symbols.

Definition 2.11 (Regular representation of a symbol). The representation of the symbol P in (2.1) is said to be *regular* if we have

$$\pi_\gamma P \not\equiv 0 \qquad (2.2)$$

for all $\gamma \in (0, \infty]$.

Example 2.12. (i) Let $\psi(\lambda, z) := \sqrt{-z^2} - \sqrt{\lambda - z^2}$ for $(\lambda, z) \in \overline{S}_\theta \times \overline{\Sigma}_\delta$ with $\theta \in (0, \pi)$, $\delta \in (0, \pi/2)$, $\theta + 2\delta < \pi$. It is obvious that $\psi \in \widetilde{S}(\overline{S}_\theta \times \overline{\Sigma}_\delta)$ but the given representation is not regular because of

$$\pi_\gamma \psi(\lambda, z) = \sqrt{-z^2} - \sqrt{-z^2} = 0, \quad \gamma < 2, \quad (\lambda, z) \in \overline{S}_\theta \times \overline{\Sigma}_\delta.$$

(ii) For θ, δ as in (i), let

$$P(\lambda, z) := \lambda^2 + \omega \lambda z + \lambda z^{3/2} - z^3 + 1, \quad (\lambda, z) \in \overline{S}_\theta \times \overline{\Sigma}_\delta$$

where $\omega = \omega(\lambda, z) := \sqrt{\lambda - z^2}$. Then we have

$$d_\gamma(P) = \begin{cases} \max\{3, \gamma + 2, 2\gamma\}, & \gamma > 0, \\ 2, & \gamma = \infty, \end{cases}$$

$$\pi_\gamma P(\lambda, z) = \begin{cases} -z^3, & \gamma \in (0, 1), \\ \lambda z \sqrt{-z^2} - z^3, & \gamma = 1, \\ \lambda z \sqrt{-z^2}, & \gamma \in (1, 2), \\ \lambda^2 + \omega \lambda z, & \gamma = 2, \\ \lambda^2, & \gamma \in (2, \infty]. \end{cases}$$

The term $\lambda z^{3/2} + 1$ is of lower order and therefore it does not appear in the principal part. With this it is obvious that the symbol P is regular in the sense of Definition 2.11.

If we have two regular representations of the same symbol, then the γ-orders and γ-principal parts coincide. This will be stated in the next lemma.

Lemma 2.13. *For two regular representations*

$$P(\lambda, z) = \sum_{\ell \in I} \tau_\ell(\lambda, z) \varphi_\ell(\lambda) \psi_\ell(z)$$

$$= \sum_{\ell \in I'} \tau'_\ell(\lambda, z) \varphi'_\ell(\lambda) \psi'_\ell(z), \quad (\lambda, z) \in L_t \times L_x,$$

of the symbol P we have $d_\gamma(P) = d'_\gamma(P)$ and $\pi_\gamma P = \pi'_\gamma P$ for all $\gamma \in (0, \infty]$. Here $d'_\gamma(P)$ and $\pi'_\gamma P$ denote the γ-order and the γ-principal part with respect to the second representation.

2.1. Inhomogeneous symbols and the Newton polygon

Proof. Let $0 < \gamma < \infty$ and $(\lambda, z) \in L_t \times L_x$ be arbitrary. Assuming $d_\gamma(P) < d'_\gamma(P)$ we obtain

$$\pi'_\gamma P(\lambda, z) = \lim_{\eta \to \infty} \eta^{d_\gamma(P) - d'_\gamma(P)} \eta^{-d_\gamma(P)} P(\eta^\gamma \lambda, \eta z) = 0 \cdot \pi_\gamma P(\lambda, z) = 0.$$

This yields $\pi'_\gamma P \equiv 0$, which contradicts (2.2).

The same argument obviously applies if $d_\gamma(P) > d'_\gamma(P)$. Therefore, we get $d_\gamma(P) = d'_\gamma(P)$ and thus $\pi_\gamma P = \pi'_\gamma P$. The case $\gamma = \infty$ can be done in exactly the same way. $\square$

Definition 2.14 (Symbol class $S(L_t \times L_x)$). We define $S(L_t \times L_x)$ as the set of all symbols $P \in \widetilde{S}(L_t \times L_x)$ for which a regular representation exists. We tacitly assume that a given representation of $P \in S(L_t \times L_x)$ is always regular. Note that the concepts of γ-order and γ-principal part are well-defined in this class.

b) Newton polygons and order functions

Before we formulate the formal definition of a Newton polygon, we start with some examples. For the operator $\partial_t - \Delta$ related to the heat equation and its symbol $P(\lambda, z) = \lambda - z_1^2 - \cdots - z_n^2$, the Newton polygon is given by the triangle with vertices $(0,0)$, $(0,1)$, and $(2,0)$. Here the point $(0,1)$ is related to the term λ which is of order 0 in z and of order 1 in λ. Similarly, the point $(2,0)$ is related to the term z^2 which is of order 2 in z and of order 0 in λ. In general, the Newton polygon represents the orders appearing in each term of $P(\lambda, z)$ where the order with respect to z gives the horizontal coordinate and the order with respect to λ the vertical coordinate of the point. We obtain a finite set of points lying in $[0, \infty)^2$, and the Newton polygon is defined as the convex hull of these points, their projections onto the axes, and the origin. It is easily seen that the Newton polygon of a single quasi-homogeneous term is always a triangle. In particular, this is the case for the heat equation.

The symbol from the Stefan problem $P(\lambda, z) = \lambda + |z|^2_- \sqrt{\lambda + |z|^2_-}$ leads to the vertices $(0,1)$, $(2, \frac{1}{2})$, and $(3,0)$. Note here that the second term $|z|^2_- \sqrt{\lambda + |z|^2_-}$ in a natural way gives rise to the two points $(2, \frac{1}{2})$ and $(3,0)$, if one remarks the equivalence (in the sense of two-sided estimates) of $\sqrt{\lambda - z^2}$ and $|\lambda|^{1/2} + |z|$.

In general, a regular representation of $P(\lambda, z)$ consists of a sum of terms of the form $\tau_\ell(\lambda, z) \varphi_\ell(\lambda) \psi_\ell(z)$ with $\tau_\ell \in S^{(\rho, N_\ell)}(L_t \times L_x)$, $\varphi_\ell \in S^{(M_\ell)}(L_t)$, and $\psi_\ell \in S^{(N_\ell)}(L_x)$. Such a term leads to the two points $(N_\ell + L_\ell, M_\ell)$ and $(L_\ell, \frac{N_\ell}{\rho} + M_\ell)$. This is part of the following formal definition.

Definition 2.15 (Newton polygon). (i) Let $\nu := \{(a_i, b_i) : i = 0, \ldots, M\} \subseteq [0, \infty)^2$ be a finite set. Then the *Newton polygon* $N(\nu)$ is defined as the convex hull of

$$\nu \cup \{(0,0)\} \cup \bigcup_{i=0}^{M} \{(a_i, 0), (0, b_i)\}.$$

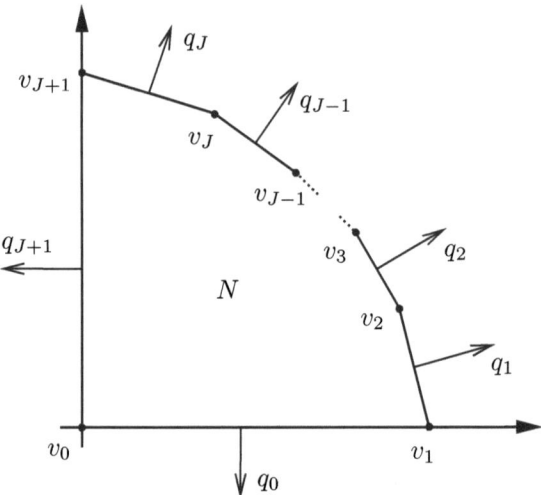

Figure 2.1: Regular Newton polygon N

A convex set $N \subseteq [0, \infty)^2$ is called a *Newton polygon* if there exists a finite set $\nu \subseteq [0, \infty)^2$ such that $N = N(\nu)$.

(ii) Let $P \in S(L_t \times L_x)$ with regular representation (2.1). We set

$$\nu(P) := \bigcup_{\ell \in I} \{(N_\ell + L_\ell, M_\ell), (L_\ell, \tfrac{N_\ell}{\rho} + M_\ell)\}.$$

Then the *Newton polygon* $N(P)$ of the symbol P is defined as $N(P) := N(\nu(P))$.

(iii) Let N be a Newton polygon. Then we denote by $v_0 := (r_0, s_0), \ldots, v_{J+1} := (r_{J+1}, s_{J+1})$ the *vertices* of N, starting at the origin and being indexed in the counter-clockwise direction. We set

$$N_V := \{v_i \colon i = 0, \ldots, J+1\}.$$

The corresponding *weight function* is then defined by

$$W_N(\lambda, z) := \sum_{(r,s) \in N_V} |z|^r |\lambda|^s, \quad (\lambda, z) \in \mathbb{C} \times \mathbb{C}^n.$$

For a finite set $\nu \subseteq [0, \infty)^2$ we also define the *weight function* $W_\nu := W_{N(\nu)}$. For a symbol $P \in S(L_t \times L_x)$ we set $W_P := W_{N(P)}$.

Example 2.16. (i) For $\nu = \{(0,0)\}$, we have $N(\nu) = \{(0,0)\}$, and the weight function equals $W_\nu(\lambda, z) = 1$ for all λ and z.

2.1. Inhomogeneous symbols and the Newton polygon

(ii) If $\nu = \{(a,b)\}$ with $a, b > 0$, then the Newton polygon is, by definition, the convex hull of the points $\{(0,0), (0,b), (a,0), (a,b)\}$ and therefore a rectangle. Here the weight function is given by

$$W_\nu(\lambda, z) = 1 + |z|^a + |\lambda|^b + |z|^a |\lambda|^b = (1 + |z|^a)(1 + |\lambda|^b), \quad (\lambda, z) \in \mathbb{C} \times \mathbb{C}^n.$$

(iii) If $\nu = \{(a,0), (0,b)\}$ with $a, b > 0$, then the Newton polygon is a triangle with vertices $(0,0), (a,0), (0,b)$. This is a typical form for classical parabolic operators like $\partial_t - \Delta$ (in this case $a = 2$ and $b = 1$). The weight function is given by

$$W_\nu(\lambda, z) = 1 + |z|^a + |\lambda|^b, \quad (\lambda, z) \in \mathbb{C} \times \mathbb{C}^n.$$

Remark 2.17. By definition, for the computation of the weight function W_N associated to the Newton polygon N, only the vertices are taken into account. This is due to the fact that the analog terms for interior points of N can be estimated by the weight function. More precisely, a convexity argument (see [GV92, Lemma 1.2.1]) shows that, for $(r_i, s_i) \in [0, \infty)^2$, $i = 0, \ldots, M$ and

$$(r, s) \in \mathrm{conv}\{(r_i, s_i) : i = 0, \ldots, M\},$$

we have the inequality

$$x^r t^s \leq \sum_{i=0}^{M} x^{r_i} t^{s_i}, \quad x, t \geq 0.$$

Definition 2.18 (Regular Newton polygon). A Newton polygon N is said to be *regular* if $r_1 \neq r_2$ and $s_J \neq s_{J+1}$ (i.e., $J > 0$ and there are no edges parallel to the axes except the trivial ones). The Newton polygon N is called *regular in time* (respectively, *regular in space*) if $r_1 \neq r_2$ (respectively, $s_J \neq s_{J+1}$).

A symbol $P \in S(L_t \times L_x)$ is called *regular/regular in time/regular in space* if the associated Newton polygon $N(P)$ is regular/regular in time/regular in space (see Figures 2.2 and 2.3).

Example 2.19. (i) The symbol $P(\lambda, z) := \lambda^2 + 3\lambda z + 5z - z^{3/2}$, $(\lambda, z) \in \overline{S}_\theta \times \overline{\Sigma}_\delta$, is regular.

(ii) The symbol $P(\lambda, z) := -2\lambda^2 + z\lambda^2 + z^2$, $(\lambda, z) \in \overline{S}_\theta \times \overline{\Sigma}_\delta$, is regular in time but it is not regular in space.

(iii) The symbol $P(\lambda, z) := 5\lambda z + z - 3$, $(\lambda, z) \in \overline{S}_\theta \times \overline{\Sigma}_\delta$, is neither regular in space nor regular in time.

For a Newton polygon N we can define the exterior normal to the edge

$$[v_j v_{j+1}] := \{t v_j + (1-t) v_{j+1} : t \in [0,1]\}$$

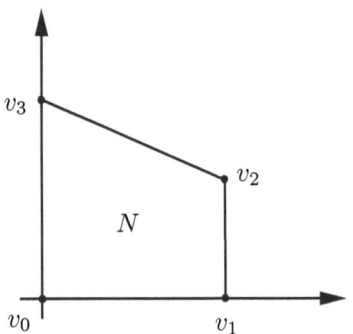
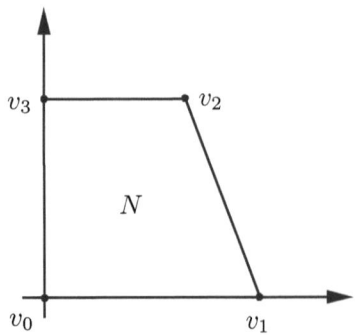

Figure 2.2: A Newton polygon that is not regular in time

Figure 2.3: A Newton polygon that is not regular in space

connecting the vertices v_j and v_{j+1} by

$$q_j := (q_{j,1}, q_{j,2}) := \left(\sqrt{1+\gamma_j^2}\right)^{-1}(1,\gamma_j), \quad j=1,\ldots,J-1,$$

$$q_J := \begin{cases} (\sqrt{1+\gamma_J^2})^{-1}(1,\gamma_J), & s_J \neq s_{J+1}, \\ (0,1), & s_J = s_{J+1} \end{cases}$$

where we have defined

$$\gamma_j := \frac{r_j - r_{j+1}}{s_{j+1} - s_j}, \quad j=1,\ldots,J-1, \quad \gamma_J := \begin{cases} \frac{r_j - r_{j+1}}{s_{J+1} - s_J}, & s_J \neq s_{J+1}, \\ \infty, & s_J = s_{J+1}. \end{cases}$$

Furthermore, we define

$$\gamma_0 := 0, \qquad q_0 := (q_{0,1}, q_{0,2}) := (0,-1),$$
$$\gamma_{J+1} := \infty, \qquad q_{J+1} := (q_{J+1,1}, q_{J+1,2}) := (-1,0),$$

and the orthogonal vectors $q_j^\perp := (-q_{j,2}, q_{j,1})$, $j = 1,\ldots, J+1$. We have $0 \leq \gamma_1 < \cdots < \gamma_J \leq \infty$. Note that in general there does not necessarily exist $j \in \{1,\ldots,J\}$ with $\gamma_j = \rho$.

Remark 2.20. Let N be a Newton polygon with vertices $N_V = \{v_j = (s_j, r_j) : j = 0,\ldots,J+1\}$. Then a simple geometric observation shows that for $(x,y) \in [0,\infty)^2$ we have $(x,y) \in N$ if and only if

$$x + \gamma y = \left\langle \binom{1}{\gamma}, \binom{x}{y} \right\rangle \leq \max_{j=1,\ldots,J+1} \left\langle \binom{1}{\gamma}, \binom{r_j}{s_j} \right\rangle = \max_{j=1,\ldots,J+1} (r_j + \gamma s_j)$$

for all $0 < \gamma < \infty$. As we have $r_j \geq r_{j+1} \geq 0$ and $0 \leq s_j \leq s_{j+1}$ for $j = 1,\ldots, J+1$, the function

$$\mu \colon (0,\infty) \to \mathbb{R}, \ \gamma \mapsto \max_{j=1,\ldots,J+1} (r_j + \gamma s_j)$$

2.1. Inhomogeneous symbols and the Newton polygon

is piecewise linear, continuous and monotonically increasing. More precisely, we have
$$\mu(\gamma) = r_j + \gamma s_j \quad \text{for } \gamma \in (\gamma_{j-1}, \gamma_j), \ j = 1, \ldots, J.$$
This is a prototype of an order function. As we will need more general order functions, we make the following definition:

Definition and Remark 2.21 (Order functions). (i) A continuous and piecewise linear function $\mu \colon [0, \infty) \to \mathbb{R}$ is called an *order function* if μ is convex or concave. More precisely, μ is an order function if there exist $M \in \mathbb{N}$ and $\gamma_\ell > 0$, $m_\ell(\mu), b_\ell(\mu) \in \mathbb{R}$ for $\ell = 0, \ldots, M$ with
$$\gamma_0 := 0 < \gamma_1 < \gamma_2 < \cdots < \gamma_M < \gamma_{M+1} := \infty$$
such that $\mu(\gamma) = m_\ell(\mu) \cdot \gamma + b_\ell(\mu)$ for $\gamma \in [\gamma_\ell, \gamma_{\ell+1})$, $\ell = 0, \ldots, M$, and if we have either
$$m_{\ell-1}(\mu) \leq m_\ell(\mu), \quad b_{\ell-1}(\mu) \geq b_\ell(\mu), \quad \ell = 1, \ldots, M$$
(i.e., μ is convex) or
$$m_{\ell-1}(\mu) \geq m_\ell(\mu), \quad b_{\ell-1}(\mu) \leq b_\ell(\mu), \quad \ell = 1, \ldots, M$$
(i.e., μ is concave). For convex order functions we have
$$\mu(\gamma) = \max_{\ell=0,\ldots,M} (b_\ell(\mu) + \gamma \cdot m_\ell(\mu)), \quad 0 \leq \gamma < \infty.$$
For concave order functions, max has to be replaced by min.

(ii) A convex order function μ is *increasing* if $m_\ell(\mu) \geq 0$ for all $\ell = 0, \ldots, M$.

A concave order function μ is *decreasing* if $m_\ell(\mu) \leq 0$ for all $\ell = 0, \ldots, M$.

(iii) A order function μ is called *strictly positive* if μ is convex and if
$$m_\ell(\mu) \geq 0 \text{ and } b_\ell(\mu) \geq 0, \quad \ell = 0, \ldots, M.$$
(In particular, in this case μ is increasing and nonnegative.)

An order function μ is called *strictly negative* if $-\mu$ is strictly positive, i.e., if we have
$$m_\ell(\mu) \leq 0 \text{ and } b_\ell(\mu) \leq 0, \quad \ell = 0, \ldots, M.$$
(In particular, in this case μ is decreasing and nonpositive.)

The following remark shows that there is a 1-1-correspondence between strictly positive order functions and Newton polygons.

Definition and Remark 2.22. (i) Let N be a Newton polygon with vertices $N_V = \{(r_j, s_j): j = 0, \ldots, J+1\} \subseteq [0, \infty)^2$ with $J \in \mathbb{N}_0$, starting at the origin and being indexed counter-clockwise. Then

$$\mu_N(\gamma) := \max_{j=1,\ldots,J+1} (r_j + \gamma s_j), \quad \gamma \geq 0,$$

is a strictly positive order function which is called the *associated* order function to N.

(ii) Let $\mu \colon [0, \infty) \to \mathbb{R}$ be a strictly positive order function. Let $\gamma_0 = 0 < \gamma_1 < \cdots < \gamma_{M+1} = \infty$ be a partition of $[0, \infty)$ such that $\mu(\gamma) = b_\ell(\mu) + \gamma m_\ell(\mu)$ for $\gamma \in [\gamma_\ell, \gamma_{\ell+1})$, $\ell = 0, \ldots, M$. We define

$$\nu(\mu) := \{(b_\ell(\mu), m_\ell(\mu)) : \ell = 0, \ldots, M\} \subseteq [0, \infty)^2.$$

Then $N(\mu) := N(\nu(\mu))$ is called the Newton polygon *associated* to μ. For $(x, y) \in [0, \infty)^2$ we have $(x, y) \in N(\mu)$ if and only if $x + \gamma y \leq \mu(\gamma)$ for all $\gamma \geq 0$.

In the situation of (i) and (ii), we obviously have $\mu_{N(\mu)} = \mu$ and $N(\mu_N) = N$.

Remark 2.23. Let $P \in S(L_t \times L_x)$. Then the Newton polygon of P can be seen as a geometrical description of the γ-principal parts and the γ-order and thus of the inhomogeneities hidden in the symbol. In fact, the following statements are easily seen:

(i) We have

$$d_\gamma(P) = \mu_{N(P)}(\gamma), \quad \gamma \geq 0,$$

i.e., the order function associated to $N(P)$ equals the γ-order of P in the sense of Definition 2.10.

(ii) For $\gamma \in (0, \infty)$, let $I_\gamma := \{\ell \in I_P : \gamma M_\ell + d_\gamma(\tau_\ell) + L_\ell = d_\gamma(P)\}$ be the index set of the γ-principal part of P (see Definition 2.10). If $N_V = \{(r_j, s_j) : j = 0, \ldots, J+1\}$ are the vertices of $N(P)$, then we have

$$I_{\gamma'} = \{\ell \in I_{\gamma_j} : M_\ell + \chi_{(\rho,\infty)}(\gamma_j) N_\ell/\rho = s_j\}$$
$$= \{\ell \in I_{\gamma_j} : \chi_{(0,\rho]}(\gamma_j) N_\ell + L_\ell = r_j\},$$
$$I_{\gamma''} = \{\ell \in I_{\gamma_j} : M_\ell + \chi_{[\rho,\infty)}(\gamma_j) N_\ell/\rho = s_{j+1}\}$$
$$= \{\ell \in I_{\gamma_j} : \chi_{(0,\rho)}(\gamma_j) N_\ell + L_\ell = r_{j+1}\}$$

for $j = 1, \ldots, J$, $0 < \gamma_j < \infty$ and all $\gamma' \in (\gamma_{j-1}, \gamma_j)$ and $\gamma'' \in (\gamma_j, \gamma_{j+1})$. Here χ_A stands for the characteristic function of the set A.

(iii) Let $N(P)$ be regular. If there exists $j \in \{1, \ldots, J+1\}$ with $\rho \in (\gamma_{j-1}, \gamma_j)$, we have $N_\ell = 0$ (i.e., $\tau_\ell = \text{const}$) for all $\ell \in I_\rho$. Otherwise we get an edge with normal direction $(1, \rho)$. Hence $M_\ell = s_j$ and $L_\ell = r_j$ for all $\ell \in I_\rho$. If $N(P)$ is not regular in time, this holds for $j \geq 2$, and if $N(P)$ is not regular in space, this holds for $j \leq J$.

2.1. Inhomogeneous symbols and the Newton polygon

Definition 2.24. (i) Let μ be a strictly positive order function. Then we define the *associated weight function* W_μ as the weight function of the Newton polygon associated to μ, i.e., we set $W_\mu := W_{N(\mu)}$.

(ii) Let μ be a convex order function. We set
$$\alpha_\mu := |\min\{0, m_0(\mu)\}| \geq 0,$$
$$\beta_\mu := |\min\{0, b_M(\mu)\}| \geq 0$$
and define the *strictly positive order function* μ_+ *associated to* μ by
$$\mu_+(\gamma) := \mu(\gamma) + \alpha_\mu \cdot \gamma + \beta_\mu, \quad \gamma \geq 0.$$
With this, the *weight function associated to* μ is defined as
$$W_\mu(\lambda, z) := \frac{W_{\nu(\mu_+)}(\lambda, z)}{W_{\{(\beta_\mu, \alpha_\mu)\}}(\lambda, z)}, \quad (\lambda, z) \in \mathbb{C} \times \mathbb{C}^n.$$

(iii) If μ is a concave order function then $\mu' := -\mu$ is a convex order function. In this case we set
$$W_\mu(\lambda, z) := \frac{1}{W_{\mu'}(\lambda, z)}, \quad (\lambda, z) \in \mathbb{C} \times \mathbb{C}^n.$$

Remark 2.25. Note that the order function $\mu(\gamma) := m\gamma + b$, $m, b \in \mathbb{R}$, is a convex order function as well as a concave order function. One easily validates that the weight function is the same if we interpret μ either as convex or concave.

Example 2.26. (i) The function
$$\mu(\gamma) := \max\left\{2, \frac{3}{2} + \frac{1}{2}\gamma, 1 + \frac{3}{4}\gamma\right\} = \begin{cases} 2, & \gamma \in [0, 1), \\ \frac{3}{2} + \frac{1}{2}\gamma, & \gamma \in [1, 2), \\ 1 + \frac{3}{4}\gamma, & \gamma \geq 2 \end{cases}$$
is strictly positive with weight function $W_\mu(\lambda, z) = 1 + |z|^2 + |\lambda|^{1/2}|z|^{3/2} + |\lambda|^{3/4}|z| + |\lambda|^{3/4}$.

(ii) Let $P \in S(L_t \times L_x)$ and $\mu(\gamma) := d_\gamma(P)$, $\gamma > 0$. Then μ is a strictly positive order function and $W_P = W_\mu$.

(iii) We define the concave order function $\mu(\gamma) := \min\{0, -\gamma + 2\}$. For the convex order function $-\mu(\gamma) = \max\{0, \gamma - 2\}$ we have $\alpha_{-\mu} = 0$ and $\beta_{-\mu} = 2$. Hence, we have
$$\mu'_+(\gamma) := \mu'(\gamma) + 2 = \max\{2, \gamma\}.$$
According to Definition 2.24 (ii) we obtain the associated weight function by
$$W_\mu(\lambda, z) = \frac{W_{\{(\beta_{-\mu}, \alpha_{-\mu})\}}(\lambda, z)}{W_{\nu(\mu'_+)}(\lambda, z)} = \frac{1 + |z|^2}{1 + |\lambda| + |z|^2}, \quad (\lambda, z) \in \mathbb{C} \times \mathbb{C}^n.$$

In the following chapters we will often need estimates from above or below for symbols. This is why we introduce the concept of upper and lower order functions.

Definition 2.27 (Upper and lower order functions). Let μ be an order function and let $P \in H_P(\mathring{L}_t \times \mathring{L}_x)$ be such that there exist $C = C(P,\mu) > 0$ and $\lambda_0 = \lambda_0(P,\mu) \geq 0$ with
$$|P(\lambda, z)| \leq C \cdot W_\mu(\lambda, z) \qquad (2.3)$$
for all $(\lambda, z) \in \mathring{L}_t \times \mathring{L}_x$ with $|\lambda| \geq \lambda_0$. Then μ is called an *upper* order function of P. If (2.3) holds with '$\geq$' instead of '$\leq$', then μ is called a *lower* order function of P.

Example 2.28. (i) The symbol $P(\lambda, z) := \lambda^2 - 3\lambda + \lambda z + \lambda z^2 - 3z^3$ for $(\lambda, z) \in \overline{S}_\theta \times \overline{\Sigma}_\delta$ fulfills
$$\begin{aligned}|P(\lambda, z)| &\leq 3\left(|\lambda|^2 + |\lambda| + |\lambda||z| + |\lambda||z|^2 + |z|^3\right) \\ &\leq 3\left(|\lambda|^2 + |\lambda| + |\lambda||z| + |\lambda||z|^2 + |z|^3 + 1\right) \\ &\leq 5\left(|\lambda|^2 + |\lambda||z|^2 + |z|^3 + 1\right), \quad (\lambda, z) \in \overline{S}_\theta \times \overline{\Sigma}_\delta\end{aligned}$$
by Remark 2.17 and due to $(0,1), (1,1) \in \mathrm{conv}\{(0,2),(2,1),(3,0),(0,0)\}$. Defining
$$\mu(\gamma) := \max\{2\gamma, \gamma + 2, 3\}, \quad \gamma \geq 0$$
we get
$$W_\mu(\lambda, z) = |\lambda|^2 + |\lambda||z|^2 + |z|^3 + 1, \quad (\lambda, z) \in \overline{S}_\theta \times \overline{\Sigma}_\delta.$$
So we obtain that μ is a convex upper order function of P where $\lambda_0(P,\mu) = 0$.

(ii) Let $\omega(\lambda, z) := \sqrt{\lambda + |z|^2_-}$, $(\lambda, z) \in \overline{S}_\theta \times \overline{\Sigma}_\delta^n$. We have $\omega \in S^{(2,1)}(\overline{S}_\theta \times \overline{\Sigma}_\delta^n)$, and therefore Remark 2.4 yields a constant $C > 0$ such that
$$\begin{aligned}|\omega(\lambda, z)| &\geq C(|\lambda|^{1/2} + |z|) = \frac{C}{2}(|\lambda|^{1/2} + 2|z| + |\lambda|^{1/2}) \\ &\geq \frac{C}{2}(|\lambda|^{1/2} + |z| + 1)\end{aligned}$$
holds for all $(\lambda, z) \in \overline{S}_\theta \times \overline{\Sigma}_\delta^n$ with $|\lambda| \geq 1$. Hence, $\mu(\gamma) := \max\{\frac{\gamma}{2}, 1\}$, $\gamma \geq 0$, is a lower order function of ω with $\lambda_0(\omega, \mu) = 1$.

Lemma 2.29. Let μ be an upper or lower order function for $P \in H_P(S_\theta \times \Sigma_\delta^n)$, $\theta \in (0, \pi)$, $\delta \in (0, \pi/2)$. Then there exists a constant $\alpha = \alpha(\lambda_0, \theta) > 0$ with $|\lambda + \alpha| \geq \lambda_0$ for all $\lambda \in S_\theta$. With this we define the shifted symbol by $P_\alpha(\lambda, z) := P(\lambda + \alpha, z)$ for all $(\lambda, z) \in S_\theta \times \Sigma_\delta^n$. Then μ is also an upper or lower order function for P_α. In this situation we can even choose $\lambda_0(P_\alpha, \mu) = 0$ in Definition 2.27.

2.1. Inhomogeneous symbols and the Newton polygon

Proof. Due to Remark 2.17 and $c|\lambda| \leq |\lambda + \alpha| \leq |\lambda| + \alpha$, $\lambda \in S_\theta$, it is easy to show that there exist $C_1, C_2 > 0$ such that

$$C_1 W_\mu(\lambda, z) \leq W_\mu(\lambda + \alpha, z) \leq C_2 W_\mu(\lambda, z) \tag{2.4}$$

for all $(\lambda, z) \in S_\theta \times \Sigma_\delta^n$. Now the assertion directly follows from (2.4) and the choice of α. $\square$

Lemma 2.30. *Let $P \in S(L_t \times L_x)$, and let μ be a strictly positive order function with*

$$d_\gamma(P) \leq \mu(\gamma), \quad \gamma \geq 0. \tag{2.5}$$

Then μ is an upper order function of P. In particular, this yields

$$|P(\lambda, z)| \leq C \cdot W_P(\lambda, z), \quad (\lambda, z) \in L_t \times L_x.$$

Proof. We have

$$|P(\lambda, z)| \leq \sum_{\ell \in I} |\tau_\ell(\lambda, z)||\varphi_\ell(\lambda)||\psi_\ell(z)|$$

$$\leq C \cdot \sum_{\ell \in I} \left(|\lambda|^{M_\ell + N_\ell/\rho}|z|^{L_\ell} + |\lambda|^{M_\ell}|z|^{L_\ell + N_\ell}\right).$$

By virtue of (2.5) and Remark 2.20 we can show $(L_\ell, M_\ell + \frac{N_\ell}{\rho}), (L_\ell + N_\ell, M_\ell) \in N(\mu)$ for all $\ell \in I$. The claim then follows from Remark 2.17 and Example 2.26 (ii). $\square$

Example 2.31. Recall the symbol $\psi(\lambda, z) := \sqrt{-z^2} - \sqrt{\lambda - z^2}$ in Example 2.12 (ii). A naive analysis of the order structure of ψ shows that $\widetilde{\mu}(\gamma) := d_\gamma(\psi) = \max\{1, \gamma/2\}$ is an upper order function of ψ. In the following we show that there exists an upper order function μ which characterizes the order structure of ψ better then $\widetilde{\mu}$.

Let $\psi'(\lambda, z) := \sqrt{-z^2} + \sqrt{\lambda - z^2}$ for $(\lambda, z) \in \overline{S}_\theta \times \overline{\Sigma}_\delta$. For sufficiently small δ we get

$$\psi' \in S^{(2,1)}(\overline{S}_\theta \times \overline{\Sigma}_\delta)$$

and therefore

$$|\psi'(\lambda, z)| \geq \frac{1}{C}\left(|\lambda|^{1/2} + |z|\right) \geq \frac{1}{2C}\left(|\lambda|^{1/2} + |z| + |\lambda|^{1/2}\right)$$

$$\geq \frac{1}{2C}\left(|\lambda|^{1/2} + |z| + 1\right), \quad (\lambda, z) \in \overline{S}_\theta \times \overline{\Sigma}_\delta, \ |\lambda| \geq 1$$

due to Remark 2.4. Considering $\psi(\lambda, z) \cdot \psi'(\lambda, z) = -\lambda$ we get

$$|\psi(\lambda, z)| = \left|\frac{\lambda}{\psi'(\lambda, z)}\right| \leq C \frac{|\lambda|}{|\lambda|^{1/2} + |z|} \leq 2C \frac{1 + |\lambda|}{|\lambda|^{1/2} + |z| + 1}$$

for all $(\lambda, z) \in \overline{S}_\theta \times \overline{\Sigma}_\delta$ with $|\lambda| \geq 1 =: \lambda_0$. This yields $|\psi(\lambda,z)| \leq C' \cdot W_\mu(\lambda,z)$ for the concave order function $\mu(\gamma) := \min\{\gamma - 1, \frac{1}{2}\gamma\}$, $\gamma \geq 0$. Hence μ is an upper order function of ψ.

We have seen that the inhomogeneous structure of a symbol P can be represented by the geometry of the Newton polygon $N(P)$ or, equivalently, by the order function $\mu_{N(P)}$. Note that we have $\mu_{N(P)}(\gamma) = d_\gamma(P)$, see Remark 2.23 (i). We have defined weight functions related to Newton polygons as well as weight functions related to order functions. Whereas the Newton polygon is a very intuitive notion and gives a geometrical description, the order functions have some advantages: First, we can define upper and lower order functions to deal with one-sided estimates. Secondly, we can also include order functions which are not strictly positive, see Definition 2.24.

The most important advantage, however, lies in the fact that the order function concept will allow us to "calculate" with Newton polygons: It is not obvious how the sum and the difference of Newton polygons can be defined, but we can make arithmetic operations with the related order functions. This will be of importance in later chapters where, e.g., we will deal with matrices of operators where each entry has its own Newton polygon structure. Moreover, both the Sobolev spaces and the operators will have inhomogeneous structure, so we are naturally facing the question of "adding" Newton polygons.

Therefore, for the remainder of this subsection, we will investigate the relation between sums and differences of order functions on one hand and the associated weight functions and Newton polygons on the other hand. We recall that the weight function corresponding to a single point $\{(\beta, \alpha)\}$ is given by $W_{\{(\beta,\alpha)\}}(\lambda, z) = (1 + |\lambda|^\alpha)(1 + |z|)^\beta$, see Example 2.16. The order function is in this case given by $\mu_{N(\{(\beta,\alpha)\})}(\gamma) = \alpha\gamma + \beta$.

We start with the relation between the sum of order functions and the product of the associated weight functions. For this, we provide the following lemma.

Lemma 2.32. *Let μ be a strictly positive order function and let $\alpha, \beta \geq 0$ such that*

$$\mu': \gamma \mapsto \mu(\gamma) - (\beta + \alpha\gamma)$$

is also strictly positive. In this situation there exists a constant $C > 0$ such that

$$W_\mu \leq W_{\{(\beta,\alpha)\}} \cdot W_{\mu'} \leq C \cdot W_\mu.$$

Proof. Due to the definition of W_μ and $W_{\mu'}$ the left inequality is obvious. The second inequality follows from $(b_\ell(\mu) + j\beta, m_\ell(\mu) + k\alpha) \in N(\mu)$, $j, k = 0, 1$, and Remark 2.17. $\square$

In the following, we will consider finitely many order functions simultaneously. For this we remark that, by refinement of the partition, we may tacitly assume that the number M in Definition 2.21 is the same for all order functions.

2.1. Inhomogeneous symbols and the Newton polygon

Lemma 2.33. (i) *Let μ_1 and μ_2 be strictly positive order functions. Then $\mu_1 + \mu_2$ is also a strictly positive order function and there exist $C, C' > 0$ such that*

$$C' \cdot W_{\mu_1+\mu_2}(\lambda, z) \leq W_{\mu_1}(\lambda, z) \cdot W_{\mu_2}(\lambda, z) \leq C \cdot W_{\mu_1+\mu_2}(\lambda, z)$$

for all $(\lambda, z) \in \mathbb{C} \times \mathbb{C}^n$.

(ii) *The two-sided estimate in (i) also holds for convex order functions μ_1 and μ_2.*

Proof. (i) The monotone structure of the strictly positive order functions ensures that the sum of strictly positive order functions is also strictly positive. For $(\lambda, z) \in \mathbb{C} \times \mathbb{C}^n$ we trivially get

$$W_{\mu_1}(\lambda, z) \cdot W_{\mu_2}(\lambda, z) = W_{\nu(\mu_1)}(\lambda, z) \cdot W_{\nu(\mu_2)}(\lambda, z)$$
$$\geq W_{\nu(\mu_1+\mu_2)}(\lambda, z) = W_{\mu_1+\mu_2}(\lambda, z)$$

because of $\nu(\mu_1 + \mu_2) = \{(b_\ell(\mu_1) + b_\ell(\mu_2), m_\ell(\mu_1) + m_\ell(\mu_2)) : \ell = 0, \ldots, M\}$. Due to Remark 2.17 it is sufficient to show

$$(b_\ell(\mu_1) + b_p(\mu_2), m_\ell(\mu_1) + m_p(\mu_2)) \in N(\mu_1 + \mu_2) \tag{2.6}$$

for all $\ell, p = 0, \ldots, M$. Let $\gamma \in [\gamma_k, \gamma_{k+1})$ be arbitrary. Then we derive

$$\left\langle \begin{pmatrix} 1 \\ \gamma \end{pmatrix}, \begin{pmatrix} b_\ell(\mu_1) + b_p(\mu_2) \\ m_\ell(\mu_1) + m_p(\mu_2) \end{pmatrix} \right\rangle = [m_\ell(\mu_1)\gamma + b_\ell(\mu_1)] + [m_p(\mu_2)\gamma + b_p(\mu_2)]$$
$$\leq [m_k(\mu_1)\gamma + b_k(\mu_1)] + [m_k(\mu_2)\gamma + b_k(\mu_2)]$$
$$= (\mu_1 + \mu_2)(\gamma).$$

Remark 2.20 then yields (2.6).

(ii) The sum of convex functions is convex, too. Let $\alpha_k := \alpha_{\mu_k}$ and $\beta_k := \beta_{\mu_k}$, $k = 1, 2$. Defining the strictly positive order functions

$$\mu_{1,+}(\gamma) := \mu_1(\gamma) + \beta_1 + \alpha_1\gamma, \qquad \alpha_{12} := \alpha_1 + \alpha_2,$$
$$\mu_{2,+}(\gamma) := \mu_2(\gamma) + \beta_2 + \alpha_2\gamma, \qquad \beta_{12} := \beta_1 + \beta_2,$$

we get

$$W_{\mu_1} \cdot W_{\mu_2} = \frac{W_{\mu_1,+} \cdot W_{\mu_2,+}}{W_{\{(\beta_1,\alpha_1)\}} \cdot W_{\{(\beta_2,\alpha_2)\}}}.$$

Using Lemma 2.33 (i) for the nominator and the denominator we obtain the two-sided estimate

$$C_1 \cdot W_{\mu_1} \cdot W_{\mu_2} \leq \frac{W_{\mu_{1,+}+\mu_{2,+}}}{W_{\{(\beta_{12},\alpha_{12})\}}} \leq C_2 \cdot W_{\mu_1} \cdot W_{\mu_2} \tag{2.7}$$

for $C_1, C_2 > 0$. Due to Lemma 2.32 we get

$$C' \cdot W_{\mu_1 + \mu_2} \leq \frac{W_{\mu'_1 + \mu'_2}}{W_{\{(\beta_{12}, \alpha_{12})\}}} \leq C'' \cdot W_{\mu_1 + \mu_2} \qquad (2.8)$$

for certain constants $C', C'' > 0$. The assertion now follows from (2.7) and (2.8). $\square$

Lemma 2.34. (i) *Let μ' and μ'' be convex order functions with $\mu'(\gamma) \leq \mu''(\gamma)$ for all $\gamma \geq 0$. Then there exists $C > 0$ with*

$$\frac{W_{\mu'}(\lambda, z)}{W_{\mu''}(\lambda, z)} \leq C, \quad (\lambda, z) \in \mathbb{C} \times \mathbb{C}^n.$$

(ii) *Let μ be a convex order function and let $\{\mu_n\}_{n=1,\ldots,m}$ be order functions with*

$$\sum_{n=1}^{m} \mu_n(\gamma) \leq \mu(\gamma), \quad \gamma > 0.$$

Then there exists $C > 0$ such that

$$\frac{\prod_{n=1}^{m} W_{\mu_n}(\lambda, z)}{W_\mu(\lambda, z)} \leq C, \quad (\lambda, z) \in \mathbb{C} \times \mathbb{C}^n.$$

Note that in general $\sum_{n=1}^{m} \mu_n$ is not an order function.

Proof. (i) Let $\alpha := \max\{\alpha_{\mu'}, \alpha_{\mu''}\}$ and $\beta := \max\{\beta_{\mu'}, \beta_{\mu''}\}$ and define

$$\mu_1(\gamma) := \mu'(\gamma) + \beta + \alpha \cdot \gamma, \quad \mu_2(\gamma) := \mu''(\gamma) + \beta + \alpha \cdot \gamma.$$

Then Lemma 2.32 yields

$$C \frac{W_{\mu_1}(\lambda, z)}{W_{\mu_2}(\lambda, z)} \leq \frac{W_{\mu'}(\lambda, z)}{W_{\mu''}(\lambda, z)} \leq C' \frac{W_{\mu_1}(\lambda, z)}{W_{\mu_2}(\lambda, z)}, \quad (\lambda, z) \in \mathbb{C} \times \mathbb{C}^n \qquad (2.9)$$

with $C, C' > 0$. The order functions μ_1 and μ_2 are strictly positive with $\mu_1(\gamma) \leq \mu_2(\gamma)$, $\gamma \geq 0$. Remark 2.20 shows $N(\mu_1) \subseteq N(\mu_2)$ and therefore the assertion follows from (2.9) with Remark 2.17.

(ii) Let

$$M_+ := \{n \in \{1, \ldots, m\} \colon \mu_n \text{ is a convex order function}\},$$
$$M_- := \{n \in \{1, \ldots, m\} \colon \mu_n \text{ is a concave order function}\} \setminus M_+.$$

Then we have

$$\frac{\prod_{n=1}^{m} W_{\mu_n}}{W_\mu} = \frac{\prod_{n \in M_+} W_{\mu_n}}{\left(\prod_{j \in M_-} W_{-\mu_j}\right) W_\mu}.$$

2.1. Inhomogeneous symbols and the Newton polygon

Using Lemma 2.33 we get

$$\frac{\prod_{n=1}^{m} W_{\mu_n}(\lambda, z)}{W_{\mu}(\lambda, z)} \leq C \cdot \frac{W_{\mu'}(\lambda, z)}{W_{\mu''}(\lambda, z)}, \quad (\lambda, z) \in \mathbb{C} \times \mathbb{C}^n$$

where $\mu' := \sum_{n \in M_+} \mu_n$ and $\mu'' := \mu - \sum_{j \in M_-} \mu_j$. Note that μ' and μ'' are both convex. As a consequence of the assumptions we have $\mu'(\gamma) \leq \mu''(\gamma)$ for all $\gamma \geq 0$. Therefore we can prove part (ii) by (i). □

Lemma 2.35. *Let μ_1 and μ_2 be strictly positive order functions. Then we have*

$$(b_\ell(\mu_1) + b_\kappa(\mu_2), m_\ell(\mu_1) + m_\kappa(\mu_2)) \in N(\mu_1 + \mu_2)$$

for all $\ell, \kappa \in \{0, \ldots, M\}$.

Proof. Let $i \in \{0, \ldots, M\}$ and $\gamma \in [\gamma_i, \gamma_{i+1})$. Definition 2.21 (ii) implies

$$[m_\ell(\mu_1) + m_\kappa(\mu_2)]\gamma + b_\ell(\mu_1) + b_\kappa(\mu_2)$$
$$= m_\ell(\mu_1)\gamma + b_\ell(\mu_1) + m_\kappa(\mu_2)\gamma + b_\kappa(\mu_2)$$
$$\leq m_i(\mu_1)\gamma + b_i(\mu_1) + m_i(\mu_2)\gamma + b_i(\mu_2) = (\mu_1 + \mu_2)(\gamma).$$

Hence Remark 2.20 completes the proof. □

Definition 2.36 (Support and index of an order function). Let μ be an order function. Then the *support* of μ is defined by

$$\operatorname{supp} \mu := \{i \in \{1, \ldots, M\} \colon (b_{i-1}(\mu), m_{i-1}(\mu)) \neq (b_i(\mu), m_i(\mu))\}.$$

The constant $I = I(\mu) := \#(\operatorname{supp} \mu)$ is called the *index* of μ. We define $i_0 := 0$ and $i_{I+1} := M + 1$. If $I \neq 0$, let $\operatorname{supp} \mu = \{i_1, \ldots, i_I\}$ with $i_0 < i_1 < \cdots < i_I < i_{I+1}$. Note that we have

$$(b_i(\mu), m_i(\mu)) = (b_j(\mu), m_j(\mu)), \quad i_k \leq i, j < i_{k+1}$$

for all $k = 0, \ldots, I$. The order function μ is said to be of *trivial index* if $I(\mu) = 0$, i.e., there exist $m, b \in \mathbb{R}$ with $\mu(\gamma) = m\gamma + b$ for all $\gamma > 0$.

Example 2.37. Let $\gamma_1 := 4$, $\gamma_2 := 6$,

$$\mu_1(\gamma) := \begin{cases} \gamma + 2, & \gamma \in [0, \gamma_1), \\ 3/2\gamma, & \gamma \in [\gamma_1, \gamma_2), \\ 3/2\gamma, & \gamma \in [\gamma_2, \infty), \end{cases} \quad \mu_2(\gamma) := \begin{cases} 1/2\gamma + 2, & \gamma \in [0, \gamma_1), \\ \gamma, & \gamma \in [\gamma_1, \gamma_2), \\ 2\gamma - 6, & \gamma \in [\gamma_2, \infty). \end{cases}$$

Then μ_1 is of index 1 and μ_2 is of index 2. Furthermore, we have $i_1(\mu_1) = 1$, $i_1(\mu_2) = 1$, and $i_2(\mu_2) = 2$.

In Lemma 2.35 we have seen that the tuples $(b_\ell(\mu_1) + b_\kappa(\mu_2), m_\ell(\mu_1) + m_\kappa(\mu_2))$ are always contained in the Newton polygon of $\mu_1 + \mu_2$. The next lemma provides more information on the position of these tuples in $N(\mu_1 + \mu_2)$.

Lemma 2.38. Let μ_1 and μ_2 be strictly positive order functions and let

$$\ell \in \{0, \ldots, M\}, \quad k \in \{0, \ldots, I(\mu_1)\}, \quad i_k(\mu_1) \leq \ell < i_{k+1}(\mu_1).$$

Define

$$p_1 := \max\{p \in \{0, \ldots, I(\mu_2) + 1\} : i_p(\mu_2) \leq i_k(\mu_1)\},$$
$$p_2 := \min\{p \in \{0, \ldots, I(\mu_2) + 1\} : i_p(\mu_2) \geq i_{k+1}(\mu_1)\},$$

and $N := N(\mu_1 + \mu_2)$. Furthermore, we define Γ as the set of all $x \in \mathbb{R}^2$ lying on a non-horizontal and non-vertical line between two vertices of N. Let N_V be the set of vertices of N (see Definition 2.15 (iii)).

(i) For all $\kappa \in \{0, \ldots, M\}$ with $i_{p_1}(\mu_2) \leq \kappa < i_{p_2}(\mu_2)$ we have

$$(b_\ell(\mu_1) + b_\kappa(\mu_2), m_\ell(\mu_1) + m_\kappa(\mu_2)) = (b_j(\mu_1 + \mu_2), m_j(\mu_1 + \mu_2)) \in N_V$$

where

$$j := \begin{cases} i_k(\mu_1), & \kappa < i_k(\mu_1), \\ \kappa, & i_k(\mu_1) \leq \kappa < i_{k+1}(\mu_1), \\ i_{k+1}(\mu_1) - 1, & \kappa \geq i_{k+1}(\mu_1). \end{cases}$$

(ii) If $p_1 \neq 0$ and $i_{p_1}(\mu_2) = i_k(\mu_1)$, then we have

$$(b_\ell(\mu_1) + b_\kappa(\mu_2), m_\ell(\mu_1) + m_\kappa(\mu_2)) \in \Gamma \setminus N_V$$

for all $\kappa \in \{0, \ldots, M\}$ with $i_{p_1-1}(\mu_2) \leq \kappa < i_{p_1}(\mu_2)$.

More precisely, $(b_\ell(\mu_1) + b_\kappa(\mu_2), m_\ell(\mu_1) + m_\kappa(\mu_2))$ lies on the edge between the vertices $(b_{j-1}(\mu_1 + \mu_2), m_{j-1}(\mu_1 + \mu_2))$ and $(b_j(\mu_1 + \mu_2), m_j(\mu_1 + \mu_2))$ where $j := i_k(\mu_1)$.

(iii) If $p_2 \neq I(\mu_2) + 1$ and $i_{p_2}(\mu_2) = i_{k+1}(\mu_1)$, then we have

$$(b_\ell(\mu_1) + b_\kappa(\mu_2), m_\ell(\mu_1) + m_\kappa(\mu_2)) \in \Gamma \setminus N_V$$

for all $\kappa \in \{0, \ldots, M\}$ with $i_{p_2}(\mu_2) \leq \kappa < i_{p_2+1}(\mu_2)$.

More precisely, $(b_\ell(\mu_1) + b_\kappa(\mu_2), m_\ell(\mu_1) + m_\kappa(\mu_2))$ lies on the edge between the vertices $(b_{j-1}(\mu_1 + \mu_2), m_{j-1}(\mu_1 + \mu_2))$ and $(b_j(\mu_1 + \mu_2), m_j(\mu_1 + \mu_2))$ where $j := i_{k+1}(\mu_1)$.

(iv) For all $\kappa \in \{0, \ldots, M\}$ not covered by the conditions of (i)-(iii) we have

$$(b_\ell(\mu_1) + b_\kappa(\mu_2), m_\ell(\mu_1) + m_\kappa(\mu_2)) \in N(\mu_1 + \mu_2) \setminus \Gamma.$$

2.2. N-parameter-ellipticity and N-parabolicity

Proof. (i) For κ and j as above we get the two equalities $(b_\ell(\mu_1), m_\ell(\mu_1)) = (b_j(\mu_1), m_j(\mu_1))$ and $(b_\kappa(\mu_2), m_\kappa(\mu_2)) = (b_j(\mu_2), m_j(\mu_2))$. Then we have

$$(\mu_1 + \mu_2)(\gamma) = [m_j(\mu_1) + m_j(\mu_2)]\gamma + b_j(\mu_1) + b_j(\mu_2)$$
$$= [m_\ell(\mu_1) + m_\kappa(\mu_2)]\gamma + b_\ell(\mu_1) + b_\kappa(\mu_2), \quad \gamma \in (\gamma_j, \gamma_{j+1}).$$

This yields $(b_\ell(\mu_1) + b_\kappa(\mu_2), m_\ell(\mu_1) + m_\kappa(\mu_2)) \in N_V$.

(ii) Let $i_{p_1-1}(\mu_2) \leq \kappa < i_{p_1}(\mu_2)$, $p_1 \neq 0$, and $i_{p_1}(\mu_2) = i_k(\mu_1) =: j > 0$. Then we simply obtain $(b_j(\mu_1), m_j(\mu_1)) = (b_\ell(\mu_1), m_\ell(\mu_1))$, $(b_{j-1}(\mu_2), m_{j-1}(\mu_2)) = (b_\kappa(\mu_2), m_\kappa(\mu_2))$, and

$$(\mu_1 + \mu_2)(\gamma_j) = [m_j(\mu_1) + m_j(\mu_2)]\gamma_j + b_j(\mu_1) + b_j(\mu_2)$$
$$= [m_\ell(\mu_1) + m_\kappa(\mu_2)]\gamma_j + b_\ell(\mu_1) + b_\kappa(\mu_2)$$

due to $m_{j-1}(\mu_2)\gamma_j + b_{j-1}(\mu_2) = m_j(\mu_2)\gamma_j + b_j(\mu_2)$.

(iii) Let $i_{p_2}(\mu_2) \leq \kappa < i_{p_2+1}(\mu_2)$, $p_2 \neq I(\mu_2) + 1$, and $i_{p_2}(\mu_2) = i_{k+1}(\mu_1) =: j < M + 1$. Then we have

$$(b_{j-1}(\mu_1), m_{j-1}(\mu_1)) = (b_\ell(\mu_1), m_\ell(\mu_1)),$$
$$(b_j(\mu_2), m_j(\mu_2)) = (b_\kappa(\mu_2), m_\kappa(\mu_2)),$$

and

$$(\mu_1 + \mu_2)(\gamma_j) = [m_j(\mu_1) + m_j(\mu_2)]\gamma_j + b_j(\mu_1) + b_j(\mu_2)$$
$$= [m_\ell(\mu_1) + m_\kappa(\mu_2)]\gamma_j + b_\ell(\mu_1) + b_\kappa(\mu_2)$$

due to $m_{j-1}(\mu_1)\gamma_j + b_{j-1}(\mu_1) = m_j(\mu_1)\gamma_j + b_j(\mu_1)$.

(iv) This can be seen by the same arguments as in the proof of the parts (i)-(iii). □

The last two lemmas help us to understand the arithmetic of Newton polygons. In particular, the last characterization turns out to be helpful when we consider compatibility conditions in Section 2.3.

2.2 N-parameter-ellipticity and N-parabolicity

> *Motivation.* We have seen that for a symbol $P \in S(L_t \times L_x)$ the inhomogeneous symbol structure can be expressed in different ways: By the Newton polygon $N(P)$, by the associated order function $\mu_{N(P)}$, or by the weight function $W_{N(P)}$ of the Newton polygon. By Lemma 2.30, we always have the inequality
> $$|P(\lambda, z)| \leq C W_{N(P)}(\lambda, z), \quad (\lambda, z) \in L_t \times L_x,$$

i.e., $\mu_{N(P)}$ is an upper order function for P.

In this section, we will consider parabolic and parameter-elliptic symbols. For a classical (quasi-homogeneous) symbol P the traditional definition of parameter-ellipticity is given by

$$P(\lambda, z) \neq 0, \quad (\lambda, z) \in (L_t \times L_x) \setminus \{(0,0)\} \qquad (2.10)$$

(see [AV64]). Due to quasi-homogeneity, this condition immediately implies a uniform estimate from below for the symbol. For instance, for the symbol $P(\lambda, z) = \lambda - z_1^2 - \cdots - z_n^2$ we obtain $|P(\lambda, z)| \geq C(|\lambda| + |z|^2)$ in suitably chosen sectors. These estimates are the basis for the solution theory of the corresponding parabolic equations.

For more general symbols $P \in S(L_t \times L_x)$, condition (2.10) is not sufficient for an estimate from below. Therefore, N-parabolicity will be defined by the condition that $\mu_{N(P)}$ is both an upper and a lower order function for P (Subsection a)). Here we first consider scalar symbols, the generalization to mixed-order systems in Section 2.3 will be done by an estimate on the determinant of the symbol matrix.

Generally, it is not easy to verify the condition that $\mu_{N(P)}$ is a lower order function for P. Therefore, it is crucial for applications to find a simple criterion for N-parameter-ellipticity. This will be done in Subsection c) where an easy-to-handle condition on the principal parts will be formulated. The proof is based on a decomposition of the co-variable space in Subsection b).

The results in this section generalize results for more specific symbol classes in [GV92, 2.1 Theorem] and [DSS08, Theorem 3.1].

a) N-parameter-elliptic symbols and $S_N(L_t \times L_x)$

Definition 2.39 (N-parameter-elliptic/N-parabolic). (i) Let $P \in S(L_t \times L_x)$, let $\mu_P(\gamma) = d_\gamma(P)$ be the canonical order function, and let $W_P = W_{\mu_P}$ be the weight function associated to P. Then P is called *N-parameter-elliptic* in $\mathring{L}_t \times \mathring{L}_x$ if μ_P is an upper as well as a lower order function of P, i.e., there exist $C, C' > 0$, and $\lambda_0 = \lambda_0(P) \geq 0$ such that the two-sided estimate

$$C' \cdot W_P(\lambda, z) \leq |P(\lambda, z)| \leq C \cdot W_P(\lambda, z) \qquad (2.11)$$

holds for all $(\lambda, z) \in \mathring{L}_t \times \mathring{L}_x$ with $|\lambda| \geq \lambda_0$.

We define $S_N(L_t \times L_x)$ as the set of all symbols $P \in S(L_t \times L_x)$ which are N-parameter-elliptic in $\mathring{L}_t \times \mathring{L}_x$.

(ii) The symbol P is called N-parameter-elliptic of *angle* θ if it is N-parameter-elliptic in $S_\theta \times \Sigma_\delta^n$ for some $\delta > 0$.

(iii) The symbol P is called *N-parabolic* if it is N-parameter-elliptic with angle $\theta \in (\pi/2, \pi)$.

2.2. N-parameter-ellipticity and N-parabolicity

Remark 2.40. Note that for every $P \in S(L_t \times L_x)$ we have the one-sided estimate

$$|P(\lambda, z)| \leq C \cdot W_P(\lambda, z)$$

by Lemma 2.30. The condition of N-parameter-ellipticity means that this estimate is two-sided which will be the basis for estimates of the inverse symbol.

The condition of N-parameter-ellipticity can also be formulated for more general symbols:

Definition 2.41. The symbol $P \in H_P(\mathring{L}_t \times \mathring{L}_x)$ is called *N-parameter-elliptic* in $\mathring{L}_t \times \mathring{L}_x$ if there exists a strictly positive order function $\mu(P)$ such that $\mu(P)$ is an upper as well as a lower order function of P. The notions of Definition 2.39 (ii) and (iii) carry over to this more general case.

The next proposition is useful for the handling of quotients of holomorphic functions. Here we can take advantage of estimates involving weight functions and order functions. This result can be used to analyze the structure of the components of an inverse matrix, where the entries are given by quotients of determinants, cf. Section 2.3 b).

Proposition 2.42. *Let $P^{(1)}, P^{(2)} \in H_P(\mathring{L}_t \times \mathring{L}_x)$ be symbols and let μ, $\{\mu_n^{(j)}\}_{n=1,\ldots,m}$ ($j = 1, \ldots, M$) be order functions. If*

(i) *$P^{(2)}$ is N-parameter-elliptic in $\mathring{L}_t \times \mathring{L}_x$ with respect to the order function $\mu(P^{(2)})$,*

(ii) *there exists $C > 0$ such that*

$$|P^{(1)}(\lambda, z)| \leq C \cdot \sum_{j=1}^{M} \prod_{n=1}^{m} W_{\mu_n^{(j)}}(\lambda, z)$$

for all $(\lambda, z) \in \mathring{L}_t \times \mathring{L}_x$ with $|\lambda| \geq \lambda_0$,

(iii) *$\mu(\gamma) + \sum_{n=1}^{m} \mu_n^{(j)}(\gamma) \leq [\mu(P^{(2)})](\gamma)$ for all $j = 1, \ldots, M$ and $\gamma \geq 0$,*

then there exists $\sigma_0 \geq 0$ such that $-\mu$ is an upper order function of the quotient

$$\frac{P_\sigma^{(1)}}{P_\sigma^{(2)}} \in H_P(\mathring{L}_t \times \mathring{L}_x), \quad \sigma \geq \sigma_0$$

where $P_\sigma^{(k)} := P^{(k)}(\sigma + \cdot, \cdot)$, $k = 1, 2$.

Proof. The function $P^{(2)}$ is N-parameter-elliptic so we obtain

$$\left| W_\mu(\lambda, z) \cdot \frac{P^{(1)}(\lambda, z)}{P^{(2)}(\lambda, z)} \right| \leq \tilde{C} \sum_{j=1}^{M} \left| \frac{W_\mu(\lambda, z) \cdot \prod_{n=1}^{m} W_{\mu_n^{(j)}}(\lambda, z)}{W_{\mu(P^{(2)})}(\lambda, z)} \right|$$

for all $(\lambda, z) \in \mathring{L}_t \times \mathring{L}_x$ with $|\lambda| \geq \lambda_0$. Using assumption (iii), Lemma 2.34 (ii), and (2.4) we get the assertion. $\square$

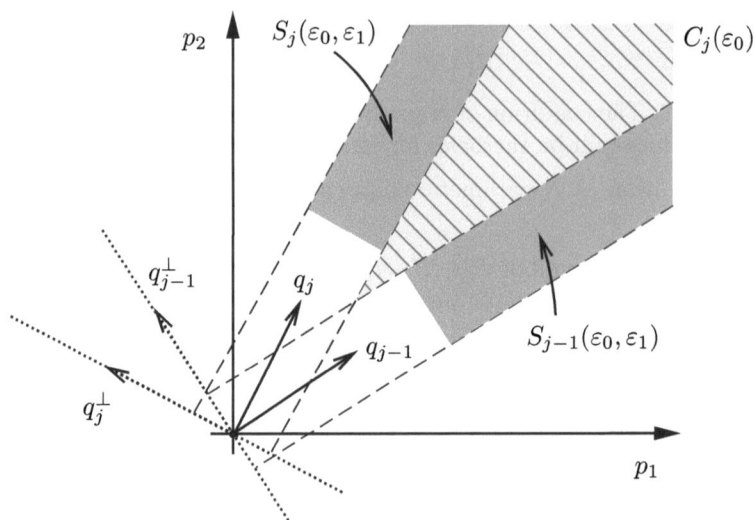

Figure 2.4: Illustration of the partition

b) Partition of the co-variable space

To prove the characterization by non-vanishing principal parts for symbols in $S_N(L_t \times L_x)$ we have to consider logarithmic partitions of the co-variable space. The used logarithmic partition was introduced in the book of S. Gindikin and L.R. Volevich [GV92]. The idea is to construct a partition where the principal parts of the symbol dominate the behavior of the whole symbol.

Definition 2.43 (Partition of the co-variable space). Let N be an arbitrary Newton polygon and let q_j and $q_j^\perp$ be defined as in Subsection 2.1 b). According to [GV92, Chapter 4, 2.3], we define for given $\varepsilon_0, \varepsilon_1 \in (0,1)$ the following sets:

$$C_j(\varepsilon_0) := \{p \in \mathbb{R}^2 : \langle q_{j-1}^\perp, p \rangle \geq \log(1/\varepsilon_0) \text{ and } \langle q_j^\perp, p \rangle \leq \log(\varepsilon_0)\},$$
$$S_\ell(\varepsilon_0, \varepsilon_1) := \{p \in \mathbb{R}^2 : \log(\varepsilon_0) \leq \langle q_\ell^\perp, p \rangle \leq \log(1/\varepsilon_0) \text{ and } \langle q_\ell, p \rangle \geq \log(1/\varepsilon_1)\}$$

and

$$\widetilde{G}_j(\varepsilon_0) := \{(\xi, \eta) \in (0, \infty)^2 : (\log(\xi), \log(\eta)) \in C_j(\varepsilon_0)\},$$
$$G_\ell(\varepsilon_0, \varepsilon_1) := \{(\xi, \eta) \in (0, \infty)^2 : (\log(\xi), \log(\eta)) \in S_\ell(\varepsilon_0, \varepsilon_1)\}$$

for $j = 1, \ldots J+1$, $\ell = 1, \ldots, J$. Here, C_j has the shape of a cone and S_ℓ has the shape of a half-strip (cf. Figure 2.4). Note that the vertices in [GV92] are indexed in the clockwise direction.

2.2. N-parameter-ellipticity and N-parabolicity

Remark 2.44. If N is regular in space, then common logarithmic identities yield the representations

$$\widetilde{G}_1(\varepsilon_0) := \left\{(\xi,\eta) \in (0,\infty)^2 : \eta \leq \varepsilon_0^{1/q_{1,1}}\xi^{\gamma_1} \text{ and } \xi \geq \varepsilon_0^{-1}\right\},$$

$$\widetilde{G}_j(\varepsilon_0) := \left\{(\xi,\eta) \in (0,\infty)^2 : \varepsilon_0^{-1/q_{j-1,1}}\xi^{\gamma_{j-1}} \leq \eta \leq \varepsilon_0^{1/q_{j,1}}\xi^{\gamma_j}\right\},$$

$$\widetilde{G}_{J+1}(\varepsilon_0) := \left\{(\xi,\eta) \in (0,\infty)^2 : \varepsilon_0^{-1/q_{J,1}}\xi^{\gamma_J} \leq \eta \text{ and } \eta \geq \varepsilon_0^{-1}\right\},$$

$$G_\ell(\varepsilon_0,\varepsilon_1) := \left\{(\xi,\eta) \in (0,\infty)^2 : \varepsilon_0^{1/q_{\ell,1}}\xi^{\gamma_\ell} \leq \eta \leq \varepsilon_0^{-1/q_{\ell,1}}\xi^{\gamma_\ell} \text{ and } \xi\eta^{\gamma_\ell} \geq \varepsilon_1^{-1/q_{\ell,1}}\right\}$$

for $j = 2,\ldots J$, $\ell = 1,\ldots, J$.

If N is not regular in space, then

$$\widetilde{G}_1(\varepsilon_0) := \left\{(\xi,\eta) \in (0,\infty)^2 : \eta \leq \varepsilon_0^{1/q_{1,1}}\xi^{\gamma_1} \text{ and } \xi \geq \varepsilon_0^{-1}\right\},$$

$$\widetilde{G}_j(\varepsilon_0) := \left\{(\xi,\eta) \in (0,\infty)^2 : \varepsilon_0^{-1/q_{j-1,1}}\xi^{\gamma_{j-1}} \leq \eta \leq \varepsilon_0^{1/q_{j,1}}\xi^{\gamma_j}\right\},$$

$$\widetilde{G}_J(\varepsilon_0) := \left\{(\xi,\eta) \in (0,\infty)^2 : \varepsilon_0^{-1/q_{J-1,1}}\xi^{\gamma_{J-1}} \leq \eta \text{ and } \xi \geq \varepsilon_0^{-1}\right\},$$

$$\widetilde{G}_{J+1}(\varepsilon_0) := \left\{(\xi,\eta) \in (0,\infty)^2 : \xi \leq \varepsilon_0 \text{ and } \eta \geq \varepsilon_0^{-1}\right\},$$

$$G_\ell(\varepsilon_0,\varepsilon_1) := \left\{(\xi,\eta) \in (0,\infty)^2 : \varepsilon_0^{1/q_{\ell,1}}\xi^{\gamma_\ell} \leq \eta \leq \varepsilon_0^{-1/q_{\ell,1}}\xi^{\gamma_\ell} \text{ and } \xi\eta^{\gamma_\ell} \geq \varepsilon_1^{-1/q_{\ell,1}}\right\},$$

$$G_J(\varepsilon_0,\varepsilon_1) := \left\{(\xi,\eta) \in (0,\infty)^2 : \varepsilon_0 \leq \xi \leq \varepsilon_0^{-1} \text{ and } \eta \geq \varepsilon_1^{-1}\right\}$$

for $j = 2,\ldots J-1$, $\ell = 1,\ldots, J-1$.

In the next lemma we state some properties of this partition. These results can also be found in [GV92, Chapter 4, 2.3/2.4/2.5]. We give the proofs to provide the versions we need and to illustrate the advantages of the partition.

Lemma 2.45 (Properties of the partition). *Let $\nu \subseteq [0,\infty)^2$ be finite and let $N(\nu)$ be defined as in Subsection 2.1 b). For arbitrary $\varepsilon > 0$ the following statements hold:*

(i) *There exists a constant $\widehat{\varepsilon}_0 = \widehat{\varepsilon}_0(\nu,\varepsilon) > 0$ such that*

$$\xi^{r'}\eta^{s'} \leq \varepsilon \cdot \xi^{r_j}\eta^{s_j}$$

for all $j = 1,\ldots, J+1$, $0 < \varepsilon_0 \leq \widehat{\varepsilon}_0$, $(\xi,\eta) \in \widetilde{G}_j(\varepsilon_0)$ and all $(r',s') \in \nu\setminus\{v_j\}$.

(ii) *Let L be the line passing through v_j and v_{j+1} for a fixed $j = 1,\ldots, J$. For every $\alpha, \beta \in L$ there exists $\theta = \theta(\alpha,\beta) \in \mathbb{R}$ such that*

$$\xi^{\alpha_1}\eta^{\alpha_2} \leq \varepsilon_0^{-|\theta|}\xi^{\beta_1}\eta^{\beta_2}$$

for all $\varepsilon_0, \varepsilon_1 > 0$ and $(\xi,\eta) \in G_j(\varepsilon_0,\varepsilon_1)$.

(iii) *For all $\varepsilon_0 > 0$ there exists a constant $\varepsilon_1 = \varepsilon_1(\nu, \varepsilon, \varepsilon_0) > 0$ such that we have*

$$\xi^{r'}\eta^{s'} \leq \varepsilon \cdot \xi^{r}\eta^{s}$$

for all $j = 1, \ldots, J$, $(\xi, \eta) \in G_j(\varepsilon_0, \varepsilon_1)$, $(r, s) \in \nu \cap [v_j v_{j+1}]$ and $(r', s') \in \nu \setminus [v_j v_{j+1}]$.

(iv) *For all $\varepsilon_0, \varepsilon_1 > 0$ there exists $\mu_0 = \mu_0(\varepsilon_0, \varepsilon_1) > 0$ with*

$$\{(\xi, \eta) \in (0, \infty)^2 : \eta \geq \mu_0\} \subseteq \bigcup_{\ell=1}^{J} G_\ell(\varepsilon_0, \varepsilon_1) \cup \bigcup_{j=1}^{J+1} \widetilde{G}_j(\varepsilon_0).$$

Here $G_1(\varepsilon_0, \varepsilon_1)$ and $\widetilde{G}_1(\varepsilon_0)$ are redundant in this context if N is not regular in time.

Proof. (i) Let $j = 1, \ldots, J+1$ and $(r', s') \in \nu \setminus \{v_j\}$ and define

$$L_j^{(1)} := \{\alpha \in \mathbb{R}^2 : \alpha = v_j + tq_j^\perp, t > 0\},$$
$$L_j^{(2)} := \{\alpha \in \mathbb{R}^2 : \alpha = v_j - tq_{j-1}^\perp, t > 0\}.$$

The convexity of $N(\nu)$ implies that there are $\delta_1, \delta_2 \in L_j^{(1)}$ and $\delta_3, \delta_4 \in L_j^{(2)}$ such that $\nu \setminus \{v_j\}$ is contained in the convex hull of $\delta_1, \ldots, \delta_4$. According to the definition of $L_j^{(1)}$ and $L_j^{(2)}$ there are $t_1, \ldots, t_4 > 0$ with $\delta_k = v_j + t_k q_j^\perp$ and $\delta_i = v_j - t_i q_{j-1}^\perp$ for $k = 1, 2$ and $i = 3, 4$. Then we define $\chi := \min\{t_k : k = 1, \ldots, 4\}$ and choose $\widehat{\varepsilon}_{0,j} > 0$ sufficiently small such that $(\widehat{\varepsilon}_{0,j})^\chi \leq \varepsilon$.

Let $0 < \varepsilon_0 \leq \widehat{\varepsilon}_{0,j}$ and $(\xi, \eta) \in \widetilde{G}_j(\varepsilon_0)$. Then there exist $\lambda_1, \ldots, \lambda_4 \in [0, 1]$ with $(r', s') = \sum_{k=1}^{4} \lambda_k \delta_k$ and $\sum_{k=1}^{4} \lambda_k = 1$. Due to the definition of $\widetilde{G}_j(\varepsilon_0)$ we deduce

$$\xi^{r'}\eta^{s'} = \prod_{k=1}^{4} \left(\xi^{\delta_{k,1}} \eta^{\delta_{k,2}}\right)^{\lambda_k}$$

$$= \xi^{r_j}\eta^{s_j} \prod_{k=1}^{2} \left(\xi^{-t_k q_{j,2}} \eta^{t_k q_{j,1}}\right)^{\lambda_k} \prod_{i=3}^{4} \left(\xi^{t_i q_{j-1,2}} \eta^{-t_i q_{j-1,1}}\right)^{\lambda_i}$$

$$= \xi^{r_j}\eta^{s_j} \prod_{k=1}^{2} [\exp(q_{j,1} \log(\eta) - q_{j,2} \log(\xi))]^{t_k \lambda_k}$$

$$\cdot \prod_{i=3}^{4} [\exp(q_{j-1,2} \log(\xi) - q_{j-1,1} \log(\eta))]^{t_i \lambda_i}$$

$$\leq \xi^{r_j}\eta^{s_j} \prod_{k=1}^{4} \varepsilon_0^{t_k \lambda_k} \leq \varepsilon_0^{\chi} \cdot \xi^{r_j}\eta^{s_j} \leq \varepsilon \cdot \xi^{r_j}\eta^{s_j}.$$

To complete the proof we define $\widehat{\varepsilon}_0 := \min\{\widehat{\varepsilon}_{0,j} : j = 1, \ldots, J+1\}$.

2.2. N-parameter-ellipticity and N-parabolicity

(ii) It is obvious that $\alpha - \beta$ is orthogonal to q_j. Therefore, there exists $\theta \in \mathbb{R}$ with $\alpha - \beta = \theta q_j^\perp$. Hence

$$\xi^{\alpha_1}\eta^{\alpha_2} \cdot \xi^{-\beta_1}\eta^{-\beta_2} = \exp((\alpha_1 - \beta_1)\log(\xi) + (\alpha_2 - \beta_2)\log(\eta))$$
$$= \exp(\theta(-q_{j,2}\log(\xi) + q_{j,1}\log(\eta))), \quad (\xi,\eta) \in (0,\infty)^2.$$

Due to the definition of $G_j(\varepsilon_0, \varepsilon_1)$ we have $\theta(-q_{j,2}\log(\xi) + q_{j,1}\log(\eta)) \leq -|\theta|\log(\varepsilon_0)$ for all tuples $(\xi,\eta) \in G_j(\varepsilon_0, \varepsilon_1)$, which implies

$$\xi^{\alpha_1}\eta^{\alpha_2} \cdot \xi^{-\beta_1}\eta^{-\beta_2} \leq \varepsilon_0^{-|\theta|}, \quad (\xi,\eta) \in G_j(\varepsilon_0, \varepsilon_1).$$

(iii) Let $\varepsilon_0 > 0$, $j = 1, \ldots, J$, $(r,s) \in \nu \cap [v_j v_{j+1}]$, and $(r',s') \in \nu \setminus [v_j v_{j+1}]$ be arbitrary. Then there exists $t = t(r', s') > 0$ and $\delta = \delta(r', s')$ on the line passing through v_j, v_{j+1} with $(r', s') = \delta - tq_j$. We choose $\varepsilon_{1,j} > 0$ sufficiently small such that $\varepsilon_{1,j}^{t(r'',s'')}\varepsilon_0^{-|\theta(r'',s'')|} \leq \varepsilon$ for all $(r'', s'') \in \nu \setminus [v_j v_{j+1}]$ and θ from (ii). The definition of $G_j(\varepsilon_0, \varepsilon_{1,j})$ and (ii) imply

$$\xi^{r'}\eta^{s'} \leq \xi^{\delta_1}\eta^{\delta_2}\exp(-t(q_{j,1}\log(\xi) + q_{j,2}\log(\eta))) \leq \varepsilon_{1,j}^{t}\varepsilon_0^{-|\theta|}\xi^r\eta^s \leq \varepsilon \cdot \xi^r\eta^s$$

for all $(\xi,\eta) \in G_j(\varepsilon_0, \varepsilon_{1,j})$. To complete the proof we define

$$\varepsilon_1 := \min\{\varepsilon_{1,j}: j = 1, \ldots, J\}.$$

(iv) For $y > 0$ we define the half space $\mathbb{R}^2(y) := \{q \in \mathbb{R}^2: q_2 \geq y\}$. In the following we want to show that there exists $y_0 > 0$ with

$$\mathbb{R}^2(y_0) \subseteq \bigcup_{j=1}^{J+1} C_j(\varepsilon_0) \cup \bigcup_{\ell=1}^{J} S_\ell(\varepsilon_0, \varepsilon_1).$$

Assume that there exists a sequence $(Q_n)_{n\in\mathbb{N}} = (q_1^{(n)}, q_2^{(n)})_{n\in\mathbb{N}} \subseteq \mathbb{R}^2$ with $\lim_{n\to\infty} q_2^{(n)} = \infty$ and

$$Q_n \notin \Big(\bigcup_{j=1}^{J+1} C_j(\varepsilon_0)\Big) \cup \Big(\bigcup_{\ell=1}^{J} S_\ell(\varepsilon_0, \varepsilon_1)\Big). \tag{2.12}$$

The sequence $(Q_n/|Q_n|)_{n\in\mathbb{N}}$ is bounded and therefore there exists a subsequence of $(Q_n)_{n\in\mathbb{N}}$, which is also denoted by $(Q_n)_{n\in\mathbb{N}}$, such that $\arg(Q_n) \in (-\pi, \pi]$ is convergent. Due to $|Q_n| \to \infty$ there is an $n_0 \in \mathbb{N}$ with $\langle q_j, Q_n \rangle \geq \log(1/\varepsilon_1)$ for all $n \geq n_0$ and $j = 1, \ldots, J$. So we have to consider two cases: If there exists $j = 1, \ldots, J$ with $\arg(q_j) = \lim_n \arg(Q_n)$, we obtain $Q_n \in S_j$ for all $n \geq n_0$. If there exists $j = 0, \ldots J$ with $\lim_n \arg(Q_n) \in (\arg(q_j), \arg(q_{j+1}))$, we get $Q_n \in C_{j+1}$ for sufficiently large n. Both cases contradict (2.12).

If $N(\nu)$ is not regular in time, it is easy to see that

$$\mathbb{R}^2(y) \cap (C_1(\varepsilon_0) \cup S_1(\varepsilon_0, \varepsilon_1)) = \emptyset$$

for sufficiently large $y > 0$. Therefore, $C_1(\varepsilon_0)$ and $S_1(\varepsilon_0, \varepsilon_1)$ are redundant for the covering of $\mathbb{R}^2(y)$. □

The last lemma shows that in $\widetilde{G}_j(\varepsilon_0)$ the vertex v_j of the Newton polygon dominates whereas in $G_\ell(\varepsilon_0, \varepsilon_1)$ the edge $[v_\ell v_{\ell+1}]$ dominates. By assertion (iv), we have found a partition of the space of all (ξ, η) into subdomains where in each subdomain we can identify a leading edge or vertex of the Newton polygon. This partition is the key for the connection between inequalities as in the definition of N-parabolicity and the fact that principal parts of the symbol do not vanish, as we will see in the next subsection.

c) Equivalent characterization of $S_N(L_t \times L_x)$

First, we prove that an N-parameter-elliptic symbol has non-vanishing principal parts. Later, we will see in our main result of this section that this is even equivalent.

As we will consider regular and non-regular Newton polygons simultaneously, we introduce the following notation:

Definition 2.46. Let N be an arbitrary Newton polygon. Then we set

$$\kappa_1(N) := \begin{cases} 1, & \text{if } N \text{ is regular in time,} \\ 2, & \text{if } N \text{ is not regular in time,} \end{cases}$$

$$\kappa_2(N) := \begin{cases} J+1, & \text{if } N \text{ is regular in space,} \\ J, & \text{if } N \text{ is not regular in space.} \end{cases}$$

Proposition 2.47. *Let $P(\lambda, z) = \sum_{\ell \in I} \tau_\ell(\lambda, z) \varphi_\ell(\lambda) \psi_\ell(z)$, $(\lambda, z) \in L_t \times L_x$, be a regular representation of the symbol $P \in S_N(L_t \times L_x)$. Then we have*

$$\pi_\gamma P(\lambda, z) \neq 0 \quad (2.13)$$

for all $(\lambda, z) \in (L_t \setminus \{0\}) \times (L_x \setminus \{0\})$, and $\gamma \in (0, \infty)$. For the ∞-principal part we get

$$\pi_\infty P(\lambda, z) \neq 0 \quad (2.14)$$

for all $(\lambda, z) \in (L_t \setminus \{0\}) \times L_x$.

Proof. Due to $P \in C(L_t \times L_x, \mathbb{C})$ we even have $|P(\lambda, z)| \geq C \cdot W_P(\lambda, z)$ for all $(\lambda, z) \in L_t \times L_x$ with $|\lambda| \geq \lambda_0$. For $(\lambda, z) \in (L_t \setminus \{0\}) \times (L_x \setminus \{0\})$, and $\gamma > 0$ the definition of the γ-principal part yields

$$|\pi_\gamma P(\lambda, z)| = \lim_{\eta \to \infty} \eta^{-d_\gamma(P)} |P(\eta^\gamma \lambda, \eta z)| \geq C \lim_{\eta \to \infty} \eta^{-d_\gamma(P)} W_P(\eta^\gamma \lambda, \eta z).$$

2.2. N-parameter-ellipticity and N-parabolicity

According to Remark 2.23 (i) we get

$$\lim_{\eta \to \infty} \eta^{-d_\gamma(P)} W_P(\eta^\gamma \lambda, \eta z) = |\lambda|^{s_j}|z|^{r_j} + |\lambda|^{s_{j+1}}|z|^{r_{j+1}}$$

for $\gamma = \gamma_j$, $j = \kappa_1(N(P)), \ldots, \kappa_2(N(P)) - 1$, and

$$\lim_{\eta \to \infty} \eta^{-d_\gamma(P)} W_P(\eta^\gamma \lambda, \eta z) = |\lambda|^{s_j}|z|^{r_j}$$

for $\gamma \in (\gamma_{j-1}, \gamma_j)$, $j = \kappa_1(N(P)), \ldots, \kappa_2(N(P))$. In both cases the limit is positive for all $(\lambda, z) \in (L_t \setminus \{0\}) \times (L_x \setminus \{0\})$ which proves the assertion for $\gamma \in (0, \infty)$. We have to consider the case $\gamma = \infty$ separately. For all $(\lambda, z) \in (L_t \setminus \{0\}) \times L_x$ we get

$$|\pi_\infty P(\lambda, z)| = \lim_{\eta \to \infty} \eta^{-d_\infty(P)} |P(\eta\lambda, z)| \geq C \lim_{\eta \to \infty} \eta^{-d_\infty(P)} W_P(\eta\lambda, z)$$

$$= C \cdot \begin{cases} |\lambda|^{s_{J+1}}, & \text{if } N(P) \text{ is space-regular,} \\ |\lambda|^{s_{J+1}}(1 + |z|^{r_J}), & \text{if } N(P) \text{ is not space-regular} \end{cases}$$

due to $d_\infty(P) = s_{J+1}$. So we have $|\pi_\infty P(\lambda, z)| > 0$ for all $(\lambda, z) \in (L_t \setminus \{0\}) \times L_x$. □

If $N(P)$ is regular in space, then the assumption $\pi_\infty P(\lambda, z) \neq 0$, $(\lambda, z) \in (L_t \setminus \{0\}) \times L_x$, already follows from $\pi_\gamma P(\lambda, z) \neq 0$ for all $(\lambda, z) \in (L_t \setminus \{0\}) \times (L_x \setminus \{0\})$ and $\gamma > 0$. This is stated in the next proposition.

Proposition 2.48. *Let $P \in S(L_t \times L_x)$ be regular in space. Then we have*

$$\pi_\infty P(\lambda, z) = \pi_\gamma P(\lambda, z)$$

for all $\gamma \geq \max\{\rho, \gamma_J\}$ and $(\lambda, z) \in L_t \times L_x$. If we have

$$\pi_\gamma P(\lambda, z) \neq 0$$

for all $(\lambda, z) \in (L_t \setminus \{0\}) \times (L_x \setminus \{0\})$, and $\gamma \in (0, \infty)$, then we also have

$$\pi_\infty P(\lambda, z) \neq 0$$

for all $(\lambda, z) \in (L_t \setminus \{0\}) \times L_x$.

Proof. For $\gamma \geq \max\{\rho, \gamma_J\}$ we have $d_\gamma(P) = s_{J+1}\gamma = d_\infty(P)\gamma$. For all $\ell \in I_\gamma$ we have $L_\ell = 0$, which yields $\psi_\ell(z) = \psi_\ell(0) = \text{const}$. With this we can show $\pi_\infty P(\lambda, z) = \pi_\gamma P(\lambda, z) \neq 0$ for all $(\lambda, z) \in (L_t \setminus \{0\}) \times (L_x \setminus \{0\})$. The principal part $\pi_\gamma P$ does not depend on z and thus we obtain

$$\pi_\infty P(\lambda, 0) = \pi_\gamma P(\lambda, 0) = \pi_\gamma P(\lambda, \zeta) \neq 0$$

for $\lambda \in L_t \setminus \{0\}$ and arbitrary $\zeta \in L_x \setminus \{0\}$. □

Remark 2.49. The statement in (2.14) generalizes the fact that the symbol P is resolved with respect to the highest power of λ (cf. [GV92, Ch. 2, 1.2.])

Example 2.50. Let $\theta \in (0, \pi)$ and $\delta > 0$, and let $|z|_- := (-\sum_{k=1}^n z_k^2)^{1/2}$ for $z \in \overline{\Sigma}_\delta^n$.

(i) Define $P_1(\lambda, z) := \lambda|z|_-^2 + |z|_-^4$. Then we have $\pi_\infty P_1(\lambda, z) = \lambda|z|_-^2$, which yields $\pi_\infty P_1(\lambda, 0) = 0$ for all λ. Hence we have $P_1 \notin S_N(\overline{S}_\theta \times \overline{\Sigma}_\delta^n)$ due to (2.14).

(ii) Define $P_2(\lambda, z) := \lambda + \lambda|z|_-^2 + |z|_-^4$. Then the symbol P_2 is not regular in space and we have $\pi_\infty P_2(\lambda, z) = \lambda(1 + |z|_-^2)$. Therefore (2.14) is fulfilled for P_2. Using Theorem 2.56 below, we will even see that $P_2 \in S_N(\overline{S}_\theta \times \overline{\Sigma}_\delta^n)$.

(iii) Let $P_3(\lambda, z) := \lambda z + z^2 - 1$, $(\lambda, z) \in \overline{S}_\theta \times \overline{\Sigma}_\delta$. Then P_3 is not regular in space and we have
$$\pi_\infty P_3(\lambda, z) = \lambda z, \quad (\lambda, z) \in \overline{S}_\theta \times \overline{\Sigma}_\delta.$$
This shows $\pi_\infty P_3(\lambda, 0) = 0$ for all $\lambda \in \overline{S}_\theta$. According to Proposition 2.47, P_3 is not N-parameter-elliptic.

The next result shows that in each subdomain $G_j(\varepsilon_0, \varepsilon_1)$ of the partition, a ρ-homogeneous function is close to its γ_j-principal part.

Lemma 2.51. *For a Newton polygon N and arbitrary $\psi \in S^{(\rho, M)}(L_t \times L_x)$, $M \geq 0$, and $\vartheta, \varepsilon_0, \varepsilon_1 > 0$ there exists $\mu_1 = \mu_1(\vartheta, \varepsilon_0, \psi) > 0$ such that*
$$|\psi(\lambda, z) - \pi_{\gamma_j}\psi(\lambda, z)| \leq \vartheta|\pi_{\gamma_j}\psi(\lambda, z)|$$
holds for all $j = \kappa_1(N), \ldots, J$ and $(\lambda, z) \in (L_t \setminus \{0\}) \times (L_x \setminus \{0\})$ with $(|\lambda|, |z|) \in G_j(\varepsilon_0, \varepsilon_1)$ and $|\lambda| \geq \mu_1$.

Proof. First, we define the function
$$\Gamma(\lambda, z) := \frac{\psi(\lambda, z) - \pi_{\gamma_j}\psi(\lambda, z)}{\pi_{\gamma_j}\psi(\lambda, z)}$$
for all $(\lambda, z) \in (L_t \setminus \{0\}) \times (L_x \setminus \{0\})$. For a fixed $j \in \{\kappa_1(N), \ldots, J\}$ we have to distinguish three cases:

(I) $\gamma_j < \rho$: In this case we have
$$|\Gamma(\lambda, z)| \leq C_\psi |z|^{-M} |\psi(\lambda, z) - \psi(0, z)| = C_\psi \left|\psi\left(\frac{\lambda}{|z|^\rho}, \frac{z}{|z|}\right) - \psi\left(0, \frac{z}{|z|}\right)\right|$$
according to Lemma 2.7 and Remark 2.4. The definition of $G_j(\varepsilon_0, \varepsilon_1)$ yields the estimate
$$\frac{|\lambda|}{|z|^\rho} \leq \varepsilon_0^{-\rho/(\gamma_j q_{j,1})} \cdot |\lambda|^{1-\rho/\gamma_j} \tag{2.15}$$

2.2. N-parameter-ellipticity and N-parabolicity

for all $(\lambda, z) \in (L_t \setminus \{0\}) \times (L_x \setminus \{0\})$ with $(|\lambda|, |z|) \in G_j(\varepsilon_0, \varepsilon_1)$. The continuity of ψ implies that ψ is uniformly continuous on the compact set $K_t \times K_x \subseteq (L_t \times L_x) \setminus \{(0,0)\}$ where

$$K_t := \{\lambda' \in L_t : |\lambda'| \leq 1\}, \quad K_x := \{z' \in L_x : |z'| = 1\}.$$

Hence we have $|\psi(\lambda'_1, z'_1) - \psi(\lambda'_2, z'_2)| \leq \vartheta C_\psi$ for all $(\lambda'_1, z'_1), (\lambda'_2, z'_2) \in K_t \times K_x$ with $|\lambda'_1 - \lambda'_2| + |z'_1 - z'_2| \leq \delta(\vartheta)$. By (2.15) we can choose $\mu(j, \varepsilon_0, \vartheta, \psi) > 0$ sufficiently large such that

$$\frac{|\lambda|}{|z|^\rho} \leq \min\{\delta(\vartheta), 1\}$$

for all $(\lambda, z) \in (L_t \setminus \{0\}) \times (L_x \setminus \{0\})$ with $(|\lambda|, |z|) \in G_j(\varepsilon_0, \varepsilon_1)$ and $|\lambda| \geq \mu(j, \varepsilon_0, \vartheta, \psi)$. So the uniform continuity implies $|\Gamma(\lambda, z)| \leq \vartheta$ for all $(\lambda, z) \in (L_t \setminus \{0\}) \times (L_x \setminus \{0\})$ with $(|\lambda|, |z|) \in G_j(\varepsilon_0, \varepsilon_1)$ and $|\lambda| \geq \mu(j, \varepsilon_0, \vartheta)$.

(II) $\gamma_j > \rho$: Using the same arguments as in (I) we get

$$|\Gamma(\lambda, z)| \leq C_\psi \cdot \left| \psi\left(\frac{\lambda}{|\lambda|}, \frac{z}{|\lambda|^{1/\rho}}\right) - \psi\left(\frac{\lambda}{|\lambda|}, 0\right)\right|,$$

$$\frac{|z|}{|\lambda|^{1/\rho}} \leq \begin{cases} \varepsilon_0^{-1/(\gamma_j q_{j,1})} \cdot |\lambda|^{1/\gamma_j - 1/\rho}, & \gamma_j < \infty, \\ \varepsilon_0^{-1} |\lambda|^{-1/\rho}, & \gamma_j = \infty \end{cases}$$

for all $(\lambda, z) \in (L_t \setminus \{0\}) \times (L_x \setminus \{0\})$ with $(|\lambda|, |z|) \in G_j(\varepsilon_0, \varepsilon_1)$. As in (I) there exists a large $\mu(j, \varepsilon_0, \vartheta, \psi) > 0$ such that we have $|\Gamma(\lambda, z)| \leq \vartheta$ for all $(\lambda, z) \in (L_t \setminus \{0\}) \times (L_x \setminus \{0\})$ with $(|\lambda|, |z|) \in G_j(\varepsilon_0, \varepsilon_1)$ and $|\lambda| \geq \mu(j, \varepsilon_0, \vartheta, \psi)$.

(III) $\gamma_j = \rho$: Here nothing has to be shown because of $\pi_{\gamma_j} \psi = \psi$.

To complete the proof we define

$$\mu_1 := \max\{\mu(j, \varepsilon_0, \vartheta, \psi) : j \in \{\kappa_1(N), \ldots, J\}, \gamma_j \neq \rho\}. \qquad \square$$

The analog result of the last lemma also holds when the weight γ lies in the open interval (γ_{j-1}, γ_j). Now we obtain an estimate in $\widetilde{G}_j(\varepsilon_0)$.

Lemma 2.52. *Let N be a Newton polygon and let $\psi \in S^{(\rho,M)}(L_t \times L_x)$, $M \geq 0$.*

(i) *For all $j = \kappa_1(N), \ldots, \kappa_2(N)$ with $\rho \notin (\gamma_{j-1}, \gamma_j)$ and $\vartheta > 0$ there exists $\varepsilon_{0,j} = \varepsilon_{0,j}(\psi) > 0$ such that we get*

$$|\psi(\lambda, z) - \pi_\gamma \psi(\lambda, z)| \leq \vartheta |\pi_\gamma \psi(\lambda, z)|$$

for all $\gamma \in (\gamma_{j-1}, \gamma_j)$, $\varepsilon_0 \leq \varepsilon_{0,j}$ and $(\lambda, z) \in (L_t \setminus \{0\}) \times (L_x \setminus \{0\})$ with $(|\lambda|, |z|) \in \widetilde{G}_j(\varepsilon_0)$ and $|\lambda| \geq 1$.

(ii) *If N is not regular in space, i.e., $\kappa_2(N) = J$, then for all $\vartheta > 0$ there exists $\varepsilon_{0,J+1} > 0$ such that we get*

$$|\psi(\lambda, z) - \psi(\lambda, 0)| \leq \vartheta |\psi(\lambda, 0)|$$

for all $\varepsilon_0 \leq \varepsilon_{0,J+1}$ and $(\lambda, z) \in (L_t \setminus \{0\}) \times (L_x \setminus \{0\})$ with $(|\lambda|, |z|) \in \widetilde{G}_{J+1}(\varepsilon_0)$ and $|\lambda| \geq 1$.

Proof. (i) Let $j = \kappa_1(N), \ldots, \kappa_2(N)$ with $\rho \notin (\gamma_{j-1}, \gamma_j)$ as well as $\vartheta > 0$ be fixed. As in Lemma 2.51 we define

$$\Gamma(\lambda, z) := \frac{\psi(\lambda, z) - \pi_\gamma \psi(\lambda, z)}{\pi_\gamma \psi(\lambda, z)}, \quad (\lambda, z) \in (L_t \setminus \{0\}) \times (L_x \setminus \{0\})$$

for all $\gamma \in (\gamma_{j-1}, \gamma_j)$. According to Remark 2.4 there exists a constant $C_\psi > 0$ such that we get $|\psi(\lambda, z)| \geq C_\psi(|z|^M + |\lambda|^{M/\rho})$ for all $(L_t \times L_x) \setminus \{(0,0)\}$. The function ψ is uniformly continuous on the compact set

$$\mathcal{K} := \{(\lambda', z') \in (L_t \times L_x) \setminus \{(0,0)\} : (|\lambda'| \leq 1 \text{ and } |z'| = 1)$$
$$\text{or } (|\lambda'| = 1 \text{ and } |z'| \leq 1)\}.$$

Therefore, there exists a constant $\delta = \delta(\vartheta) \in (0,1)$ with $|\psi(\lambda'_1, z'_1) - \psi(\lambda'_2, z'_2)| \leq \vartheta C_\psi$ for all $(\lambda'_1, z'_1), (\lambda'_2, z'_2) \in \mathcal{K}$ with $|\lambda'_1 - \lambda'_2| + |z'_1 - z'_2| \leq \delta(\vartheta)$. We have to consider two cases:

(I) $\gamma_j \leq \rho$: Due to $0 < \rho < \infty$ we have $j < \kappa_2(N)$. Define $\varepsilon_{0,j} := \delta^{(\gamma_j q_{j,1})/\rho}$ and let $\varepsilon_0 \leq \varepsilon_{0,j}$. As in Lemma 2.51 we get

$$|\Gamma(\lambda, z)| \leq C_\psi \left| \psi\left(\frac{\lambda}{|z|^\rho}, \frac{z}{|z|}\right) - \psi\left(0, \frac{z}{|z|}\right) \right|, \quad (\lambda, z) \in (L_t \setminus \{0\}) \times (L_x \setminus \{0\})$$

and $|\lambda| |z|^{-\rho} \leq \varepsilon_0^{\rho/(\gamma_j q_{j,1})} \cdot |\lambda|^{1-\rho/\gamma_j} \leq \varepsilon_0^{\rho/(\gamma_j q_{j,1})} \leq \delta$ for all $(\lambda, z) \in (L_t \setminus \{0\}) \times (L_x \setminus \{0\})$ with $(|\lambda|, |z|) \in \widetilde{G}_j(\varepsilon_0)$ and $|\lambda| \geq 1$.

(II) $\gamma_j > \rho$: We have $j > \kappa_1(N)$. Hence we have $\gamma_{j-1} \notin \{0, \infty\}$. Define $\varepsilon_{0,j} := \delta^{\gamma_{j-1} q_{j-1,1}}$ and let $\varepsilon_0 \leq \varepsilon_{0,j}$. As in the previous lemma we obtain

$$|\Gamma(\lambda, z)| \leq C_\psi \cdot \left| \psi\left(\frac{\lambda}{|\lambda|}, \frac{z}{|\lambda|^{1/\rho}}\right) - \psi\left(\frac{\lambda}{|\lambda|}, 0\right) \right|$$

for all $(\lambda, z) \in (L_t \setminus \{0\}) \times (L_x \setminus \{0\})$ and

$$\frac{|z|}{|\lambda|^{1/\rho}} \leq \varepsilon_0^{1/(\gamma_{j-1} q_{j-1,1})} \cdot |\lambda|^{1/\gamma_{j-1} - 1/\rho} \leq \varepsilon_0^{1/(\gamma_{j-1} q_{j-1,1})} \leq \delta$$

for all $(\lambda, z) \in (L_t \setminus \{0\}) \times (L_x \setminus \{0\})$ with $(|\lambda|, |z|) \in \widetilde{G}_j(\varepsilon_0)$ and $|\lambda| \geq 1$.

2.2. N-parameter-ellipticity and N-parabolicity

Both cases imply $|\Gamma(\lambda, z)| \leq \vartheta$ for all $(\lambda, z) \in (L_t \setminus \{0\}) \times (L_x \setminus \{0\})$ with $(|\lambda|, |z|) \in \widetilde{G}_j(\varepsilon_0)$ ($\varepsilon_0 \leq \varepsilon_{0,j}$) and $|\lambda| \geq 1$, which proves the assertion.

(ii) Let N be not regular in space. Then we define $\varepsilon_{0,J+1} := \delta^{1+1/\rho}$ and obtain

$$|\Gamma(\lambda, z)| \leq C_\psi^{-1} \cdot \left|\psi\left(\frac{\lambda}{|\lambda|}, \frac{z}{|\lambda|^{1/\rho}}\right) - \psi\left(\frac{\lambda}{|\lambda|}, 0\right)\right|, \quad (\lambda, z) \in (L_t \setminus \{0\}) \times (L_x \setminus \{0\})$$

as in part (II) of the proof in (i). Due to $|z||\lambda|^{-1/\rho} \leq \varepsilon_0^{1+1/\rho} \leq \delta$ for all $\varepsilon_0 \leq \varepsilon_{0,J+1}$ and all $(\lambda, z) \in (L_t \setminus \{0\}) \times (L_x \setminus \{0\})$ with $(|\lambda|, |z|) \in \widetilde{G}_{J+1}(\varepsilon_0)$ we get $|\Gamma(\lambda, z)| \leq \vartheta$. $\square$

We do not consider the situation when $\rho \in (\gamma_{j-1}, \gamma_j)$ for $\psi \in S^{(\rho, M)}(L_t \times L_x)$ because it is not necessary in our variant of the proof of the main result in this section.

Lemma 2.53. *Let $P(\lambda, z) = \sum_{\ell \in I} \tau_\ell(\lambda, z) \varphi_\ell(\lambda) \psi_\ell(z)$, $(\lambda, z) \in L_t \times L_x$, be a regular representation of the symbol $P \in S(L_t \times L_x)$. If we have*

$$\pi_\gamma P(\lambda, z) \neq 0, \quad \pi_\infty P(\lambda, 0) \neq 0 \qquad (2.16)$$

for all $(\lambda, z) \in (L_t \setminus \{0\}) \times (L_x \setminus \{0\})$, and $\gamma \in (0, \infty]$, then for every $j = \kappa_1(N(P)), \ldots, J$ there exists a constant $C_j > 0$ such that

$$|\pi_{\gamma_j} P(\lambda, z)| \geq C_j \left(|z|^{r_j} |\lambda|^{s_j} + |z|^{r_{j+1}} |\lambda|^{s_{j+1}}\right)$$

for all $(\lambda, z) \in (L_t \setminus \{0\}) \times (L_x \setminus \{0\})$.

Proof. Let $j \in \{\kappa_1(N(P)), \ldots, \kappa_2(N(P)) - 1\}$ be fixed. For this proof we will use the notations $\mathfrak{v}_1(\ell) := (N_\ell + L_\ell, M_\ell)$ and $\mathfrak{v}_2(\ell) := (L_\ell, \frac{N_\ell}{\rho} + M_\ell)$ (cf. Definition 2.15 for the definition of $N(P)$). Once more we have to distinguish three cases:

(I) $\gamma_j > \rho$: According to Definition 2.10 we have

$$\pi_{\gamma_j} P(\lambda, z) = \sum_{\ell \in I_{\gamma_j}} \tau_\ell(\lambda, 0) \varphi_\ell(\lambda) \psi_\ell(z), \quad (\lambda, z) \in (L_t \times L_x).$$

For $\ell \in I_{\gamma_j}$ a geometrical argument shows $\mathfrak{v}_2(\ell) \in [v_j v_{j+1}]$, which implies

$$L_\ell \geq r_{j+1} \quad \text{and} \quad N_\ell/\rho + M_\ell \geq s_j.$$

We define

$$\widetilde{P}(\lambda, z) := \sum_{\ell \in I_{\gamma_j}} \left[\tau_\ell\left(\frac{\lambda}{|\lambda|}, 0\right) \varphi_\ell\left(\frac{\lambda}{|\lambda|}\right) \psi_\ell\left(\frac{z}{|z|}\right)\right] |\lambda|^{M_\ell + N_\ell/\rho - s_j} |z|^{L_\ell - r_{j+1}}$$

for $(\lambda, z) \in (L_t \setminus \{0\}) \times (L_x \setminus \{0\})$ and get $\pi_{\gamma_j} P(\lambda, z) = |\lambda|^{s_j} |z|^{r_{j+1}} \widetilde{P}(\lambda, z)$.

In the following we want to prove that there exists a constant $C > 0$ such that the estimate

$$|\widetilde{P}(\lambda, z)| \geq C \left(|\lambda|^{s_{j+1}-s_j} + |z|^{r_j-r_{j+1}}\right) \tag{2.17}$$

holds for all $(\lambda, z) \in (L_t \setminus \{0\}) \times (L_x \setminus \{0\})$. The symbol $\widetilde{P}$ is γ_j-homogeneous of degree $r_j - r_{j+1}$ and therefore (2.17) is equivalent to

$$|\widetilde{P}(\lambda, z)| \geq C, \quad (\lambda, z) \in \Omega, \tag{2.18}$$

where $\Omega := \{(\mu, \zeta) \in (L_t \setminus \{0\}) \times (L_x \setminus \{0\}) \colon |\mu|^{s_{j+1}-s_j} + |\zeta|^{r_j-r_{j+1}} = 1\}$. Assuming that (2.18) is false we find a sequence $(\mu_m, \zeta^{(m)})_{m\in\mathbb{N}} \subseteq \Omega$ with $\widetilde{P}(\mu_m, \zeta^{(m)}) \to 0$, $m \to \infty$. Due to compactness, we can choose a convergent subsequence (which will not be relabeled) such that $(\mu_m/|\mu_m|)_{m\in\mathbb{N}}$ and $(\zeta^{(m)}/|\zeta^{(m)}|)_{m\in\mathbb{N}}$ are convergent. If we define $\lim_{m\to\infty}(\mu_m, \zeta^{(m)}) =: (\mu_0, \zeta_0) \in \overline{\Omega}$, $\lim_{m\to\infty} \mu_m/|\mu_m| =: \hat{\mu} \neq 0$ and $\lim_{m\to\infty} \zeta^{(m)}/|\zeta^{(m)}| =: \hat{\zeta} \neq 0$, we have to handle three cases:

(1) $\mu_0 \neq 0$, $\zeta_0 \neq 0$: Since $\widetilde{P}$ is continuous on Ω we get $\widetilde{P}(\mu_m, \zeta^{(m)}) \to \widetilde{P}(\mu_0, \zeta_0)$, $m \to \infty$, and therefore $\widetilde{P}(\mu_0, \zeta_0) = 0$. Thus $\pi_{\gamma_j} P(\mu_0, \zeta_0) = 0$, contradicting (2.16).

(2) $\mu_0 = 0$: In this case we obviously have $|\zeta_0| = 1$, $\zeta_0 = \hat{\zeta}$ and

$$0 = \lim_{m\to\infty} \widetilde{P}(\mu_m, \zeta^{(m)}) = \sum_{\substack{\ell \in I_{\gamma_j}, \\ M_\ell + N_\ell/\rho = s_j}} \tau_\ell(\hat{\mu}, 0)\, \varphi_\ell(\hat{\mu})\, \psi_\ell(\zeta_0).$$

Let $\gamma \in (\max\{\gamma_{j-1}, \rho\}, \gamma_j)$. Then we get the γ-principal part

$$\pi_\gamma P(\hat{\mu}, \zeta_0) = \sum_{\ell \in I_\gamma} \tau_\ell(\hat{\mu}, 0)\varphi_\ell(\hat{\mu})\psi_\ell(\zeta_0) = \sum_{\substack{\ell \in I_{\gamma_j}, \\ M_\ell + N_\ell/\rho = s_j}} \tau_\ell(\hat{\mu}, 0)\varphi_\ell(\hat{\mu})\psi_\ell(\zeta_0)$$

according to Remark 2.23 (iii). But this implies $\pi_\gamma P(\hat{\mu}, \zeta_0) = 0$, which contradicts (2.16).

(3) $\zeta_0 = 0$: Here we have $|\mu_0| = 1$, $\mu_0 = \hat{\mu}$ and

$$0 = \lim_{m\to\infty} \widetilde{P}(\mu_m, \zeta^{(m)}) = \sum_{\substack{\ell \in I_{\gamma_j}, \\ M_\ell + N_\ell/\rho = s_{j+1}}} \tau_\ell(\mu_0, 0)\, \varphi_\ell(\mu_0)\psi_\ell(\hat{\zeta}).$$

For $\gamma \in (\gamma_j, \gamma_{j+1})$ we obtain the γ-principal part

$$\pi_\gamma P(\mu_0, \hat{\zeta}) = \sum_{\ell \in I_\gamma} \tau_\ell(\mu_0, 0)\varphi_\ell(\mu_0)\psi_\ell(\hat{\zeta})$$

$$= \sum_{\substack{\ell \in I_{\gamma_j}, \\ M_\ell + N_\ell/\rho = s_{j+1}}} \chi_\ell(\vartheta_0)\tau_\ell(\mu_0, 0)\varphi_\ell(\mu_0)\psi_\ell(\hat{\zeta})$$

2.2. N-parameter-ellipticity and N-parabolicity

according to Remark 2.23 (ii). This implies $\pi_\gamma P(\mu_0, \hat{\zeta}) = 0$, which again contradicts (2.16).

So we have proved (2.17) for $\gamma_j > \rho$, which directly leads to

$$|\pi_{\gamma_j} P(\lambda, z)| \geq C_j \left(|z|^{r_j} |\lambda|^{s_j} + |z|^{r_{j+1}} |\lambda|^{s_{j+1}} \right)$$

for all $(\lambda, z) \in (L_t \setminus \{0\}) \times (L_x \setminus \{0\})$.

(II) $\gamma_j < \rho$: It is clear that $\gamma_j \neq \infty$. We essentially follow the procedure of (I), therefore we omit some details. For all $\ell \in I_{\gamma_j}$ we have $\mathfrak{v}_1(\ell) \in [v_j v_{j+1}]$, which implies $L_\ell + N_\ell \geq r_{j+1}$ and $M_\ell \geq s_j$. If we define

$$\widetilde{P}(\lambda, z) := \sum_{\ell \in I_{\gamma_j}} \left[\tau_\ell \left(0, \frac{z}{|z|}\right) \varphi_\ell \left(\frac{\lambda}{|\lambda|}\right) \psi_\ell \left(\frac{z}{|z|}\right) \right] |\lambda|^{M_\ell - s_j} |z|^{L_\ell + N_\ell - r_{j+1}}$$

for all $(\lambda, z) \in (L_t \setminus \{0\}) \times (L_x \setminus \{0\})$, we get $\pi_{\gamma_j} P(\lambda, z) = |\lambda|^{s_j} |z|^{r_{j+1}} \widetilde{P}(\lambda, z)$ again. Assume that there exists a sequence $(\mu_m, \zeta^{(m)})_{m \in \mathbb{N}} \subseteq \Omega$ as in (I) with the same properties as in (I). So we have to consider the same three cases as in (I):

(1) $\mu_0 \neq 0$, $\zeta_0 \neq 0$: This directly yields $\widetilde{P}(\mu_m, \zeta^{(m)}) \to \widetilde{P}(\mu_0, \zeta_0)$, $m \to \infty$, and therefore $\pi_{\gamma_j} P(\mu_0, \zeta_0) = 0$, contradicting (2.16).

(2) $\mu_0 = 0$: Similar to (I) we get

$$0 = \lim_{m \to \infty} \widetilde{P}(\mu_m, \zeta^{(m)}) = \sum_{\substack{\ell \in I_{\gamma_j}, \\ L_\ell + N_\ell = r_j}} \tau_\ell(0, \zeta_0) \varphi_\ell(\hat{\mu}) \psi_\ell(\zeta_0).$$

Let $\gamma \in (\gamma_{j-1}, \gamma_j)$. Then we get the γ-principal part

$$\pi_\gamma P(\hat{\mu}, \zeta_0) = \sum_{\substack{\ell \in I_{\gamma_j}, \\ L_\ell + N_\ell = r_j}} \tau_\ell(0, \zeta_0) \varphi_\ell(\hat{\mu}) \psi_\ell(\zeta_0)$$

according to Remark 2.23 (ii). But this yields $\pi_\gamma P(\hat{\mu}, \zeta_0) = 0$, which contradicts (2.16).

(3) $\zeta_0 = 0$: Here we have

$$0 = \lim_{m \to \infty} \widetilde{P}(\mu_m, \zeta^{(m)}) = \sum_{\substack{\ell \in I_{\gamma_j}, \\ L_\ell + N_\ell = r_{j+1}}} \tau_\ell(0, \hat{\zeta}) \varphi_\ell(\mu_0) \psi_\ell(\hat{\zeta}).$$

With $\gamma \in (\gamma_j, \min\{\gamma_{j+1}, \rho\})$ we get

$$\pi_\gamma P(\mu_0, \hat{\zeta}) = \sum_{\substack{\ell \in I_{\gamma_j}, \\ L_\ell + N_\ell = r_{j+1}}} \tau_\ell(0, \hat{\zeta}) \varphi_\ell(\mu_0) \psi_\ell(\hat{\zeta})$$

according to Remark 2.23 (ii). This implies $\pi_\gamma P(\mu_0, \hat{\zeta}) = 0$, which again contradicts (2.16). So we get

$$|P(\lambda, z)| \geq C \left(|z|^{r_{j+1}} |\lambda|^{s_{j+1}} + |z|^{r_j} |\lambda|^{s_j}\right)$$

for all $(\lambda, z) \in (L_t \setminus \{0\}) \times (L_x \setminus \{0\})$.

(III) $\gamma_j = \rho$: In this case the γ_j-principal part is given by

$$\pi_\rho P(\lambda, z) = \sum_{\ell \in I_\rho} \tau_\ell(\lambda, z) \varphi_\ell(\lambda) \psi_\ell(z).$$

For all $\ell \in I_\rho$ we have $\mathfrak{v}_1(\ell), \mathfrak{v}_2(\ell) \in [v_j v_{j+1}]$, which implies $L_\ell \geq r_{j+1}$ and $M_\ell \geq s_j$. We define

$$\widetilde{P}(\lambda, z) := \sum_{\ell \in I_\rho} \left[\tau_\ell(\lambda, z) \varphi_\ell\left(\frac{\lambda}{|\lambda|}\right) \psi_\ell\left(\frac{z}{|z|}\right) \right] |\lambda|^{M_\ell - s_j} |z|^{L_\ell - r_{j+1}}$$

for all $(\lambda, z) \in (L_t \setminus \{0\}) \times (L_x \setminus \{0\})$ and get $\pi_\rho P(\lambda, z) = |\lambda|^{s_j} |z|^{r_{j+1}} \widetilde{P}(\lambda, z)$. Assume that there exists a sequence $(\mu_m, \zeta^{(m)})_{m \in \mathbb{N}} \subseteq \Omega$ as in (I) with the same properties and nomenclatures. Then we consider the same three cases as in (I):

(1) $\mu_0 \neq 0$, $\zeta_0 \neq 0$: This yields $\pi_{\gamma_j} P(\mu_0, \zeta_0) = 0$.

(2) $\mu_0 = 0$: As in (I) we get

$$0 = \lim_{m \to \infty} \widetilde{P}(\mu_m, \zeta^{(m)}) = \sum_{\substack{\ell \in I_\rho, \\ L_\ell + N_\ell = r_j}} \tau_\ell(0, \zeta_0) \varphi_\ell(\hat{\mu}) \psi_\ell(\zeta_0).$$

Let $\gamma \in (\gamma_{j-1}, \rho)$. Then we get the γ-principal part

$$\pi_\gamma P(\hat{\mu}, \zeta_0) = \sum_{\substack{\ell \in I_\rho, \\ L_\ell + N_\ell = r_j}} \tau_\ell(0, \zeta_0) \varphi_\ell(\hat{\mu}) \psi_\ell(\zeta_0),$$

which yields $\pi_\gamma P(\hat{\mu}, \zeta_0) = 0$.

(3) $\zeta_0 = 0$: Here we have

$$0 = \lim_{m \to \infty} \widetilde{P}(\mu_m, \zeta^{(m)}) = \sum_{\substack{\ell \in I_{\gamma_j}, \\ M_\ell + N_\ell/\rho = s_{j+1}}} \tau_\ell(\mu_0, 0) \varphi_\ell(\mu_0) \psi_\ell(\hat{\zeta}).$$

With $\gamma \in (\gamma_j, \gamma_{j+1})$ we get

$$\pi_\gamma P(\mu_0, \hat{\zeta}) = \sum_{\substack{\ell \in I_{\gamma_j}, \\ M_\ell + N_\ell/\rho = s_{j+1}}} \tau_\ell(\mu_0, 0) \varphi_\ell(\mu_0) \psi_\ell(\hat{\zeta}).$$

This implies $\pi_\gamma P(\mu_0, \hat{\zeta}) = 0$.

2.2. N-parameter-ellipticity and N-parabolicity

All three cases contradict (2.16).

At last we consider the case where $N(P)$ is not regular in space, i.e., $\gamma_J = \infty$. For $\ell \in I_\infty$ we have $N_\ell/\rho + M_\ell = s_{J+1} = s_J$ and $L_\ell \in [0, r_J]$. So we have to prove that there exists $C_J > 0$ with

$$|\pi_\infty P(\lambda, z)| \geq C_J |\lambda|^{s_J}(|z|^{r_J} + 1) \tag{2.19}$$

for all $(\lambda, z) \in (L_t \setminus \{0\}) \times (L_x \setminus \{0\})$. We define

$$\widetilde{P}(\lambda, z) := |\lambda|^{-s_J}(|z|^{r_J} + 1)^{-1} \pi_\infty P(\lambda, z)$$

$$= \sum_{\ell \in I_\infty} \left[\tau_\ell\left(\frac{\lambda}{|\lambda|}, 0\right) \varphi_\ell\left(\frac{\lambda}{|\lambda|}\right) \psi_\ell\left(\frac{z}{|z|}\right) \right] \frac{|z|^{L_\ell}}{|z|^{r_J} + 1}.$$

Assuming that (2.19) is not true we can find a sequence $(\mu_m, \zeta^{(m)})_{m \in \mathbb{N}} \subseteq (L_t \setminus \{0\}) \times (L_x \setminus \{0\})$ with $\widetilde{P}(\mu_m, \zeta^{(m)}) \to 0$, $m \to \infty$. Without loss of generality we can assume that the sequences $(\mu_m/|\mu_m|)_{m \in \mathbb{N}}$ and $(\zeta^{(m)}/|\zeta^{(m)}|)_{m \in \mathbb{N}}$ are convergent with

$$(\hat{\mu}, \hat{\zeta}) := \lim_{m \to \infty}(\mu_m/|\mu_m|, \zeta^{(m)}/|\zeta^{(m)}|).$$

We have to consider two cases:

(i) $\lim_{m \to \infty}|\zeta^{(m)}| = \infty$: We obtain

$$0 = \lim_{m \to \infty} \widetilde{P}(\mu_m, \zeta^{(m)}) = \sum_{\substack{\ell \in I_\infty, \\ L_\ell = r_J}} \tau_\ell(\hat{\mu}, 0)\, \varphi_\ell(\hat{\mu})\, \psi_\ell(\hat{\zeta}) = \pi_\gamma P(\hat{\mu}, \hat{\zeta})$$

for $\gamma \in (\max\{\gamma_{J-1}, \rho\}, \infty)$ according to Remark 2.23 (ii). This contradicts (2.16).

(ii) Due to (i) the sequence $(|\zeta^{(m)}|)_{m \in \mathbb{N}}$ has to be bounded. Therefore, we can find a convergent subsequence of $(\zeta^{(m)})_{m \in \mathbb{N}}$ which is also denoted by $(\zeta^{(m)})_{m \in \mathbb{N}}$. We define $\zeta' := \lim_{m \to \infty}|\zeta^{(m)}|$.

If $\zeta' = 0$, then we get

$$0 = \lim_{m \to \infty} \widetilde{P}(\mu_m, \zeta^{(m)}) = \sum_{\substack{\ell \in I_\infty, \\ L_\ell = 0}} \tau_\ell(\hat{\mu}, 0)\, \varphi_\ell(\hat{\mu})\, \psi_\ell(\hat{\zeta}) = \pi_\infty P(\hat{\mu}, 0)$$

since we have $\psi_\ell = \text{const}$ for $L_\ell = 0$ and $\psi_\ell(0) = 0$ for $L_\ell > 0$. This contradicts (2.16).

If $\zeta' \neq 0$, then we have $\zeta^{(m)} \to \hat{\zeta}/\zeta' \in L_x \setminus \{0\}$, $m \to \infty$, which yields

$$0 = \lim_{m \to \infty} \widetilde{P}(\mu_m, \zeta^{(m)}) = (\zeta' + 1)^{-1} \sum_{\ell \in I_\infty} \tau_\ell(\hat{\mu}, 0)\, \varphi_\ell(\hat{\mu})\, \psi_\ell(\hat{\zeta}/\zeta')$$

$$= (\zeta' + 1)^{-1} \pi_\infty P(\hat{\mu}, \hat{\zeta}/\zeta').$$

This also contradicts (2.16). □

Lemma 2.54. Let $P(\lambda, z) = \sum_{\ell \in I} \tau_\ell(\lambda, z) \varphi_\ell(\lambda) \psi_\ell(z)$, $(\lambda, z) \in L_t \times L_x$, be a regular representation of the symbol $P \in S(L_t \times L_x)$. Let

$$\pi_\gamma P(\lambda, z) \neq 0 \qquad (2.20)$$

for all $(\lambda, z) \in (L_t \setminus \{0\}) \times (L_x \setminus \{0\})$, and $\gamma \in (0, \infty]$. Then for every $j = \kappa_1(N(P)), \ldots, \kappa_2(N(P))$ such that $\rho \notin (\gamma_{j-1}, \gamma_j)$ there exists a constant $C_j > 0$ such that

$$|\pi_\gamma P(\lambda, z)| \geq C_j |\lambda|^{s_j} |z|^{r_j}$$

for all $\gamma \in (\gamma_{j-1}, \gamma_j)$ and $(\lambda, z) \in (L_t \setminus \{0\}) \times (L_x \setminus \{0\})$.

Proof. Let $j \in \{\kappa_1(N(P)), \ldots, \kappa_2(N(P))\}$ be fixed. We have to consider two cases:

(I) $\gamma_j > \rho$: Since $\rho \notin (\gamma_{j-1}, \gamma_j)$ we have $j > \kappa_1(N(P))$ and $\gamma_{j-1} \geq \rho$. Let $\gamma \in (\gamma_{j-1}, \gamma_j)$ be arbitrary. Then Remark 2.23 (ii) yields

$$\pi_\gamma P(\lambda, z) = \sum_{\substack{\ell \in I_{\gamma_{j-1}} \\ L_\ell = r_j}} \tau_\ell(\lambda, 0) \varphi_\ell(\lambda) \psi_\ell(z).$$

We have $\pi_\gamma P(\lambda, z) = |\lambda|^{s_j} |z|^{r_j} \cdot \pi_\gamma P\left(\frac{\lambda}{|\lambda|}, \frac{z}{|z|}\right)$ for all $(\lambda, z) \in (L_t \setminus \{0\}) \times (L_x \setminus \{0\})$. The continuity of $\pi_\gamma P$ and the compactness of

$$\Omega := \{(\mu, \zeta) \in (L_t \setminus \{0\}) \times (L_x \setminus \{0\}) : |\mu| = |\zeta| = 1\}$$

imply that there exists $C > 0$ such that

$$|\pi_\gamma P(\mu, \zeta)| \geq C > 0, \quad (\mu, \zeta) \in \Omega.$$

This yields $|\pi_\gamma P(\lambda, z)| \geq C |\lambda|^{s_j} |z|^{r_j}$ for all $(\lambda, z) \in (L_t \setminus \{0\}) \times (L_x \setminus \{0\})$.

(II) $\gamma_j \leq \rho$: Obviously, we have $j \neq \kappa_2(N(P))$ in this case. For $\gamma \in (\gamma_{j-1}, \gamma_j)$ Remark 2.23 (iii) yields

$$\pi_\gamma P(\lambda, z) = \sum_{\substack{\ell \in I_{\gamma_j} \\ M_\ell = s_j}} \tau_\ell(0, z) \varphi_\ell(\lambda) \psi_\ell(z).$$

Then we have $\pi_\gamma P(\lambda, z) = |\lambda|^{s_j} |z|^{r_j} \cdot \pi_\gamma P\left(\frac{\lambda}{|\lambda|}, \frac{z}{|z|}\right)$ for all $(\lambda, z) \in (L_t \setminus \{0\}) \times (L_x \setminus \{0\})$. Using the same arguments as in (I) we obtain a constant $C > 0$ such that we have

$$|\pi_\gamma P(\lambda, z)| \geq C |\lambda|^{s_j} |z|^{r_j}$$

for all $(\lambda, z) \in (L_t \setminus \{0\}) \times (L_x \setminus \{0\})$. $\square$

2.2. N-parameter-ellipticity and N-parabolicity

Lemma 2.55. Let $P(\lambda, z) = \sum_{\ell \in I} \tau_\ell(\lambda, z) \varphi_\ell(\lambda) \psi_\ell(z)$, $(\lambda, z) \in L_t \times L_x$, be a regular representation of the symbol $P \in S(L_t \times L_x)$. Let

$$\pi_\gamma P(\lambda, z) \neq 0 \qquad (2.21)$$

for all $(\lambda, z) \in (L_t \setminus \{0\}) \times (L_x \setminus \{0\})$, and $\gamma \in (0, \infty]$. Then for $j = \kappa_1(N(P)), \ldots, \kappa_2(N(P))$ such that $\rho \in (\gamma_{j-1}, \gamma_j)$ there exists a constant $C_j > 0$ such that

$$|\pi_\rho P(\lambda, z)| \geq C_j |\lambda|^{s_j} |z|^{r_j}$$

for all $(\lambda, z) \in (L_t \setminus \{0\}) \times (L_x \setminus \{0\})$.

Proof. Due to Remark 2.23 (iii) it is obvious that τ_ℓ is constant for all $\ell \in I_\rho$. If we define $a_\ell := \tau_\ell(\lambda, z) \neq 0$, we get

$$\pi_\rho P(\lambda, z) = \sum_{\ell \in I_\rho} a_\ell \varphi_\ell(\lambda) \psi_\ell(z) = |\lambda|^{s_j} |z|^{r_j} \pi_\rho P\left(\frac{\lambda}{|\lambda|}, \frac{z}{|z|}\right)$$

for all $(\lambda, z) \in (L_t \setminus \{0\}) \times (L_x \setminus \{0\})$. The same arguments as in Lemma 2.54 yield the assertion. $\square$

Next, we formulate our main result of this section. It gives a sufficient condition for N-parameter-ellipticity of symbols. Some special versions can be found in [Don74, Theorem 3.3] and [GV92, Theorem 2.1], where the authors consider polynomials, or in [DSS08, Theorem 3.1], where the authors consider

$$\tau_\ell(\lambda, z) = \left(\sqrt{\lambda + z^2}\right)^{N_\ell}, \quad \varphi_\ell(\lambda) = \lambda^{M_\ell}, \quad \psi_\ell(z) = z^{L_\ell}$$

for $(\lambda, z) \in (\overline{S}_\theta \times \overline{S}_\delta) \setminus \{(0,0)\} \subseteq \mathbb{C}^2$. Here we want to generalize these results to our situation by using the same approach and techniques as in the works of R. Denk, S. Gindikin, J. Saal, J. Seiler, and L.R. Volevich, cf. [GV92] and [DSS08]. We also want to mention that in [DSS08] regular symbols are considered.

Theorem 2.56. Let $P \in S(L_t \times L_x)$ be a symbol satisfying

$$\pi_\gamma P(\lambda, z) \neq 0, \quad \pi_\infty P(\lambda, 0) \neq 0$$

for all $(\lambda, z) \in (L_t \setminus \{0\}) \times (L_x \setminus \{0\})$, and $\gamma \in (0, \infty]$. Then P is N-parameter-elliptic in $\mathring{L}_t \times \mathring{L}_x$.

Proof. For any $\gamma \in (0, \infty]$ and $(\lambda, z) \in (L_t \setminus \{0\}) \times (L_x \setminus \{0\})$ we obtain

$$|P(\lambda, z)| \geq |\pi_\gamma P(\lambda, z)| - \left|\pi_\gamma P(\lambda, z) - \sum_{\ell \in I_\gamma} \tau_\ell(\lambda, z) \varphi_\ell(\lambda) \psi_\ell(z)\right| \qquad (2.22)$$

$$- \left|\sum_{\ell \in I \setminus I_\gamma} \tau_\ell(\lambda, z) \varphi_\ell(\lambda) \psi_\ell(z)\right|$$

$$\geq |\pi_\gamma P(\lambda, z)| - \Lambda_1(\gamma, \lambda, z) - \Lambda_2(\gamma, \lambda, z) \qquad (2.23)$$

where
$$\Lambda_1(\gamma,\lambda,z) := \sum_{\ell \in I_\gamma} |\pi_\gamma \tau_\ell(\lambda,z) - \tau_\ell(\lambda,z)|\, |\varphi_\ell(\lambda)|\, |\psi_\ell(z)|,$$
$$\Lambda_2(\gamma,\lambda,z) := \sum_{\ell \in I \setminus I_\gamma} |\tau_\ell(\lambda,z)|\, |\varphi_\ell(\lambda)|\, |\psi_\ell(z)|.$$

Let $C > 0$ be such that
$$\frac{1}{C}\left(|\lambda|^{N_\ell/\rho} + |z|^{N_\ell}\right) \leq |\tau_\ell(\lambda,z)| \leq C\left(|\lambda|^{N_\ell/\rho} + |z|^{N_\ell}\right), \qquad (2.24)$$
$$|\varphi_\ell(\lambda)| \leq C|\lambda|^{M_\ell}, \quad |\psi_\ell(z)| \leq C|z|^{L_\ell} \qquad (2.25)$$

for all $\ell \in I$ and $(\lambda,z) \in L_t \times L_x$. Due to Lemmas 2.53, 2.54, and 2.55 there exists a constant $C' > 0$ such that
$$|\pi_{\gamma_j} P(\lambda,z)| \geq C'\left(|z|^{r_j}|\lambda|^{s_j} + |z|^{r_{j+1}}|\lambda|^{s_{j+1}}\right),\; j = \kappa_1(N(P)),\ldots,J, \qquad (2.26)$$
$$|\pi_\gamma P(\lambda,z)| \geq C'|\lambda|^{s_k}|z|^{r_k},\; k = \kappa_1(N(P)),\ldots,\kappa_2(N(P)),\; \gamma \in (\gamma_{k-1},\gamma_k)$$
$$\text{if } \rho \notin (\gamma_{k-1},\gamma_k), \qquad (2.27)$$
$$|\pi_\rho P(\lambda,z)| \geq C'|\lambda|^{s_\ell}|z|^{r_\ell},\; \ell = \kappa_1(N(P)),\ldots,\kappa_2(N(P)),\; \text{if } \rho \in (\gamma_{\ell-1},\gamma_\ell), \qquad (2.28)$$

for all $(\lambda,z) \in (L_t \setminus \{0\}) \times (L_x \setminus \{0\})$.

The concept of the proof consists in using the partition of Remark 2.44 and in proving estimates for $|P(\lambda,z)|$ on each subdomain $G_j, \widetilde{G}_j$ separately. According to Lemma 2.45 (iv) we can ignore G_1 and $\widetilde{G}_1$ if P is not regular in time. Therefore, we only consider $j \geq \kappa_1(N(P))$.

1. Determination of the partition:
We define $\varepsilon := C'/(4C^4 \#I)$ (where $\#I$ stands for the cardinality of I) and choose $\widehat{\varepsilon}_0 = \widehat{\varepsilon}_0(\nu(P),\varepsilon)$ according to Lemma 2.45 (i). Then define $\vartheta := \varepsilon_0/4$. Due to Lemma 2.52 we can choose $\varepsilon_0 \leq \widehat{\varepsilon}_0$ sufficiently small such that
$$|\tau_\ell(\lambda,z) - \pi_\gamma \tau_\ell(\lambda,z)| \leq \vartheta|\pi_\gamma \tau_\ell(\lambda,z)| \qquad (2.29)$$

for all $\ell \in I$, $j = \kappa_1(N(P)),\ldots,\kappa_2(N(P))$ with $\rho \notin (\gamma_{j-1},\gamma_j)$, $\gamma \in (\gamma_{j-1},\gamma_j)$, and all non-vanishing tuples $(\lambda,z) \in (L_t \setminus \{0\}) \times (L_x \setminus \{0\})$ with $(|\lambda|,|z|) \in \widetilde{G}_j(\varepsilon_0)$ and $|\lambda| \geq 1$. If $N(P)$ is not regular in space, then we also choose $\varepsilon_0 \leq \widehat{\varepsilon}_0$ sufficiently small such that
$$|\tau_\ell(\lambda,z) - \tau_\ell(\lambda,0)| \leq \vartheta|\tau_\ell(\lambda,0)| \qquad (2.30)$$

for all $(\lambda,z) \in (L_t \setminus \{0\}) \times (L_x \setminus \{0\})$ with $(|\lambda|,|z|) \in \widetilde{G}_{J+1}(\varepsilon_0)$ and $|\lambda| \geq 1$.

Next, we choose $\varepsilon_1 = \varepsilon_1(\nu(P),\varepsilon,\varepsilon_0) > 0$ such that Lemma 2.45 (iii) holds. Lemma 2.51 allows us to define $\mu_1 > 0$ sufficiently large such that
$$|\tau_\ell(\lambda,z) - \pi_{\gamma_j}\tau_\ell(\lambda,z)| \leq \vartheta|\pi_{\gamma_j}\tau_\ell(\lambda,z)| \qquad (2.31)$$

2.2. N-parameter-ellipticity and N-parabolicity

for all $\ell \in I$, $j = \kappa_1(N(P)), \ldots, J$, and $(\lambda, z) \in (L_t \setminus \{0\}) \times (L_x \setminus \{0\})$ with $(|\lambda|, |z|) \in G_j(\varepsilon_0, \varepsilon_1)$ and $|\lambda| \geq \mu_1$.

For the sake of readability we define

$$\widetilde{U}_j := \left\{ (\lambda, z) \in L_t \times L_x : (|\lambda|, |z|) \in \widetilde{G}_j(\varepsilon_0) \right\}, \quad j = \kappa_1(N(P)), \ldots, J+1,$$
$$U_\ell := \left\{ (\lambda, z) \in L_t \times L_x : (|\lambda|, |z|) \in G_\ell(\varepsilon_0, \varepsilon_1) \right\}, \quad \ell = \kappa_1(N(P)), \ldots, J.$$

According to Lemma 2.45 (iv) there exists $\mu_0 = \mu_0(\varepsilon_0, \varepsilon_1)$ with

$$\{(\lambda, z) \in L_t \times (L_x \setminus \{0\}) : |\lambda| \geq \mu_0\} \subseteq \bigcup_{j=\kappa_1(N(P))}^{J+1} \widetilde{U}_j \cup \bigcup_{\ell=\kappa_1(N(P))}^{J} U_\ell.$$

In the following we set $\lambda_0 := \max\{1, \mu_1, \mu_0\}$ and verify that it satisfies the assertion.

2. Estimate on U_j:

Let $(\lambda, z) \in U_j$ with $j = \kappa_1(N(P)), \ldots, J$ and $|\lambda| \geq \lambda_0$. We apply (2.23) with $\gamma := \gamma_j$ and prove estimates for $\Lambda_k(\gamma_j, \lambda, z)$, $k = 1, 2$.

(I) According to (2.31) we have

$$\Lambda_1(\gamma_j, \lambda, z) \leq \vartheta \cdot \sum_{\ell \in I_{\gamma_j}} |\pi_{\gamma_j} \tau_\ell(\lambda, z)| \, |\varphi_\ell(\lambda)| |\psi_\ell(z)|.$$

Using (2.24), (2.25), and Remark 2.17 we easily obtain

$$|\pi_{\gamma_j} \tau_\ell(\lambda, z)| \, |\varphi_\ell(\lambda)| |\psi_\ell(z)|$$
$$\leq 2C^4 \begin{cases} |\lambda|^{\mathfrak{s}_1(\ell)} |z|^{\mathfrak{r}_1(\ell)}, & \gamma_j < \rho, \\ |\lambda|^{\mathfrak{s}_1(\ell)} |z|^{\mathfrak{r}_1(\ell)} + |\lambda|^{\mathfrak{s}_2(\ell)} |z|^{\mathfrak{r}_2(\ell)}, & \gamma_j = \rho, \\ |\lambda|^{\mathfrak{s}_2(\ell)} |z|^{\mathfrak{r}_2(\ell)}, & \gamma_j > \rho \end{cases}$$
$$\leq 4C^4 \left(|z|^{r_j} |\lambda|^{s_j} + |z|^{r_{j+1}} |\lambda|^{s_{j+1}} \right)$$

since $\mathfrak{v}_1(\ell), \mathfrak{v}_2(\ell) \in [v_j v_{j+1}]$ for all $\ell \in I_{\gamma_j}$. Here we have used the notation

$$\mathfrak{v}_1(\ell) = (\mathfrak{r}_1(\ell), \mathfrak{s}_1(\ell)) := (N_\ell + L_\ell, M_\ell),$$
$$\mathfrak{v}_2(\ell) = (\mathfrak{r}_2(\ell), \mathfrak{s}_2(\ell)) := (L_\ell, \tfrac{N_\ell}{\rho} + M_\ell),$$

see Definition 2.15 in Subsection 2.1 b) for the definition of $N(P)$. Hence we get

$$\Lambda_1(\gamma_j, \lambda, z) \leq 4C^4 \vartheta \cdot (\#I_{\gamma_j}) \left(|z|^{r_j} |\lambda|^{s_j} + |z|^{r_{j+1}} |\lambda|^{s_{j+1}} \right)$$
$$\leq \frac{C'}{4} \left(|z|^{r_j} |\lambda|^{s_j} + |z|^{r_{j+1}} |\lambda|^{s_{j+1}} \right). \tag{2.32}$$

(II) Since $\mathfrak{v}_1(\ell), \mathfrak{v}_2(\ell) \in \nu(P) \setminus [v_j v_{j+1}]$ for all $\ell \in I \setminus I_{\gamma_j}$ we obtain

$$\Lambda_2(\gamma_j, \lambda, z) \leq C^4 \cdot \sum_{\ell \in I \setminus I_{\gamma_j}} \left(|\lambda|^{\mathfrak{s}_1(\ell)} |z|^{\mathfrak{r}_1(\ell)} + |\lambda|^{\mathfrak{s}_2(\ell)} |z|^{\mathfrak{r}_2(\ell)} \right)$$

$$\leq \varepsilon C^4 \cdot \#(I \setminus I_{\gamma_j}) \left(|z|^{r_j} |\lambda|^{s_j} + |z|^{r_{j+1}} |\lambda|^{s_{j+1}} \right)$$

$$\leq \frac{C'}{4} \left(|z|^{r_j} |\lambda|^{s_j} + |z|^{r_{j+1}} |\lambda|^{s_{j+1}} \right) \tag{2.33}$$

by Lemma 2.45 (iii).

Using (2.26), (2.32), and (2.33) we get $|P(\lambda, z)| \geq C'/2 \cdot (|z|^{r_j} |\lambda|^{s_j} + |z|^{r_{j+1}} |\lambda|^{s_{j+1}})$. Once more we apply Lemma 2.45 (iii) and get

$$W_P(\lambda, z) = \sum_{(r,s) \in N_V(P)} |z|^r |\lambda|^s$$

$$= (|z|^{r_j} |\lambda|^{s_j} + |z|^{r_{j+1}} |\lambda|^{s_{j+1}}) + \sum_{(r,s) \in N_V(P) \setminus \{v_j, v_{j+1}\}} |z|^r |\lambda|^s$$

$$\leq (1 + J\varepsilon) \left(|z|^{r_j} |\lambda|^{s_j} + |z|^{r_{j+1}} |\lambda|^{s_{j+1}} \right)$$

since $(\lambda, z) \in U_j$ and $\#N_V(P) = J + 2$. Altogether it follows that

$$|P(\lambda, z)| \geq C \cdot W_P(\lambda, z), \quad (\lambda, z) \in U_j, \ |\lambda| \geq \lambda_0$$

where $C := C'/(2(1 + J\varepsilon))$.

3. Estimate on $\widetilde{U}_j$:

First, let $(\lambda, z) \in \widetilde{U}_j$ with $j = \kappa_1(N(P)), \ldots, \kappa_2(N(P))$, $\rho \notin (\gamma_{j-1}, \gamma_j)$ and $|\lambda| \geq \lambda_0$. Next, we apply (2.23) with an arbitrary $\gamma \in (\gamma_{j-1}, \gamma_j)$ and prove estimates for $\Lambda_k(\gamma, \lambda, z)$, $k = 1, 2$, again.

(I) With (2.24), (2.25), and Remark 2.23 (ii) one obtains

$$|\pi_\gamma \tau_\ell(\lambda, z)| \, |\varphi_\ell(\lambda)| \, |\psi_\ell(z)| \leq 2C^4 \cdot |\lambda|^{s_j} |z|^{r_j}$$

for all $\ell \in I_\gamma$. Using (2.29) we get

$$\Lambda_1(\gamma, \lambda, z) \leq 2C^4 \vartheta(\#I_\gamma) \cdot |\lambda|^{s_j} |z|^{r_j} \leq \frac{C'}{8} \cdot |\lambda|^{s_j} |z|^{r_j}. \tag{2.34}$$

(II) We have $\mathfrak{v}_1(\ell), \mathfrak{v}_2(\ell) \in \nu(P) \setminus \{v_j\}$ for all $\ell \in I \setminus I_\gamma$. Thus we obtain

$$\Lambda_2(\gamma, \lambda, z) \leq C^4 \cdot \sum_{\ell \in I \setminus I_\gamma} \left(|\lambda|^{\mathfrak{s}_1(\ell)} |z|^{\mathfrak{r}_1(\ell)} + |\lambda|^{\mathfrak{s}_2(\ell)} |z|^{\mathfrak{r}_2(\ell)} \right)$$

$$\leq 2\varepsilon C^4 \cdot \#(I \setminus I_\gamma) \cdot |\lambda|^{s_j} |z|^{r_j}$$

$$\leq \frac{C'}{2} \cdot |\lambda|^{s_j} |z|^{r_j} \tag{2.35}$$

by Lemma 2.45 (i).

2.2. N-parameter-ellipticity and N-parabolicity

Employing (2.27), (2.34), and (2.35) we get $|P(\lambda,z)| \geq \frac{3}{8}C' \cdot |z|^{r_j}|\lambda|^{s_j}$. Lemma 2.45 (i) then implies

$$W_P(\lambda,z) \leq (1+\varepsilon(J+1)) \cdot |z|^{r_j}|\lambda|^{s_j}$$

since $(\lambda,z) \in \widetilde{U}_j$. We get

$$|P(\lambda,z)| \geq \widetilde{C} \cdot W_P(\lambda,z), \quad (\lambda,z) \in \widetilde{U}_j, \ |\lambda| \geq \lambda_0$$

where $\widetilde{C} := 3/[8(1+\varepsilon(J+1))] \cdot C'$.

In the following we have to consider the case when there exists no $i \in \{\kappa_1(N(P)), \ldots, \kappa_2(N(P))\}$ with $\rho = \gamma_i$. Let $(\lambda,z) \in \widetilde{U}_j$ with $j = \kappa_1(N(P)), \ldots, \kappa_2(N(P))$, $\rho \in (\gamma_{j-1}, \gamma_j)$ and $|\lambda| \geq \lambda_0$. Here we use (2.23) with $\gamma := \rho$. We already have $\Lambda_1(\gamma, \lambda, z) = 0$ and in the same way as in the previous case we obtain

$$\Lambda_2(\gamma,\lambda,z) \leq \frac{C'}{2} \cdot |z|^{r_j}|\lambda|^{s_j}$$

and $W_P(\lambda,z) \leq (1+\varepsilon(J+1)) \cdot |z|^{r_j}|\lambda|^{s_j}$. Now (2.28) yields

$$|P(\lambda,z)| \geq \widetilde{C} \cdot W_P(\lambda,z), \quad (\lambda,z) \in \widetilde{U}_j, \ |\lambda| \geq \lambda_0$$

where $\widetilde{C} := 1/[2(1+\varepsilon(J+1))] \cdot C'$.

4. Estimate on $\widetilde{U}_{J+1}$ with $\kappa_2(N(P)) = J$: If $N(P)$ is not regular in space (i.e., $\kappa_2(N(P)) = J$) we have to modify the argumentation slightly to obtain an estimate in $\widetilde{U}_{J+1}$. Let $(\lambda,z) \in \widetilde{U}_{J+1}$. As in the very beginning we decompose the symbol

$$|P(\lambda,z)| \geq |\pi_\infty P(\lambda,0)| - \Lambda_1(\lambda,z) - \Lambda_2(\lambda,z)$$

where

$$\Lambda_1(\lambda,z) := \sum_{\substack{\ell \in I_\infty, \\ L_\ell = 0}} |\tau_\ell(\lambda,0) - \tau_\ell(\lambda,z)| \, |\varphi_\ell(\lambda)| |\psi_\ell(z)|,$$

$$\Lambda_2(\lambda,z) := \sum_{\substack{\ell \in I \setminus I_\infty \\ \vee L_\ell \neq 0}} |\tau_\ell(\lambda,z)| \, |\varphi_\ell(\lambda)| |\psi_\ell(z)|.$$

We have $|\pi_\infty P(\lambda,0)| \geq C'|\lambda|^{s_{J+1}}$ by (2.26) and continuity.

(I) With (2.24) and (2.25) one obtains

$$|\tau_\ell(\lambda,0)| \, |\varphi_\ell(\lambda)| |\psi_\ell(z)| \leq 2C^4 \cdot |\lambda|^{s_{J+1}}$$

for all $\ell \in I_\infty$ with $L_\ell = 0$. Using (2.30) we get

$$\Lambda_1(\gamma,\lambda,z) \leq 2C^4 \vartheta \cdot \#I \cdot |\lambda|^{s_j}|z|^{r_j} \leq \frac{C'}{8} \cdot |\lambda|^{s_j}|z|^{r_j}.$$

(II) With Lemma 2.45 (i) and $\mathfrak{v}_1(\ell), \mathfrak{v}_2(\ell) \in \nu(P) \setminus \{v_{J+1}\}$ for all $\ell \in I$ with $\ell \notin I_\infty$ or $L_\ell \neq 0$ we obtain

$$\Lambda_2(\lambda, z) \leq C^4 \cdot \sum_{\substack{\ell \in I \setminus I_\infty \\ \vee L_\ell \neq 0}} \left(|\lambda|^{\mathfrak{s}_1(\ell)} |z|^{\mathfrak{r}_1(\ell)} + |\lambda|^{\mathfrak{s}_2(\ell)} |z|^{\mathfrak{r}_2(\ell)} \right)$$

$$\leq 2\varepsilon C^4 \cdot \#I \cdot |\lambda|^{s_{J+1}} \leq \frac{C'}{2} \cdot |\lambda|^{s_{J+1}}$$

as before.

We get
$$|P(\lambda, z)| \geq \widetilde{C} \cdot W_P(\lambda, z), \quad (\lambda, z) \in \widetilde{U}_{J+1}, \ |\lambda| \geq \lambda_0$$
where $\widetilde{C} := 3/[8(1 + \varepsilon(J+1))] \cdot C'$. □

Summarizing the results of Proposition 2.47 and Theorem 2.56, we get the following characterization of N-parameter-elliptic symbols in $S_N(L_t \times L_x)$.

Corollary 2.57 (Characterization of N-parameter-elliptic symbols). *The symbol class $S_N(L_t \times L_x)$ consists of all symbols $P \in S(L_t \times L_x)$ with non-vanishing γ-principal parts, i.e.,*

$$\pi_\gamma P(\lambda, z) \neq 0, \quad \pi_\infty P(\lambda, 0) \neq 0$$

for all $(\lambda, z) \in (L_t \setminus \{0\}) \times (L_x \setminus \{0\})$, and $\gamma \in (0, \infty]$.

2.3 H^∞-calculus of N-parabolic mixed-order systems

This section combines the results of the joint time-space H^∞-calculus developed in Section 1.3 and the concepts of order functions and N-parabolicity introduced in Sections 2.1 and 2.2. In the first part of this section, we are interested in the properties of operators $P_\sigma(\mathcal{D}_+)$ where $P \in H_P(S_\theta \times \Sigma_\delta^n)$ and $P_\sigma := P(\sigma + \cdot, \cdot)$. Let μ be an upper order function of the symbol P. Then we prove that the domain of $P_\sigma(\mathcal{D}_+)$ always contains an intersection of spaces which is related to the structure of the order function μ. To obtain this result we first have to consider a class of order reducing functions Φ_N corresponding to a Newton polygon N. As soon as we have determined the domain of $\Phi_N(\mathcal{D}_+)$, we can use this to obtain information about $D(P_\sigma(\mathcal{D}_+))$.

2.3. H^∞-calculus of N-parabolic mixed-order systems

a) The H^∞-calculus of N-parabolic symbols

Motivation. In order to apply the results on the joint bounded H^∞-calculus to inhomogeneous symbols as considered in the previous section, we will use generalized order reduction operators which provide an isomorphism between anisotropic Sobolev spaces related to the Newton polygon and basic spaces like $L_p(\mathbb{R}^{n+1}_+)$. This is done in analogy to the well-known order reduction operator for Bessel potential spaces $H^r_p(\mathbb{R}^n)$ (see Definition 1.55). Here $(1-\Delta)^{r/2}$ induces an isomorphism between $H^r_p(\mathbb{R}^n)$ and $L_p(\mathbb{R}^n)$. Note that $(1-\Delta)^{r/2} = \mathrm{op}[\Lambda_r]$ as considered in Section 1.2.

In the more general case, the order reduction operator will be of the form $\Phi_N(\mathcal{D}_+)$ where Φ_N is a suitable combination of Ψ_s and Λ_r. Here Φ_N is again defined by the structure of the Newton polygon N. With this construction, we will obtain suitable mapping properties for scalar N-parabolic operators.

In the following, we use the symbols $\Psi_s(\lambda) := (1+\lambda)^s$ and $\Lambda_r(z) := (1 - \sum_{k=1}^n z_k^2)^{r/2}$, which were introduced in Section 1.2. We start with an elementary observation.

Lemma 2.58. *Let $M \in \mathbb{N}_0$, $\theta \in (\pi/2, \pi)$, $\alpha := (\alpha_i)_{i=0,\ldots,M} \subseteq [0, 1/2]$, and $\beta := (\beta_i)_{i=0,\ldots,M} \subseteq [0, \infty)$. Then there exist $\delta_0 = \delta_0(\beta, \theta) \in (0, \pi/2)$, $\varphi = \varphi(\delta_0) \in (0, \pi/2)$ such that*

$$\Psi_{\alpha_i}(\lambda)\Lambda_{\beta_i}(z) \in S_\varphi \tag{2.36}$$

for all $(\lambda, z) \in S_\theta \times \Sigma^n_\delta$ with $\delta \in (0, \delta_0)$.

Proof. For $\lambda \in S_\theta$ we get $(1+\lambda)^{\alpha_i} \in S_{\theta/2}$ due to $\alpha_i \leq 1/2$. Let $\delta_0 > 0$ be sufficiently small such that $\varphi := \max_i\{\theta/2 + \delta_0\beta_i\} < \pi/2$. We can easily see that $(1 - \sum_k z_k^2)^{\beta_i/2} \in S_{\delta\beta_i}$ for all $z \in \Sigma^n_\delta$ with $\delta \leq \delta_0$ and $i = 0, \ldots, M$. Hence we get

$$(1+\lambda)^{\alpha_i}\left(1 - \sum_k z_k^2\right)^{\beta_i/2} \in S_{\theta/2 + \delta\beta_i} \subseteq S_\varphi$$

for all $(\lambda, z) \in S_\theta \times \Sigma^n_\delta$ with $\delta \leq \delta_0$. □

Lemma 2.59. *Let $M \in \mathbb{N}_0$ and $\nu := (\beta_i, \alpha_i)_{i=0,\ldots M} \subseteq [0, \infty)^2$ be arbitrary with $\sigma := \max_{i=0,\ldots,M} \alpha_i > 0$ and $\lceil 2\sigma \rceil := \min\{m \in \mathbb{N}_0 : 2\sigma \leq m\}$. Then there exist $C, C' > 0$ such that*

$$C \cdot W_\nu(\lambda, z) \leq \left|\sum_{i=0}^M \Psi_{\alpha_i/\lceil 2\sigma \rceil}(\lambda)\Lambda_{\beta_i/\lceil 2\sigma \rceil}(z)\right|^{\lceil 2\sigma \rceil} \leq C' \cdot W_\nu(\lambda, z) \tag{2.37}$$

for all $(\lambda, z) \in S_\theta \times \Sigma^n_\delta$ with $\delta \leq \delta_0 = \delta_0(\theta, (\beta_i/\lceil 2\sigma \rceil)_i)$ as in Lemma 2.58. The weight function W_ν is defined as in Subsection 2.1 b).

Proof. To prove the assertion we define the two continuous functions

$$\varphi_1 \colon \overline{S}_\varphi^{M+1} \to \mathbb{C}, \quad x \mapsto \sum_{i=0}^{M} |x_i|^{\lceil 2\sigma \rceil}, \quad \varphi_2 \colon \overline{S}_\varphi^{M+1} \to \mathbb{C}, \quad x \mapsto \left(\sum_{i=0}^{M} x_i\right)^{\lceil 2\sigma \rceil}$$

where $\varphi = \varphi(\delta_0)$ is given as in Lemma 2.58. Both functions are homogeneous of degree $\lceil 2\sigma \rceil$ and $\varphi_1(x), \varphi_2(x) \neq 0$, $x \neq 0$, due to $\varphi < \pi/2$. This yields $C_1 |\varphi_1(x)| \le |\varphi_2(x)| \le C_1' |\varphi_1(x)|$ for all $x \in \overline{S}_\varphi^{M+1}$. Setting $x_i := \Psi_{\alpha_i/\lceil 2\sigma \rceil}(\lambda)\Lambda_{\beta_i/\lceil 2\sigma \rceil}(z)$ for $i = 0, \ldots, M$ and $(\lambda, z) \in S_\theta \times \Sigma_\delta^n$ ($\delta \le \delta_0$) we get

$$C_1 \sum_{i=0}^{M} |\Psi_{\alpha_i}(\lambda)| \cdot |\Lambda_{\beta_i}(z)| \le \left| \sum_{i=0}^{M} \Psi_{\alpha_i/\lceil 2\sigma \rceil}(\lambda) \Lambda_{\beta_i/\lceil 2\sigma \rceil}(z) \right|^{\lceil 2\sigma \rceil}$$

$$\le C_1' \sum_{i=0}^{M} |\Psi_{\alpha_i}(\lambda)| \cdot |\Lambda_{\beta_i}(z)|$$

by virtue of Lemma 2.58.

With similar arguments we can show that there are $C_2, C_2', C_3, C_3' > 0$ such that
$$\begin{aligned} C_2(1 + |\lambda|^{\alpha_i}) &\le |\Psi_{\alpha_i}(\lambda)| \le C_2'(1 + |\lambda|^{\alpha_i}), \\ C_3(1 + |z|^{\beta_i}) &\le |\Lambda_{\beta_i}(z)| \le C_3'(1 + |z|^{\beta_i}) \end{aligned} \qquad (2.38)$$

for all $(\lambda, z) \in S_\theta \times \Sigma_\delta^n$ and $i = 0, \ldots, M$. Hence, we get

$$C \sum_{i=0}^{M} (1 + |\lambda|^{\alpha_i})(1 + |z|^{\beta_i}) \le \left| \sum_{i=0}^{M} \Psi_{\alpha_i/\lceil 2\sigma \rceil}(\lambda) \Lambda_{\beta_i/\lceil 2\sigma \rceil}(z) \right|^{\lceil 2\sigma \rceil}$$

$$\le C' \sum_{i=0}^{M} (1 + |\lambda|^{\alpha_i})(1 + |z|^{\beta_i})$$

with constants $C, C' > 0$, and $(\lambda, z) \in S_\theta \times \Sigma_\delta^n$. Trivially, we have

$$W_\nu(\lambda, z) \le \sum_{i=0}^{M} (1 + |\lambda|^{\alpha_i})(1 + |z|^{\beta_i}), \quad (\lambda, z) \in S_\theta \times \Sigma_\delta^n,$$

which yields the first inequality in the assertion. The inverse inequality can be obtained by Remark 2.17, which yields $\sum_{i=0}^{M}(1 + |\lambda|^{\alpha_i})(1 + |z|^{\beta_i}) \le cW_\nu(\lambda, z)$, $(\lambda, z) \in S_\theta \times \Sigma_\delta^n$, for some $c > 0$. □

Let us fix the situation for the remainder of this section: Let $s' \ge 0, r' \in \mathbb{R}$, $1 < p_0, p_1, q_0, q_1 < \infty$, $\mathcal{F} \in \{B_{p_0 q_0}, H_{p_0}\}$ with $s' > 0$ if $\mathcal{F} = B_{p_0 q_0}$, and $\mathcal{K} \in \{B_{p_1 q_1}, H_{p_1}\}$ (see Definition 1.71). Let X be a Banach space of class $\mathcal{HT}$ with property (α). As before, we set

$$_0\mathcal{F}^{s'}(\mathcal{K}^{r'}) := {}_0\mathcal{F}^{s'}(\mathbb{R}_+, \mathcal{K}^{r'}(\mathbb{R}^n, X)).$$

2.3. H^∞-calculus of N-parabolic mixed-order systems

We will also use the notation
$$_0\mathcal{F}_\varrho^{s'}(\mathcal{K}^{r'}) := {_0\mathcal{F}_\varrho^{s'}}(\mathbb{R}_+, \mathcal{K}^{r'}(\mathbb{R}^n, X))$$
for the spaces with exponential weight $\varrho \geq 0$ (cf. Remark 1.91). We denote the $_0\mathcal{F}^{s'}(\mathcal{K}^{r'})$-realization of $(\sigma + \partial_t, \nabla_x)$ by $\mathcal{D}_{+,\sigma}$, and the $_0\mathcal{F}_\varrho^{s'}(\mathcal{K}^{r'})$-realization of $(\sigma + \partial_t, \nabla_x)$ (see Remark 1.91) by $\mathcal{D}_{+,\sigma}^{(\varrho)}$.

In the next proposition we will present order reduction operators for intersections of spaces of mixed regularity. This result can be seen as a generalization of the results in Proposition 1.92, Lemma 1.58, and Proposition 1.79.

Proposition 2.60. *Let $J \in \mathbb{N}_0$ and let $N_V := (r_i, s_i)_{i=0,\ldots J+1} \subseteq [0, \infty)^2$ be the vertices (starting at the origin and being indexed in the counter-clockwise direction) of a Newton polygon N with $s_{\max} := \max_{i=0,\ldots,J+1} s_i$. Define*

$$\Phi_N(\lambda, z) := \begin{cases} \Lambda_{r_1}(z), & s_{\max} = 0 \ (i.e., J = 0), \\ \left(\sum_{i=0}^{J+1} \Psi_{s_i/\lceil 2\sigma \rceil}(\lambda) \Lambda_{r_i/\lceil 2\sigma \rceil}(z)\right)^{\lceil 2\sigma \rceil}, & s_{\max} > 0 \end{cases}$$

for $(\lambda, z) \in S_\theta \times \Sigma_\delta^n$, $\theta > \pi/2$ and $\delta \leq \delta_0$ with δ_0 from Lemma 2.58. Then there are constants $C, C' > 0$ such that we have the estimates

$$C \cdot W_N(\lambda, z) \leq |\Phi_N(\lambda, z)| \leq C' \cdot W_N(\lambda, z), \quad (\lambda, z) \in S_\theta \times \Sigma_\delta^n \quad (2.39)$$

for the symbol Φ_N. For all $\sigma \geq 0$ the operator $\Phi_N(\mathcal{D}_{+,\sigma})$ is invertible with

$$D(\Phi_N(\mathcal{D}_{+,\sigma})) = \bigcap_{(r,s) \in N_V} {_0\mathcal{F}^{s'+s}}(\mathcal{K}^{r'+r}),$$

$$\Phi_N(\mathcal{D}_{+,\sigma}) \in L_{\mathrm{Isom}}\left(\bigcap_{(r,s) \in N_V} {_0\mathcal{F}^{s'+s}}(\mathcal{K}^{r'+r}), {_0\mathcal{F}^{s'}}(\mathcal{K}^{r'})\right). \quad (2.40)$$

Proof. The claimed estimates for the symbol Φ_N follow directly from Lemma 2.59 and (2.38). Note that $W_N(\lambda, z) = 1 + |z|^{r_1}$ if $s_{\max} = 0$.

(I) First, we consider the case $\sigma = 0$ and $s_{\max} = 0$. So we have $\Phi_N = \Lambda_{r_1}$ and the assertion follows from Proposition 1.92.

Next, we consider the case $\sigma = 0$ and $s_{\max} > 0$. In this case we get $\Phi_N^{-1} \in H^\infty(S_\theta \times \Sigma_\delta^n)$ by (2.39) as well as $D(\Phi_N(\mathcal{D}_+)) = R(\Phi_N^{-1}(\mathcal{D}_+))$ and $(\Phi_N(\mathcal{D}_+))^{-1} = (\Phi_N)^{-1}(\mathcal{D}_+)$. Then Theorem 1.26 (ii) yields

$$\Phi_N(\mathcal{D}_+) \supseteq \left(\sum_{i=0}^{J+1} \Psi_{s_i/\lceil 2\sigma \rceil}(\mathcal{D}_+) \Lambda_{r_i/\lceil 2\sigma \rceil}(\mathcal{D}_+)\right)^{\lceil 2\sigma \rceil}$$

due to $\lceil 2\sigma \rceil \in \mathbb{N}$. According to Proposition 1.92 it follows that

$$D(\Phi_N(\mathcal{D}_+)) \supseteq \bigcap_{(r,s) \in N_V} {_0\mathcal{F}^{s'+s}}(\mathcal{K}^{r'+r}).$$

For each $i \in \{0,\ldots,J+1\}$ the function $\Psi_{s_i}\Lambda_{r_i}\Phi_N^{-1}$ is bounded. This can be seen by Lemma 2.59 with $M=0$ and $\nu := \{(r_i,s_i)\}$. We can write

$$\left(\frac{1}{\Phi_N}\right)(\mathcal{D}_+) = \Psi_{s_i}^{-1}(\mathcal{D}_+)\Lambda_{r_i}^{-1}(\mathcal{D}_+)\underbrace{\left(\frac{\Psi_{s_i}\Lambda_{r_i}}{\Phi_N}\right)(\mathcal{D}_+)}_{\in H^\infty} \qquad (2.41)$$

for $i=0,\ldots,J+1$ by Proposition 1.92 and Theorem 1.26 (iii). Therefore, we obtain

$$R((1/\Phi_N)(\mathcal{D}_+)) \subseteq \bigcap_{i=0}^{J+1} R(\Psi_{s_i}^{-1}(\mathcal{D}_+)\Lambda_{r_i}^{-1}(\mathcal{D}_+)) = \bigcap_{i=0}^{J+1} {}_0\mathcal{F}^{s'+s_i}(\mathcal{K}^{r'+r_i}).$$

So we get $D(\Phi_N(\mathcal{D}_+)) = \bigcap_{(r,s)\in N_V} {}_0\mathcal{F}^{s'+s}(\mathcal{K}^{r'+r})$. By virtue of (2.41) and Proposition 1.92 we easily see that

$$(1/\Phi_N)(\mathcal{D}_+) \in L\Big({}_0\mathcal{F}^{s'}(\mathcal{K}^{r'}), \bigcap_{(r,s)\in N_V} {}_0\mathcal{F}^{s'+s}(\mathcal{K}^{r'+r})\Big)$$

and therefore

$$\Phi_N(\mathcal{D}_+) \in L\Big(\bigcap_{(r,s)\in N_V} {}_0\mathcal{F}^{s'+s}(\mathcal{K}^{r'+r}), {}_0\mathcal{F}^{s'}(\mathcal{K}^{r'})\Big).$$

(II) For $\sigma \geq 0$ we obtain

$$\Phi_N(\mathcal{D}_{+,\sigma})u = \mathscr{M}_\sigma \Phi_N(\mathcal{D}_{+,0}^{(\sigma)})\mathscr{M}_\sigma^{-1} u, \quad u \in {}_0\mathcal{F}^{s'}(\mathcal{K}^{r'})$$

by Theorem 1.89 (iii) where $\mathcal{D}_{+,0}^{(\sigma)}$ is the ${}_0\mathcal{F}^{s'}(\mathcal{K}^{r'})$-realization of (∂_t, ∇_x) (see Remark 1.70 and Remark 1.91). From (I) we already know that $\Phi_N(\mathcal{D}_{+,0}^{(\sigma)})$ is an isomorphism and thus the assertion follows from the isomorphism results for $\mathscr{M}_\sigma$ given in Remark 1.70. □

The next result generalizes the embedding in [DSS08, Lemma 4.3] to our spaces of mixed scales. In particular, this result holds for mixed L_p-L_q-scales. In contrast to [DSS08] we directly develop a Banach space valued theory where the Banach space has to be of class $\mathcal{HT}$ with property (α).

Lemma 2.61 (Embedding results I). (i) *For all $s,r \geq 0$ and $\sigma \in [0,1]$ the embedding*

$$ {}_0\mathcal{F}^{s'+s}(\mathcal{K}^{r'}) \cap {}_0\mathcal{F}^{s'}(\mathcal{K}^{r'+r}) \hookrightarrow {}_0\mathcal{F}^{s'+\sigma s}(\mathcal{K}^{r'+(1-\sigma)r})$$

holds.

2.3. H^∞-calculus of N-parabolic mixed-order systems

(ii) Let $\nu := (\beta_i, \alpha_i)_{i=0,\ldots,M} \subseteq [0,\infty)^2$ be an arbitrary set and let N_V be the vertices of the Newton polygon $N(\nu)$. Then we have

$$\bigcap_{i=0,\ldots,M} {}_0\mathcal{F}^{s'+\alpha_i}(\mathcal{K}^{r'+\beta_i}) = \bigcap_{(r,s)\in N_V} {}_0\mathcal{F}^{s'+s}(\mathcal{K}^{r'+r})$$

with equivalent norms.

In the proof, we will frequently use the following elementary fact: Let X_k, Y_k be Banach spaces and $T_k \in L(X_k, Y_k)$ for $k = 0,1$. If we have $T_0 x = T_1 x$ for all $x \in X_0 \cap X_1$ then

$$T_1|_{X_0 \cap X_1} \in L(X_0 \cap X_1, Y_0 \cap Y_1).$$

Proof of Lemma 2.61. (i) For $s = 0$ the assertion is obvious. Let $s > 0$ and define the Newton polygons

$$N_1 := N(\{(r,0),(0,s)\}), \quad N_2 := N(\{((1-\sigma)r, \sigma s)\}).$$

Let $\Phi_1 := \Phi_{N_1}$ and $\Phi_2 := \Phi_{N_2}$ be the corresponding shift functions given in Proposition 2.60. Remark 2.17 and (2.39) then yield

$$|\Phi_2(\lambda, z)| \leq C(1 + |\lambda|^{\sigma s} + |z|^{(1-\sigma)r} + |\lambda|^{\sigma s}|z|^{(1-\sigma)r})$$
$$\leq C(1 + |\lambda|^s + |z|^r) \leq C'|\Phi_1(\lambda, z)|, \quad (\lambda, z) \in S_\theta \times \Sigma_\delta^n.$$

Now Lemma 1.30 (i) and Proposition 2.60 yield

$${}_0\mathcal{F}^{s'+s}(\mathcal{K}^{r'}) \cap {}_0\mathcal{F}^{s'}(\mathcal{K}^{r'+r}) = D(\Phi_1(\mathcal{D}_+))$$
$$\subseteq D(\Phi_2(\mathcal{D}_+)) = {}_0\mathcal{F}^{s'+\sigma s}(\mathcal{K}^{r'+(1-\sigma)r})$$

as well as the estimate

$$\|\Phi_2(\mathcal{D}_+)u\|_{{}_0\mathcal{F}^{s'}(\mathcal{K}^{r'})} \leq C \cdot \|\Phi_1(\mathcal{D}_+)u\|_{{}_0\mathcal{F}^{s'}(\mathcal{K}^{r'})} \quad (2.42)$$

for all $u \in {}_0\mathcal{F}^{s'+s}(\mathcal{K}^{r'}) \cap {}_0\mathcal{F}^{s'}(\mathcal{K}^{r'+r})$. Then the assertion follows easily by virtue of (2.42) and (2.40).

(ii) (I) For all $i \in \{0,\ldots,M\}$ there exist $(s_1(i), r_1(i)), (s_2(i), r_2(i)) \in N_V$ and $\sigma \in [0,1]$ such that $\sigma r_1(i) + (1-\sigma)r_2(i) = \beta_i$ and $\sigma s_1(i) + (1-\sigma)s_2(i) \geq \alpha_i$. According to (i) we have

$$\bigcap_{k=1,2} {}_0\mathcal{F}^{s'+s_k(i)}(\mathcal{K}^{r'+r_k(i)}) \hookrightarrow {}_0\mathcal{F}^{s'+\sigma s_1(i)+(1-\sigma)s_2(i)}(\mathcal{K}^{r'+\beta_i})$$

$$\hookrightarrow {}_0\mathcal{F}^{s'+\alpha_i}(\mathcal{K}^{r'+\beta_i}) \quad (2.43)$$

and therefore

$$\bigcap_{i=0}^{M} \bigcap_{k=1,2} {}_0\mathcal{F}^{s'+s_k(i)}(\mathcal{K}^{r'+r_k(i)}) \hookrightarrow \bigcap_{i=0}^{M} {}_0\mathcal{F}^{s'+\alpha_i}(\mathcal{K}^{r'+\beta_i}).$$

It is trivial that

$$\bigcap_{(r,s)\in N_V} {}_0\mathcal{F}^{s'+s}(\mathcal{K}^{r'+r}) \hookrightarrow \bigcap_{i=0}^{M}\bigcap_{k=1,2} {}_0\mathcal{F}^{s'+s_k(i)}(\mathcal{K}^{r'+r_k(i)}). \quad (2.44)$$

Using (2.43) and (2.44) we then conclude that

$$\bigcap_{(r,s)\in N_V} {}_0\mathcal{F}^{s'+s}(\mathcal{K}^{r'+r}) \hookrightarrow \bigcap_{i=0}^{M} {}_0\mathcal{F}^{s'+\alpha_i}(\mathcal{K}^{r'+\beta_i}).$$

(II) Let $I := \{i \in \{0,\ldots,M\}\colon (\alpha_i,\beta_i) \in N_V\}$ and $I' := \{0,\ldots,M\}\setminus I$. We obviously have

$$\bigcap_{i=0}^{M} {}_0\mathcal{F}^{s'+\alpha_i}(\mathcal{K}^{r'+\beta_i}) \hookrightarrow \bigcap_{i\in I} {}_0\mathcal{F}^{s'+\alpha_i}(\mathcal{K}^{r'+\beta_i}) \cap \bigcap_{i\in I'} {}_0\mathcal{F}^{s'+\alpha_i}(\mathcal{K}^{r'+\beta_i})$$
$$\hookrightarrow \bigcap_{i\in I} {}_0\mathcal{F}^{s'+\alpha_i}(\mathcal{K}^{r'+\beta_i}) \cap {}_0\mathcal{F}^{s'}(\mathcal{K}^{r'})$$
$$= \bigcap_{(r,s)\in N_V} {}_0\mathcal{F}^{s'+s}(\mathcal{K}^{r'+r}). \qquad \square$$

The next two theorems enable us to derive mapping properties of an operator $P(\mathcal{D}_+)$ if the associated symbol P possesses an upper order function. Note that $\lambda_0(P,\mu) \geq 0$ is given in the definition of an upper order function (Definition 2.27).

Theorem 2.62 (H^∞-calculus for symbols in H_P, part I)**.** Let $\Omega := S_\theta \times \Sigma_\delta^n$ and $\theta > \pi/2$, $\delta > 0$. Let $P \in H_P(\Omega)$ be an arbitrary function with upper convex order function μ satisfying $\alpha_\mu \leq s'$, and let

$$\mu_+(\gamma) := \mu(\gamma) + \alpha_\mu \gamma + \beta_\mu$$

be the associated strictly positive order function, where the definitions of α_μ and β_μ are given in Definition 2.24. We set $s'' := s' - \alpha_\mu \geq 0$ and $r'' := r' - \beta_\mu \in \mathbb{R}$.

Let $\mathcal{D}_+^{(s'',r'')}$ be the ${}_0\mathcal{F}^{s''}(\mathcal{K}^{r''})$-realization of ∇_+, and for $\sigma \geq 0$ let $P_\sigma(\lambda,z) := P(\sigma+\lambda,z)$. Then we have, for all $\sigma \geq \lambda_0(P,\mu)$,

$$D(P_\sigma(\mathcal{D}_+^{(s'',r'')})) \supseteq \bigcap_{\ell=0}^{M} {}_0\mathcal{F}^{s'+m_\ell(\mu)}(\mathcal{K}^{r'+b_\ell(\mu)})$$
$$= \bigcap_{(r,s)\in N_V(\mu_+)} {}_0\mathcal{F}^{s''+s}(\mathcal{K}^{r''+r}) =: \mathcal{V} \quad (2.45)$$

and the restriction of the maximal realization of $P_\sigma(\mathcal{D}_+)$ to $\mathcal{V}$ induces the bounded linear operator

$$P_\sigma(\mathcal{D}_+)|_\mathcal{V} \in L(\mathcal{V}, {}_0\mathcal{F}^{s'}(\mathcal{K}^{r'})).$$

If the case $s' = \alpha_\mu$, we have to choose $\mathcal{F}_\ell = H_{p_0}$ to ensure that ${}_0\mathcal{F}^{s''}(\mathcal{K}^{r''})$ is defined as in Definition 1.71.

2.3. H^∞-calculus of N-parabolic mixed-order systems

Proof. Let $\sigma \geq \lambda_0(P,\mu)$ be sufficiently large. By assumption and Lemma 2.29 we have

$$|P_\sigma(\lambda, z)| \leq C \cdot W_\mu(\lambda, z) = C \cdot \frac{W_{\mu_+}(\lambda, z)}{W_{\{(\beta_\mu, \alpha_\mu)\}}(\lambda, z)} \quad (2.46)$$

for $(\lambda, z) \in \Omega$. Defining $N_+ := N(\mu_+)$ and $N_- := N(\{(\beta_\mu, \alpha_\mu)\})$ we obtain

$$W_{\mu_+}(\lambda, z) = W_{N_+}(\lambda, z) \leq C' \cdot |\Phi_{N_+}(\lambda, z)|, \quad (2.47)$$
$$W_{\{(\beta_\mu, \alpha_\mu)\}}(\lambda, z) = W_{N_-}(\lambda, z) \geq C'' \cdot |\Phi_{N_-}(\lambda, z)|, \quad (\lambda, z) \in \Omega \quad (2.48)$$

by Proposition 2.60. Using (2.46)-(2.48) we get

$$|P_\sigma(\lambda, z)| \leq C \cdot |\Phi_{N_-}^{-1}(\lambda, z)\Phi_{N_+}(\lambda, z)|, \quad (\lambda, z) \in \Omega.$$

Due to Lemma 1.30 (i), Proposition 2.60, and Theorem 1.26 (ii) we then obtain

$$\mathcal{V} = D\bigl(\Phi_{N_+}(\mathcal{D}_+^{(s'',r'')})\bigr) = D\bigl(\Phi_{N_-}^{-1}(\mathcal{D}_+^{(s'',r'')})\Phi_{N_+}(\mathcal{D}_+^{(s'',r'')})\bigr)$$
$$\subseteq D\bigl((\Phi_{N_-}^{-1}\Phi_{N_+})(\mathcal{D}_+^{(s'',r'')})\bigr) \subseteq D\bigl(P_\sigma(\mathcal{D}_+^{(s'',r'')})\bigr)$$

and for $u \in \mathcal{V}$ we get

$$\|P_\sigma(\mathcal{D}_+^{(s'',r'')})u\|_{{}_0\mathcal{F}^{s'}(\mathcal{K}^{r'})}$$
$$= \left\|\left(\frac{P_\sigma}{\Phi_{N_+}\Phi_{N_-}^{-1}}\right)(\mathcal{D}_+^{(s'',r'')}) \underbrace{\Phi_{N_-}^{-1}(\mathcal{D}_+^{(s'',r'')})\Phi_{N_+}(\mathcal{D}_+^{(s'',r'')})u}_{\in {}_0\mathcal{F}^{s'}(\mathcal{K}^{r'})}\right\|_{{}_0\mathcal{F}^{s'}(\mathcal{K}^{r'})}$$
$$\leq C_1 \left\|\Phi_{N_-}^{-1}(\mathcal{D}_+^{(s'',r'')})\Phi_{N_+}(\mathcal{D}_+^{(s'',r'')})u\right\|_{{}_0\mathcal{F}^{s'}(\mathcal{K}^{r'})}$$
$$\leq C_2 \left\|\Phi_{N_+}(\mathcal{D}_+^{(s'',r'')})u\right\|_{{}_0\mathcal{F}^{s''}(\mathcal{K}^{r''})}$$
$$\leq C_3 \|u\|_\mathcal{V}$$

by Proposition 1.35 and Proposition 2.60. We have shown (2.45) and the claimed boundedness. □

Theorem 2.63 (H^∞-**calculus for symbols in** H_P, **part II**). *Let* $\Omega := S_\theta \times \Sigma_\delta^n$, $\theta > \pi/2$, $\delta > 0$, *and let* $P \in H_P(\Omega)$ *be an arbitrary function with upper concave order function* μ *satisfying* $\alpha_{-\mu} \leq s'$ *and*

$$\mu_+(\gamma) := -\mu(\gamma) + \alpha_{-\mu}\gamma + \beta_{-\mu},$$

where the definitions of $\alpha_{-\mu}$ *and* $\beta_{-\mu}$ *are given in Definition 2.21. We define* $s'' := s' - \alpha_{-\mu} \geq 0$, $r'' := r' - \beta_{-\mu} \in \mathbb{R}$. *Let* $\mathcal{D}_+^{(s'',r'')}$ *be the* ${}_0\mathcal{F}^{s''}(\mathcal{K}^{r''})$-*realization of* ∇_+. *Then we get, for all* $\sigma \geq \lambda_0(P,\mu)$,

$$D(P_\sigma(\mathcal{D}_+^{(s'',r'')})) \supseteq {}_0\mathcal{F}^{s'}(\mathcal{K}^{r'}) \quad (2.49)$$

and the restriction of the maximal realization of $P_\sigma(\mathcal{D}_+)$ to $_0\mathcal{F}^{s'}(\mathcal{K}^{r'})$ induces the bounded linear operator

$$P_\sigma(\mathcal{D}_+) \in L(_0\mathcal{F}^{s'}(\mathcal{K}^{r'}), \mathcal{V})$$

where

$$\mathcal{V} := \bigcap_{\ell=0}^{M} {}_0\mathcal{F}^{s'-m_\ell(\mu)}(\mathcal{K}^{r'-b_\ell(\mu)}) = \bigcap_{(r,s)\in N_\mathcal{V}(\mu_+)} {}_0\mathcal{F}^{s''+s}(\mathcal{K}^{r''+r}).$$

Note that we have to choose $\mathcal{F}_\ell = H_{p_0}$ in the case $s' = \alpha_{-\mu}$.

Proof. Let $\sigma \geq \lambda_0(P, \mu)$ be sufficiently large. In a similar way as in the proof of Theorem 2.62 we obtain

$$|P_\sigma(\lambda, z)| \leq C \cdot |\Phi_{N_-}^{-1}(\lambda, z)\Phi_{N_+}(\lambda, z)|, \quad (\lambda, z) \in \Omega$$

where $N_+ := N(\{(\beta_{-\mu}, \alpha_{-\mu})\})$, and $N_- := N(\mu_+)$. Due to Lemma 1.30 (i) and Proposition 2.60 we then obtain

$$_0\mathcal{F}^{s'}(\mathcal{K}^{r'}) = D\big(\Phi_{N_+}(\mathcal{D}_+^{(s'',r'')})\big) = D\big(\Phi_{N_-}^{-1}(\mathcal{D}_+^{(s'',r'')})\Phi_{N_+}(\mathcal{D}_+^{(s'',r'')})\big)$$
$$\subseteq D\big((\Phi_{N_-}^{-1}\Phi_{N_+})(\mathcal{D}_+^{(s'',r'')})\big) \subseteq D\big(P_\sigma(\mathcal{D}_+^{(s'',r'')})\big)$$

and $P_\sigma(\mathcal{D}_+^{(s'',r'')})u \in \mathcal{V}$ for $u \in {}_0\mathcal{F}^{s'}(\mathcal{K}^{r'})$. As in Theorem 2.62 we get

$$\|P_\sigma(\mathcal{D}_+^{(s'',r'')})u\|_\mathcal{V} = \left\|\left(\frac{P_\sigma}{\Phi_{N_+}\Phi_{N_-}^{-1}}\right)(\mathcal{D}_+^{(s'',r'')}) \underbrace{\Phi_{N_-}^{-1}(\mathcal{D}_+^{(s'',r'')})\Phi_{N_+}(\mathcal{D}_+^{(s'',r'')})u}_{\in \mathcal{V}}\right\|_\mathcal{V}$$
$$\leq C_1 \left\|\Phi_{N_-}^{-1}(\mathcal{D}_+^{(s'',r'')})\Phi_{N_+}(\mathcal{D}_+^{(s'',r'')})u\right\|_\mathcal{V}$$
$$\leq C_2 \left\|\Phi_{N_+}(\mathcal{D}_+^{(s'',r'')})u\right\|_{_0\mathcal{F}^{s''}(\mathcal{K}^{r''})}$$
$$\leq C_3 \|u\|_{_0\mathcal{F}^{s'}(\mathcal{K}^{r'})}, \quad u \in {}_0\mathcal{F}^{s'}(\mathcal{K}^{r'}).$$

So we have proved the claim. $\square$

Example 2.64. We define the convex order function

$$\mu(\gamma) := \max\left\{1, \gamma - \frac{1}{2}\right\}, \quad \gamma \geq 0.$$

Then we have $\alpha_\mu = 0$ and $\beta_\mu = 1/2$. If $P \in H_P(S_\theta \times \Sigma_\delta^n)$, $\theta > \pi/2$, is a function such that μ is an upper order function, i.e.,

$$|P(\lambda, z)| \leq C \cdot W_\mu(\lambda, z) = C \cdot \frac{1 + |\lambda| + |z|^{3/2}}{1 + |z|^{1/2}}, \quad |\lambda| \geq \lambda_0(P, \mu)$$

2.3. H^∞-calculus of N-parabolic mixed-order systems

(for example $P(\lambda, z) = z + \lambda(1-z^2)^{-1/4}$, $(\lambda, z) \in S_\theta \times \Sigma_\delta$), then Theorem 2.62 yields
$$\mathcal{V} = L_p(\mathbb{R}_+, H_q^1(\mathbb{R}^n)) \cap {}_0H_p^1(\mathbb{R}_+, H_q^{-1/2}(\mathbb{R}^n))$$
and
$$P_\sigma(\mathcal{D}_+)|_\mathcal{V} \in L(\mathcal{V}, L_p(\mathbb{R}_+, L_q(\mathbb{R}^n))), \quad \sigma \geq \lambda_0(P, \mu).$$

Here we have set $s' := r' := 0$, $\mathcal{F} := H_p$, $\mathcal{K} := H_q$, $1 < p, q < \infty$, and $\sigma \geq \lambda_0(P, \mu)$. Note that here $s'' = 0$ and $r'' = -\frac{1}{2}$.

For a special class of symbols which only depend on λ and $|\xi|$ the next result can be found in [DSS08, Theorem 3.2]. In particular, the quasi-homogeneous coefficients are more specific in [DSS08] as in our situation. Additionally, we obtain the vector-valued result.

Corollary 2.65. *If $P \in S_N(\overline{S}_\theta \times \overline{\Sigma}_\delta^n)$, $\theta > \pi/2$, $\delta > 0$, then there exists $\sigma > 0$ such that*
$$P_\sigma(\mathcal{D}_+) \in L_{\text{Isom}}\Big(\bigcap_{(r,s) \in N_V(P)} {}_0\mathcal{F}^{s'+s}(\mathcal{K}^{r'+r}), {}_0\mathcal{F}^{s'}(\mathcal{K}^{r'})\Big).$$

Proof. From the N-parabolicity we directly obtain $1/P_\sigma \in H^\infty(S_\theta \times \Sigma_\delta^n)$ for sufficiently large $\sigma > 0$. Theorem 1.26 yields that $P_\sigma(\mathcal{D}_+)$ is invertible with $[P_\sigma(\mathcal{D}_+)]^{-1} = P_\sigma^{-1}(\mathcal{D}_+)$. Now we can apply Theorem 2.63 and Lemma 2.29. We obtain
$$\Big(\frac{1}{P_\sigma}\Big)(\mathcal{D}_+) \in L\Big({}_0\mathcal{F}^{s'}(\mathcal{K}^{r'}), \bigcap_{(r,s) \in N_V(P)} {}_0\mathcal{F}^{s'+s}(\mathcal{K}^{r'+r})\Big)$$
and therefore the assertion follows by Theorem 2.62. $\square$

b) Mixed-order systems on spaces of mixed scales

> *Motivation.* The continuity result of the previous subsection is the basis for analog results on mixed-order systems, i.e., matrices of operators where the order of the operator may be different in each entry. Elliptic theory for mixed-order systems or Douglis-Nirenberg systems was developed, e.g., in the papers by Agmon, Douglis, and Nirenberg ([ADN59], [ADN64]) in the 1960's. We start with a simple example of a classical mixed-order system.
>
> The thermoelastic plate equation in the whole space can be written as $\partial_t u = Au$ with the operator matrix
> $$A = (A_{j,k})_{j,k=1,2,3} = \begin{pmatrix} 0 & 1 & 0 \\ -\Delta^2 & 0 & -\Delta \\ 0 & \Delta & \Delta \end{pmatrix}.$$

Setting formally the order of a zero entry to $-\infty$, we obtain the matrix

$$\begin{pmatrix} -\infty & 0 & -\infty \\ 4 & -\infty & 2 \\ -\infty & 2 & 2 \end{pmatrix}$$

which describes the order of each entry. Setting $s := (0, 2, 2)^T$ and $t := (2, 0, 0)^T$, we see that $\operatorname{ord} A_{j,k} \leq s_j + t_k$ holds for $j, k = 1, 2, 3$. Moreover, the matrix A coincides with its principal part, i.e., every non-zero entry $A_{j,k}$ has exactly order $s_j + t_k$. Therefore, this is an example of a classical (elliptic) Douglis-Nirenberg system.

In the situation considered in this section, the operators are matrices $\mathscr{L}(\partial_t, \nabla_x)$ of differential operators in time and space with general inhomogeneous structure. More precisely, instead of real numbers s_j, t_k we now assume that there exist order functions s_j, t_k such that $s_j + t_k$ is an upper order function for the entry $\mathscr{L}_{j,k}$. As (strictly positive) order functions are related to Newton polygons, we essentially assume a Newton polygon structure which may be different in each entry. We will simultaneously consider concave and convex order functions.

Under a suitable N-parabolicity condition, we obtain that $\mathscr{L}(\mathcal{D}_+)$ induces an isomorphism between tuples of Sobolev spaces. As before, there is a large degree of freedom in the choice of these spaces, including spaces of mixed scales. In order to obtain an isomorphism, the spaces have to satisfy a certain compatibility condition which will be made more precise below.

We start with the definition of Douglis-Nirenberg systems and N-parabolicity. For other approaches to mixed-order systems we refer to [DMV98], [DV02a], [DSS09], [DD11], and [DS11] for example. These references include approaches by pseudo-differential operators and parameter-ellipticity. Our procedure can mainly be compared with [DSS08] and [DV08].

Definition 2.66 (Douglis-Nirenberg system). Let $\mathscr{L} \in [H_P(S_\theta \times \Sigma_\delta^n)]^{m \times m}$, $m \in \mathbb{N}$, $\theta > \pi/2$ be a matrix of holomorphic functions which are polynomially bounded. The system $\mathscr{L}$ is called a mixed-order system *in the sense of Douglis-Nirenberg* if there exist order functions $s_j := \mu_j^{\mathrm{row}}$ and $t_k := \mu_k^{\mathrm{col}}$, $j, k = 1, \ldots, m$, such that $s_j + t_k$ is an upper order function of $\mathscr{L}_{j,k}$ for all $j, k = 1, \ldots, m$, i.e., the upper order structure of each component splits into rows and columns.

Similar to [Vol01, Definition 3.3] and [DV08, Definition 3.2], respectively, we introduce the concept of N-parabolic matrices. This useful property enables us to prove that a mixed-order system gives rise to an isomorphism in appropriate spaces, cf. Theorem 2.69 below.

2.3. H^∞-calculus of N-parabolic mixed-order systems

Definition 2.67 (N-parabolic mixed-order system). Let $\mathscr{L} \in [H_P(S_\theta \times \Sigma_\delta^n)]^{m \times m}$, $\theta > \pi/2$, be a mixed-order system in the sense of Douglis-Nirenberg. Then the system $\mathscr{L}$ is called an *N-parabolic* mixed-order system if

(i) $\det \mathscr{L}$ is N-parabolic,

(ii) $[\mu(\det \mathscr{L})](\gamma) = \sum_{j=1}^m (s_j(\gamma) + t_j(\gamma))$ for all $\gamma > 0$.

Note that our definition of an order function is more restrictive than in [DV08] because we assume convexity or concavity in the definition of order functions.

In Chapter 4 we will present a selection of N-parabolic mixed-order systems, which occur in the treatment of parabolic differential equations. In [Vol63], L.R. Volevich showed, by a connection to a linear optimization problem, that for classical Douglis-Nirenberg systems suitable orders s_j, t_j can always be found, see also [Vol01] and [DV08]. In the general situation considered here, it is not clear in which cases there exist convex or concave order functions fulfilling Definition 2.66 and Definition 2.67. In many applications, however, it is transparent how to choose them.

For fixed $1 < p_0, p_1 < \infty$ we consider the maximal realization $\mathscr{L}_\sigma(\mathcal{D}_+)$ of $\mathscr{L}(\sigma + \partial_t, \nabla_x)$, $\sigma \geq 0$, which is given by

$$\mathscr{L}_\sigma(\mathcal{D}_+) := ((\mathscr{L}_\sigma)_{j,k}(\mathcal{D}_+))_{j,k=1,\ldots,m},$$

cf. Definition 1.90. In general, it is not clear how to choose appropriate spaces $\mathbb{H} := \prod_{i=1}^m \mathbb{H}_i$ and $\mathbb{F} := \prod_{i=1}^m \mathbb{F}_i$ such that $\mathscr{L}_\sigma(\mathcal{D}_+)$ acts as a bounded operator between them. Here the point of interest is that $\mathscr{L}_\sigma(\mathcal{D}_+)$ is even an isomorphism between $\mathbb{H}$ and $\mathbb{F}$.

Before stating the main result of this section we want to motivate the embedding assumptions (2.51) and (2.52) appearing in Definition 2.68 below. Up to now we have only considered the mapping properties of operators $P(\mathcal{D}_+)$ on ground spaces of the form $_0\mathcal{F}^\alpha(\mathbb{R}_+, \mathcal{K}^\beta(\mathbb{R}^n, X))$. If we want to consider $P(\mathcal{D}_+)$ as an operator between intersections of spaces with different scales such as

$$_0B_{pp}^{s_1}(\mathbb{R}_+, H_p^{r_1}(\mathbb{R}^n, X)) \cap {}_0H_p^{s_2}(\mathbb{R}_+, B_{pp}^{r_2}(\mathbb{R}^n, X)), \quad s_1, s_2 > 0,\ r_1, r_2 \in \mathbb{R},\ 1 < p < \infty,$$

the scales $(\mathcal{F}_1, \mathcal{K}_1) := (B_{pp}, H_p)$ and $(\mathcal{F}_0, \mathcal{K}_0) := (H_p, B_{pp})$ have to provide some compatibility attributes. The next example demonstrates the necessity of this compatibility assumption:

Let $N := N(\{(1,0),(0,2)\})$ and $P := \Phi_N^{-1}$. Then P has the upper strictly negative order function $\mu(\gamma) = -\max\{\gamma, 2\}$ due to Proposition 2.60. Moreover, we define the spaces

$$\mathbb{H} := \mathbb{H}^{(1)} \cap \mathbb{H}^{(0)} := {}_0B_{pp}^s(\mathbb{R}_+, L_p(\mathbb{R}^n, X)) \cap L_p(\mathbb{R}_+, B_{pp}^{2s}(\mathbb{R}^n, X)),$$

$$\mathbb{F} := \mathbb{F}^{(1)} \cap \mathbb{F}^{(0)} := {}_0B_{pp}^{s+1}(\mathbb{R}_+, L_p(\mathbb{R}^n, X)) \cap L_p(\mathbb{R}_+, B_{pp}^{2(s+1)}(\mathbb{R}^n, X)), \quad s > 0.$$

Considering the order structure of P it seems to be natural that we want to have that $P(\mathcal{D}_+)$ acts as an isomorphism between $\mathbb{H}$ and $\mathbb{F}$. According to Proposition 2.60 we already have

$$P(\mathcal{D}_+)|_{\mathbb{H}^{(1)}} \in L_{\text{Isom}}\left(\mathbb{H}^{(1)}, \mathbb{F}^{(1)} \cap {}_0B^s_{pp}(\mathbb{R}_+, H^2_p(\mathbb{R}^n, X))\right),$$

$$P(\mathcal{D}_+)|_{\mathbb{H}^{(0)}} \in L_{\text{Isom}}\left(\mathbb{H}^{(0)}, {}_0H^1_p(\mathbb{R}_+, B^{2s}_{pp}(\mathbb{R}^n, X)) \cap \mathbb{F}^{(0)}\right).$$

So we deduce

$$P(\mathcal{D}_+)|_{\mathbb{H}} \in L_{\text{Isom}}\left(\mathbb{H}, \mathbb{F} \cap {}_0H^1_p(\mathbb{R}_+, B^{2s}_{pp}(\mathbb{R}^n, X)) \cap {}_0B^s_{pp}(\mathbb{R}_+, H^2_p(\mathbb{R}^n, X))\right).$$

Therefore, we need the embedding

$$\mathbb{F} \hookrightarrow {}_0H^1_p(\mathbb{R}_+, B^{2s}_{pp}(\mathbb{R}^n, X)) \cap {}_0B^s_{pp}(\mathbb{R}_+, H^2_p(\mathbb{R}^n, X))$$

to deduce $P(\mathcal{D}_+)|_{\mathbb{H}} \in L_{\text{Isom}}(\mathbb{H}, \mathbb{F})$.

We summarize the compatibility conditions in the following definition. For this, we recall that an order function μ is characterized by $M \in \mathbb{N}$, $m_\ell(\mu)$, $b_\ell(\mu)$, $\ell = 0, \ldots, M$, and a partition $0 = \gamma_0 < \gamma_1 < \cdots < \gamma_{M+1} = \infty$ such that

$$\mu(\gamma) = m_\ell(\mu)\gamma + b_\ell(\mu) \quad \text{in } [\gamma_\ell, \gamma_{\ell+1})$$

(see Definition 2.21). For strictly positive order functions, the set

$$\nu(\mu) := \{(b_\ell(\mu), m_\ell(\mu)) : \ell = 0, \ldots, M\} \subseteq [0, \infty)^2$$

generates the Newton polygon $N(\mu) := N(\nu(\mu))$ (see Definition 2.22). By a refinement of the partition, we may assume that the number M is the same for all appearing order functions s_j, t_k. The number α_μ related to the order function μ is defined in Definition 2.24.

Definition 2.68 (Compatible tuple of spaces). Let $\mathcal{L} \in [H_P(S_\theta \times \Sigma^n_\delta)]^{m \times m}$, $\theta > \pi/2$, $\delta > 0$, be an N-parabolic mixed-order system with order functions s_j, t_k, $j, k = 1, \ldots, m$. Then a tuple of spaces $\mathbb{H} = \prod_{i=1}^m \mathbb{H}_i$ and $\mathbb{F} = \prod_{j=1}^m \mathbb{F}_j$ is called *compatible* with $\mathcal{L}$ if the following conditions are satisfied:

(i) For $i, j = 1, \ldots, m$, the spaces $\mathbb{H}_i, \mathbb{F}_j$ are of the form

$$\mathbb{H}_i := \bigcap_{\ell=0}^M {}_0\mathcal{F}_\ell^{s'_\ell + m_\ell(t_i)}(\mathcal{K}_\ell^{r'_\ell + b_\ell(t_i)}), \quad \mathbb{F}_j := \bigcap_{\ell=0}^M {}_0\mathcal{F}_\ell^{s'_\ell - m_\ell(s_j)}(\mathcal{K}_\ell^{r'_\ell - b_\ell(s_j)})$$

with $\mathcal{F}_\ell \in \{B_{p_0 q_0}, H_{p_0}\}$, $\mathcal{K}_\ell \in \{B_{p_1 q_1}, H_{p_1}\}$, $s'_\ell \geq 0$, and $r'_\ell \in \mathbb{R}$ ($\ell = 0, \ldots, M$).

2.3. H^∞-calculus of N-parabolic mixed-order systems

(ii) For each $i, j = 1, \ldots, m$, the functions

$$\mu_{\mathbb{H}_i}(\gamma) := \max_\ell \left([s'_\ell + m_\ell(t_i)]\gamma + r'_\ell + b_\ell(t_i) \right), \quad \gamma \geq 0,$$

$$\mu_{\mathbb{F}_j}(\gamma) := \max_\ell \left([s'_\ell - m_\ell(s_j)]\gamma + r'_\ell - b_\ell(s_j) \right), \quad \gamma \geq 0,$$

are convex increasing order functions and

$$s'_\ell \geq \max \left\{ \max_{i=1,\ldots,m} (\delta_{i,1} - m_\ell(t_i)), \max_{j=1,\ldots,m} (\delta_{j,2} + m_\ell(s_j)) \right\} \quad (2.50)$$

holds for all $\ell = 0, \ldots, M$ where

$$\delta_{i,1} := \begin{cases} 0, & \text{if } s_j + t_i \text{ is not concave for all } j \in \{0, \ldots, m\}, \\ \max\{\alpha_{-s_j-t_i} : j \in \{0, \ldots, m\} \text{ such that } s_j + t_i \text{ is concave}\}, & \text{else,} \end{cases}$$

$$\delta_{j,2} := \begin{cases} 0, & \text{if } s_j + t_i \text{ is not convex for all } i \in \{0, \ldots, m\}, \\ \max\{\alpha_{s_j+t_i} : i \in \{0, \ldots, m\} \text{ such that } s_j + t_i \text{ is convex}\}, & \text{else.} \end{cases}$$

Note that for each $\ell \in \{0, \ldots, M\}$ we have to choose $\mathcal{F}_\ell = H_{p_0}$ if equality holds in (2.50).

(iii) Embedding conditions: For $i, j = 1, \ldots, m$ we define

$$\mathbb{H}_{ij} := \bigcap_{\substack{\ell, \kappa = 0, \\ \ell \neq \kappa}}^{M} {}_0\mathcal{F}^{\sigma_{ji}(\ell,\kappa,t)}(\mathcal{K}^{\eta_{ji}(\ell,\kappa,t)}), \quad \mathbb{F}_{ij} := \bigcap_{\substack{\ell, \kappa = 0, \\ \ell \neq \kappa}}^{M} {}_0\mathcal{F}^{\sigma_{ji}(\ell,\kappa,s)}(\mathcal{K}^{\eta_{ji}(\ell,\kappa,s)})$$

where

$$\sigma_{ji}(\ell, \kappa, t) := s'_\ell - m_\ell(s_j) + m_\kappa(s_j) + m_\kappa(t_i),$$
$$\sigma_{ji}(\ell, \kappa, s) := s'_\ell + m_\ell(t_i) - [m_\kappa(s_j) + m_\kappa(t_i)],$$
$$\eta_{ji}(\ell, \kappa, t) := r'_\ell - b_\ell(s_j) + b_\kappa(s_j) + b_\kappa(t_i),$$
$$\eta_{ji}(\ell, \kappa, s) := r'_\ell + b_\ell(t_i) - [b_\kappa(s_j) + b_\kappa(t_i)].$$

Then the embeddings

$$\mathbb{H}_i \hookrightarrow \mathbb{H}_{ij} \quad \text{if } s_j + t_i \text{ is convex}, \quad (2.51)$$
$$\mathbb{F}_j \hookrightarrow \mathbb{F}_{ij} \quad \text{if } s_j + t_i \text{ is concave} \quad (2.52)$$

hold for $i, j = 1, \ldots m$.

In this definition, part (i) states that we have general Newton polygon spaces in each component. Condition (ii) ensures that no space with negative time regularity appears. This condition should be seen as a condition on the orders s'_ℓ, t'_ℓ which should be large enough. The embedding condition (iii) is connected with the fact that we have, in general, mixed scales of Bessel potential and Besov spaces. It is satisfied for many reasonable choices of spaces as appearing in the applications below. The embedding condition will be discussed in detail in the subsequent section. For further information on the assumptions of this definition, see Remark 2.70 below. Now we can state our main result on N-parabolic mixed-order systems.

Theorem 2.69 (Main Theorem on N-parabolic mixed-order systems). Let $\mathscr{L} \in [H_P(S_\theta \times \Sigma_\delta^n)]^{m \times m}$, $\theta > \pi/2$, $\delta > 0$, be an N-parabolic mixed-order system. Let the tuples $\mathbb{H} = \prod_{i=1}^m \mathbb{H}_i$ and $\mathbb{F} = \prod_{j=1}^m \mathbb{F}_j$ be compatible with $\mathscr{L}$ as formulated in Definition 2.68. Then there exists $\sigma_0 > 0$ such that, for all $\sigma \geq \sigma_0$,

$$\mathscr{L}_\sigma(\mathcal{D}_+)\big|_{\mathbb{H}} \in L_{\mathrm{Isom}}(\mathbb{H}, \mathbb{F}) \quad \text{and} \quad \big(\mathscr{L}_\sigma(\mathcal{D}_+)\big|_{\mathbb{H}}\big)^{-1} = \mathscr{L}_\sigma^{-1}(\mathcal{D}_+)\big|_{\mathbb{F}}.$$

Proof. (I) First, we consider the H^∞-calculus of the inverse matrix symbol. According to Definition 2.67 and Lemma 2.29 there exists $\sigma_1 \geq 0$ such that $\det(\mathscr{L}_\sigma(\lambda, z)) \neq 0$ for all tuples $(\lambda, z) \in S_\theta \times \Sigma_\delta^n$ and $\sigma \geq \sigma_1$. So we can define for $\sigma \geq \sigma_1$,

$$\mathscr{L}_\sigma^{-1}(\lambda, z) := (\mathscr{L}_\sigma(\lambda, z))^{-1} = \frac{1}{\det \mathscr{L}_\sigma(\lambda, z)} \cdot \mathscr{L}_\sigma(\lambda, z)^\#, \quad (\lambda, z) \in S_\theta \times \Sigma_\delta^n,$$

where $\mathscr{L}_\sigma(\lambda, z)^\# := ((-1)^{i+j} \cdot \det((\mathscr{L}_\sigma)_{j,i}^\times(\lambda, z)))_{i,j=1,\ldots,m}$. Here $(\mathscr{L}_\sigma)_{j,i}^\times(\lambda, z)$ denotes the $(m-1) \times (m-1)$-matrix that results from deleting row j and column i in $\mathscr{L}_\sigma(\lambda, z)$. For fixed $i, j \in \{1, \ldots, m\}$ we define

$$\mu(\gamma) := -s_j(\gamma) - t_i(\gamma),$$

$$\widetilde{\mu}(\gamma) := \sum_{k=1, k \neq j}^{m+1} s_k(\gamma) + \sum_{k=1, k \neq i}^{m+1} t_k(\gamma), \quad \gamma \geq 0,$$

$$Q_{ij}(\lambda, z) := \det(\mathscr{L})_{j,i}^\times(\lambda, z), \quad (\lambda, z) \in S_\theta \times \Sigma_\delta^n.$$

According to the definition of a Douglis-Nirenberg system, the function $s_k + t_p$ is an upper order function of $\mathscr{L}_{k,p}$ for every $k, p \in \{1, \ldots, m\}$. Let $\mathcal{S}_{ij}$ be the set of all bijective functions $\pi \colon \{1, \ldots, m\} \setminus \{j\} \to \{1, \ldots, m\} \setminus \{i\}$. Then there exists $C' > 0$ such that

$$|Q_{ij}(\lambda, z)| \leq \sum_{\pi \in \mathcal{S}_{ij}} \prod_{k=1, k \neq j}^m |\mathscr{L}_{k, \pi(k)}(\lambda, z)| \leq C' \sum_{\pi \in \mathcal{S}_{ij}} \prod_{k=1, k \neq j}^m W_{s_k + t_{\pi(k)}}(\lambda, z)$$

for $(\lambda, z) \in S_\theta \times \Sigma_\delta^n$ with sufficiently large $|\lambda|$. We also have $\sum_{k=1, k \neq j}^m (s_k + t_{\pi(k)}) = \mu_2$ for all $\pi \in \mathcal{S}_{ij}$. Due to Proposition 2.42 we see that μ is an upper order function of the quotient $(Q_{ij})_\sigma / \det \mathscr{L}_\sigma$ if σ is sufficiently large. Note that we have

$$(b_\ell(\mu), m_\ell(\mu)) = \big(-(b_\ell(s_j) + b_\ell(t_i)), -(m_\ell(s_j) + m_\ell(t_i))\big), \quad \ell = 0, \ldots, M.$$

In the following we have to distinguish two cases:

2.3. H^∞-calculus of N-parabolic mixed-order systems

(a1) Let μ be a concave order function. For $\ell \in \{0,\ldots,M\}$ we define
$$s(\ell) := -(m_\ell(s_j) + m_\ell(t_i)) = m_\ell(\mu),$$
$$r(\ell) := -(b_\ell(s_j) + b_\ell(t_i)) = b_\ell(\mu),$$
$$\mu_\ell(\gamma) := s(\ell) \cdot \gamma + r(\ell) \geq \mu(\gamma), \quad \gamma \geq 0.$$

Due to Lemma 2.34 (i) μ_ℓ is also a concave upper order function of $(Q_{ij})_\sigma / \det \mathscr{L}_\sigma$. If we define
$$s'(\ell) := s'_\ell - m_\ell(s_j) \geq \alpha_{s_j + t_i},$$
$$r'(\ell) := r'_\ell - b_\ell(s_j),$$
$$\mathcal{W}_\ell := {}_0\mathcal{F}^{s'(\ell)}(\mathcal{K}^{r'(\ell)}), \quad \ell = 0,\ldots,M,$$

then we obtain $s'(\ell) \geq \alpha_{s_j+t_i} = \alpha_{-\mu} \geq \alpha_{-\mu_\ell}$. Therefore Theorem 2.63 yields for large σ,
$$[(\mathscr{L}_\sigma^{-1})_{ij}(\mathcal{D}_+)]\Big|_{\mathcal{W}_\ell} \in L\left(\mathcal{W}_\ell, {}_0\mathcal{F}^{s'_\ell + m_\ell(t_i)}(\mathcal{K}^{r'_\ell + b_\ell(t_i)})\right).$$

Due to $\mathbb{F}_j = \bigcap_{\ell=0}^M \mathcal{W}_\ell$ and $\mathbb{H}_i = \bigcap_{\ell=0}^M {}_0\mathcal{F}^{s'_\ell + m_\ell(t_i)}(\mathcal{K}^{r'_\ell + b_\ell(t_i)})$, we then get for large σ,
$$[(\mathscr{L}_\sigma^{-1})_{ij}(\mathcal{D}_+)]\Big|_{\mathbb{F}_j} \in L(\mathbb{F}_j, \mathbb{H}_i).$$

(b1) Let μ be a convex order function. If we define
$$s'(\ell) := s'_\ell + m_\ell(t_i) \geq \alpha_{-s_j - t_i},$$
$$r'(\ell) := r'_\ell + b_\ell(t_i),$$
$$\mathcal{W}_\ell := {}_0\mathcal{F}^{s'(\ell)}(\mathcal{K}^{r'(\ell)}), \quad \ell = 0,\ldots,M,$$

then we have $\alpha_\mu \leq s'(\ell)$ due to $\alpha_{-s_j - t_i} = \alpha_\mu$. Hence Theorem 2.62 yields for large σ and all $\ell = 0,\ldots,M$,
$$[(\mathscr{L}_\sigma^{-1})_{ij}(\mathcal{D}_+)]|_{\mathcal{V}_\ell(\mathbb{F})} \in L(\mathcal{V}_\ell(\mathbb{F}), \mathcal{W}_\ell),$$
$$\mathcal{V}_\ell(\mathbb{F}) := \bigcap_{\kappa=0}^M {}_0\mathcal{F}^{\sigma_{ji}(\ell,\kappa,s)}(\mathcal{K}^{\eta_{ji}(\ell,\kappa,s)}).$$

With $\mathbb{H}_i = \bigcap_{\ell=0}^M \mathcal{W}_\ell$ we then derive for large σ
$$[(\mathscr{L}_\sigma^{-1})_{ij}(\mathcal{D}_+)]|_{\bigcap_{\ell=0}^M \mathcal{V}_\ell(\mathbb{F})} \in L\left(\bigcap_{\ell=0}^M \mathcal{V}_\ell(\mathbb{F}), \mathbb{H}_i\right). \tag{2.53}$$

Here assumption (2.52) comes into play. Due to (2.52) it is obvious that
$$\bigcap_{\ell=0}^M \mathcal{V}_\ell(\mathbb{F}) = \mathbb{F}_j \cap \mathbb{F}_{ij} = \mathbb{F}_j.$$

From (2.53) we deduce for large σ,
$$[(\mathcal{L}_\sigma^{-1})_{ij}(\mathcal{D}_+)]\big|_{\mathbb{F}_j} \in L(\mathbb{F}_j, \mathbb{H}_i).$$

In both cases we derive $[\mathcal{L}_\sigma^{-1}(\mathcal{D}_+)]\big|_{\mathbb{F}} \in L(\mathbb{F}, \mathbb{H})$ for large σ.

(II) Now we determine the mapping properties of the operator $(\mathcal{L}_\sigma)_{ji}(\mathcal{D}_+)$. We already know that $-\mu = s_j + t_i$ is an upper order function of the symbol $\mathcal{L}_{ji}$. As in the previous part we have to distinguish two cases:

(a2) Let $s_j + t_i$ be a convex order function. For $\ell = 0, \ldots, M$ and large σ we obtain
$$[(\mathcal{L}_\sigma)_{ji}(\mathcal{D}_+)]\big|_{\mathcal{V}_\ell(\mathbb{H})} \in L(\mathcal{V}_\ell(\mathbb{H}), \mathcal{W}_\ell),$$
$$\mathcal{V}_\ell(\mathbb{H}) := \bigcap_{\kappa=0}^{M} {}_0\mathcal{F}^{\sigma_{ji}(\ell,\kappa,t)}(\mathcal{K}^{\eta_{ji}(\ell,\kappa,t)})$$

by virtue of Theorem 2.62 with $s'(\ell)$, $r'(\ell)$, and $\mathcal{W}_\ell$ as in (a1). Note that $s'(\ell) \geq \alpha_{s_j+t_i}$. For sufficiently large σ this directly yields
$$[(\mathcal{L}_\sigma)_{ji}(\mathcal{D}_+)]\big|_{\bigcap_{\ell=0}^M \mathcal{V}_\ell(\mathbb{H})} \in L\Big(\bigcap_{\ell=0}^M \mathcal{V}_\ell(\mathbb{H}), \mathbb{F}_j\Big)$$

because of $\mathbb{F}_j = \bigcap_{\ell=0}^M \mathcal{W}_\ell$. Due to (2.51) we can see that
$$\bigcap_{\ell=0}^M \mathcal{V}_\ell(\mathbb{H}) = \bigcap_{\ell,\kappa=0}^M {}_0\mathcal{F}^{\sigma_{ji}(\ell,\kappa,t)}(\mathcal{K}^{\eta_{ji}(\ell,\kappa,t)}) = \mathbb{H}_{ji} \cap \mathbb{H}_i = \mathbb{H}_i$$

and therefore $[(\mathcal{L}_\sigma)_{ji}(\mathcal{D}_+)]\big|_{\mathbb{H}_i} \in L(\mathbb{H}_i, \mathbb{F}_j)$ for sufficiently large σ.

(b2) Let $s_j + t_i$ be a concave order function. Let
$$s(\ell) := m_\ell(s_j) + m_\ell(t_i),$$
$$r(\ell) := b_\ell(s_j) + b_\ell(t_i),$$
$$\mu_\ell(\gamma) := s(\ell) \cdot \gamma + r(\ell) \geq (s_j + t_i)(\gamma), \quad \gamma \geq 0.$$

Then μ_ℓ is also a concave upper order function of $\mathcal{L}_{ji}$, cf. Lemma 2.34 (i). Now Theorem 2.63 with $s'(\ell)$, $r'(\ell)$, and $\mathcal{W}_\ell$ as in (b1) yields
$$[(\mathcal{L}_\sigma)_{ji}(\mathcal{D}_+)]\big|_{\mathcal{W}_\ell} \in L\Big(\mathcal{W}_\ell, {}_0\mathcal{F}^{s'_\ell - m_\ell(s_j)}(\mathcal{K}^{r'_\ell - b_\ell(s_j)})\Big)$$

due to $s'(\ell) \geq \alpha_{-\mu_1} \geq \alpha_{-\mu_\ell}$. For sufficiently large σ we then derive
$$[(\mathcal{L}_\sigma)_{ji}(\mathcal{D}_+)]\big|_{\mathbb{H}_i} \in L(\mathbb{H}_i, \mathbb{F}_j).$$

2.3. H^∞-calculus of N-parabolic mixed-order systems

Altogether we get for large σ,

$$\left[\mathscr{L}_\sigma^{-1}(\boldsymbol{D}_+)\right]\Big|_\mathbb{F} \in L(\mathbb{F},\mathbb{H}), \quad \left[\mathscr{L}_\sigma(\boldsymbol{D}_+)\right]\Big|_\mathbb{H} \in L(\mathbb{H},\mathbb{F}).$$

Therefore, we can compose these operators. Using Theorem 1.26 we obtain

$$\mathscr{L}_\sigma^{-1}(\boldsymbol{D}_+)\mathscr{L}_\sigma(\boldsymbol{D}_+)u = \left(\sum_{k,j=1}^m (\mathscr{L}_\sigma^{-1})_{ik}(\boldsymbol{D}_+)(\mathscr{L}_\sigma)_{kj}(\boldsymbol{D}_+)u_j\right)_{i=1,\ldots,m}$$

$$= \left(\sum_{j=1}^m \underbrace{\left[\sum_{k=1}^m (\mathscr{L}_\sigma^{-1})_{ik}(\mathscr{L}_\sigma)_{kj}\right]}_{=\delta_{ij}}(\boldsymbol{D}_+)u_j\right)_{i=1,\ldots,m}$$

$$= (u_i)_{i=1,\ldots,m} = u$$

for all $u \in \mathbb{H}$. In exactly the same way we derive $\mathscr{L}_\sigma(\boldsymbol{D}_+)\mathscr{L}_\sigma^{-1}(\boldsymbol{D}_+)v = v$ for all $v \in \mathbb{F}$. This yields $([\mathscr{L}_\sigma(\boldsymbol{D}_+)]|_\mathbb{H})^{-1} = [\mathscr{L}_\sigma^{-1}(\boldsymbol{D}_+)]|_\mathbb{F}$, which ends the proof. □

Remark 2.70. (i) In the next subsection, especially in Proposition 2.72 and Proposition 2.79, we will give sufficient conditions for the embedding conditions (2.52) and (2.51) in case of a "tame choice" of the scales $(\mathcal{F}_\ell, \mathcal{K}_\ell)_{\ell=0,\ldots,M}$.

(ii) If we only have positive and negative order functions $s_j + t_i$ in the situation of Theorem 2.69, then the assumptions become simpler. If we additionally have a "tame choice" of scales $(\mathcal{F}_\ell, \mathcal{K}_\ell)$, the assumptions (2.52) and (2.51) can also be dropped. This special case is sufficient for many applications, and for the sake of clarity we will state this in Corollary 2.80.

(iv) The results of Theorem 2.69 also hold in exponentially weighted spaces as can be easily seen using Remark 1.91. With the help of exponential weights, we can eliminate the translation σ in the time co-variable in the situation of Theorem 2.69. For this we define the spaces

$$\mathbb{H}_i(\varrho) := \bigcap_{\ell=0}^M {}_0\mathcal{F}_\varrho^{s'_\ell+m_\ell(t_i)}(\mathbb{R}_+, \mathcal{K}_\ell^{r'_\ell+b_\ell(t_i)}(\mathbb{R}^n, X)),$$

$$\mathbb{F}_j(\varrho) := \bigcap_{\ell=0}^M {}_0\mathcal{F}_\varrho^{s'_\ell-m_\ell(s_j)}(\mathbb{R}_+, \mathcal{K}_\ell^{r'_\ell-b_\ell(s_j)}(\mathbb{R}^n, X)),$$

$$\mathbb{H}(\varrho) := \prod_{i=0}^m \mathbb{H}_i(\sigma),$$

$$\mathbb{F}(\varrho) := \prod_{j=0}^m \mathbb{F}_j(\sigma)$$

where $\varrho \geq 0$ is an exponential weight (see Remark 1.70 and Remark 1.91). Theorem 2.69 and Remark 1.91 already yield

$$[\mathscr{L}(\mathcal{D}_{+,\sigma}^{(\varrho)})]|_{\mathbb{H}(\varrho)} = [\mathscr{L}_\sigma(\mathcal{D}_+^{(\varrho)})]|_{\mathbb{H}(\varrho)} \in L_{\text{Isom}}(\mathbb{H}(\varrho), \mathbb{F}(\varrho)), \quad \sigma \geq \sigma_0.$$

According to Theorem 2.69 and Remark 1.91 we then derive

$$\left[\mathscr{L}(\mathcal{D}_+^{(\varrho+\sigma)})\right]\bigg|_{\mathbb{H}(\varrho+\sigma)} = \left[\mathscr{M}_\sigma^{-1} \mathscr{L}_\sigma(\mathcal{D}_+^{(\varrho)}) \mathscr{M}_\sigma\right]\bigg|_{\mathbb{H}(\varrho+\sigma)}.$$

Thus, we obtain

$$\left[\mathscr{L}(\mathcal{D}_+^{(\varrho+\sigma)})\right]\bigg|_{\mathbb{H}(\varrho+\sigma)} \in L_{\text{Isom}}(\mathbb{H}(\varrho+\sigma), \mathbb{F}(\varrho+\sigma)),$$

$$\left(\left[\mathscr{L}(\mathcal{D}_+^{(\varrho+\sigma)})\right]\bigg|_{\mathbb{H}(\varrho+\sigma)}\right)^{-1} = \left[\mathscr{M}_\sigma^{-1}(\mathscr{L}_\sigma^{-1})(\mathcal{D}_+^{(\varrho)}) \mathscr{M}_\sigma\right]\bigg|_{\mathbb{F}(\varrho+\sigma)}$$

due to Remark 1.91. Note that in general $\mathscr{L}^{-1}(\mathcal{D}_+^{(\varrho+\sigma)})$ is not well-defined because we only have

$$\det \mathscr{L}(\lambda, z) \neq 0$$

for sufficiently large $|\lambda|$. We will use this formulation of the main theorem for the applications in Chapter 4.

Remark 2.71. In Section 4.7, we will need a generalization of the above results where the matrix $\mathscr{L}$ depends additionally on a parameter ϑ varying in a compact set $K \subseteq \mathbb{C}^m$. This generalization is straight-forward, provided that the symbol depends continuously on ϑ, and that the structure of the Newton polygon does not depend on ϑ. For instance, if $P[\vartheta] \in S(L_t \times L_x)$ for each $\vartheta \in K$, one condition is that $d_\gamma(P[\vartheta]) = d_\gamma(P[\vartheta'])$ for all $\vartheta, \vartheta' \in K$. With this, one can define the class $S_N[K](L_t \times L_x)$ of all symbols which are N-parameter-elliptic with compact parameter. All results above carry over to this case, where now all estimates hold uniformly in the parameter. In particular, in the situation of Theorem 2.69, the mappings $\vartheta \mapsto \mathscr{L}_\sigma[\vartheta](\mathcal{D}_+)$ as well as $\vartheta \mapsto (\mathscr{L}_\sigma[\vartheta](\mathcal{D}_+))^{-1}$ are continuous, and the norm of the inverse matrix $(\mathscr{L}_\sigma[\vartheta](\mathcal{D}_+))^{-1}$ can be estimated independently of ϑ. For details of this generalization, we refer to [Kai12].

c) Remarks on the compatibility condition

Motivation. In formulation of the main result of the previous subsection, Theorem 2.69, the compatibility conditions should be discussed in detail. As already noted, conditions (i) and (ii) in Definition 2.68 are quite natural: While in (i) the structure of the spaces is formulated in the form of a Newton polygon space (where the inhomogeneous structure may be different in each component), condition (ii) avoids negative orders in the time derivative. The

2.3. H^∞-calculus of N-parabolic mixed-order systems

> parameters r'_ℓ and s'_ℓ appearing in (ii) are degrees of freedom which should be chosen in accordance to the given mixed-order system.
>
> However, the embedding conditions (iii) in Definition 2.68 are connected with the different types of Sobolev spaces which are considered here. In many applications, the embedding conditions are satisfied automatically. This is the case, e.g., if the spaces are intersections of mixed scales of uniform type, i.e., if $\mathcal{F}_\ell$ and $\mathcal{K}_\ell$ are independent of ℓ. In more general situations one may have intersections of the form $B_{p_0 q_0}^{s_0}(H_{q_0}^{r_0}) \cap H_{p_1}^{s_1}(B_{p_1 q_1}^{r_1})$. In such cases rather deep results on the embedding of Banach space valued Sobolev spaces are needed.

Throughout this subsection, let X be a Banach space of class $\mathcal{HT}$ with property (α). For $s \geq 0$, $r \in \mathbb{R}$ and exponential weight $\varrho \geq 0$ we will again use the abbreviation

$$_0\mathcal{F}_\varrho^s(\mathcal{K}^r) := {}_0\mathcal{F}_\varrho^s(\mathbb{R}_+, \mathcal{K}^r(\mathbb{R}^n, X)).$$

The next pages are devoted to illuminating the compatibility conditions as formulated in Definition 2.68. In most cases this is only a condition on the position of the regularity tuples $(\eta_{ji}(\ell, \kappa, s), \sigma_{ji}(\ell, \kappa, s))$ and $(\eta_{ji}(\ell, \kappa, t), \sigma_{ji}(\ell, \kappa, t))$ for $i, j = 1, \ldots, m$ and $\ell, \kappa = 0, \ldots, M$.

First, we investigate (2.51) and (2.52) for non-mixed scales. The next proposition clarifies this situation.

Proposition 2.72 (Compatibility condition I). *Assume that in the situation of Definition 2.68, the tuples $\mathbb{H} = \prod_{i=1}^m \mathbb{H}_i$ and $\mathbb{F} = \prod_{j=1}^m \mathbb{F}_j$ satisfy conditions 2.68 (i) and (ii). If additionally $(\mathcal{F}_\ell, \mathcal{K}_\ell) = (\mathcal{F}, \mathcal{K})$ for all $\ell \in \{0, \ldots, M\}$, then embedding conditions (2.51) and (2.52) are always fulfilled, and $\mathbb{H}, \mathbb{F}$ is compatible to $\mathscr{L}$.*

Proof. Let $i, j \in \{1, \ldots, m\}$. Without loss of generality, let $s_j + t_i$ be convex. We define $\mu_1(\gamma) := \mu_{\mathbb{F}_j}(\gamma) + \beta_1$, $\mu_2(\gamma) := (s_j + t_i)(\gamma) + \alpha\gamma + \beta_2$ with $\beta_1 := \beta_{\mu_{\mathbb{F}_j}}$, $\alpha := \alpha_{s_j + t_i}$, and $\beta_2 := \beta_{s_j + t_i}$. Note that we have

$$\mu_{\mathbb{H}_i}(\gamma) = \mu_{\mathbb{F}_j}(\gamma) + (s_j + t_i)(\gamma)$$
$$= (\mu_1 + \mu_2)(\gamma) - \alpha\gamma - (\beta_1 + \beta_2).$$

For fixed $\ell, \kappa \in \{0, \ldots, M\}$ we define

$$\sigma := \sigma_{ji}(\ell, \kappa, t) = m_\ell(\mu_{\mathbb{F}_j}) + m_\kappa(s_j + t_i),$$
$$\eta := \eta_{ji}(\ell, \kappa, t) = b_\ell(\mu_{\mathbb{F}_j}) + b_\kappa(s_j + t_i).$$

Then $\sigma + \alpha = m_\ell(\mu_1) + m_\kappa(\mu_2)$ and $\eta + \beta_1 + \beta_2 = b_\ell(\mu_1) + b_\kappa(\mu_2)$ hold. Lemma 2.35 directly yields $(\eta + \beta_1 + \beta_2, \sigma + \alpha) \in N(\mu_1 + \mu_2)$. Using Lemma 2.61 we get

$$\bigcap_{\ell=0,\ldots,M} {}_0\mathcal{F}_\varrho^{m_\ell(\mu_1 + \mu_2)}(\mathcal{K}^{b_\ell(\mu_1 + \mu_2)}) \hookrightarrow {}_0\mathcal{F}_\varrho^{\sigma + \alpha}(\mathcal{K}^{\eta + \beta_1 + \beta_2}).$$

Due to (2.50) we have $m_\ell(\mu_1 + \mu_2) > 0$ for $\ell = 0, \ldots, M$ if $\mathcal{F} = B_{p_0 q_0}$. Let $N' := N(\{(\alpha, \beta_1 + \beta_2)\})$. Then the isomorphism $\Phi_{N'}(\mathcal{D}_+^{(\varrho)})$ yields $\mathbb{H}_i \hookrightarrow {}_0\mathcal{F}^\sigma(\mathcal{K}^\eta)$ and therefore $\mathbb{H}_i \hookrightarrow \mathbb{H}_{ij}$.

The embedding $\mathbb{F}_j \hookrightarrow \mathbb{F}_{ij}$ follows in the same way. □

In contrast to Proposition 2.72 we are also interested in spaces like
$$ {}_0 B_{pp,\varrho}^s(\mathbb{R}_+, L_p(\mathbb{R}^n, X)) \cap L_{p,\varrho}(\mathbb{R}_+, B_{pp}^{2s}(\mathbb{R}^n, X)), \quad s > 0, \ p \in (1, \infty) $$
to handle boundary value problems where the mixed-order system acts on the trace spaces. To get analog results on the assumptions (2.51) and (2.52) as in Proposition 2.72 we have to provide analog results as in Lemma 2.61 (i) of the form
$$ {}_0\mathcal{F}^{s'+s}(\mathcal{K}^{r'}) \cap {}_0\mathcal{L}^{s'}(\mathcal{M}^{r'+r}) \hookrightarrow {}_0\mathcal{J}^{s'+\sigma s}(\mathcal{I}^{r'+(1-\sigma)r}). $$

To obtain further embeddings as in Lemma 2.61 (i) we need the compatibility of the real interpolation method with the intersection of Banach spaces. There are several results for the so-called "intersection problem" if the underlying Banach spaces are quasi-linearizable, see [Pee71], [Pee74], [Tri78]. The next result goes back to P. Grisvard and can be found in [Gri72]. J. Peetre showed in [Pee74] that Grisvard's result is a special case of the more general result in [Pee74]. In fact, P. Grisvard's result on the intersection problem holds without the assumption of quasi-linearizability.

Remark 2.73 (Intersection problem, see [Gri72]). Let Z and Y be Banach spaces with $Y \hookrightarrow Z$ and let
$$ A \colon Z \supseteq D(A) \to Z $$
be a linear closed unbounded operator. Let the following conditions be fulfilled:

(i) It holds that $(-\infty, 0) \subseteq \rho(A)$ and there exists $C_0 > 0$ such that
$$ \|t(t - A)^{-1} u\|_Z \leq C_0 \|u\|_Z, \quad t < 0, \ u \in Z. $$

(ii) The resolvent $(t - A)^{-1}$ is Y-invariant for all $t < 0$ and there exists $C_1 > 0$ such that
$$ \|t(t - A)^{-1} u\|_Y \leq C_1 \|u\|_Y, \quad t < 0, \ u \in Y. $$

Then we obtain
$$ (Z, Y \cap D(A))_{\theta, p} = (Z, Y)_{\theta, p} \cap (Z, D(A))_{\theta, p} $$
with equivalent norms for all $\theta \in (0, 1)$ and $p \in (1, \infty)$.

Lemma 2.74. *For all $s', s, r \geq 0$, $r' \in \mathbb{R}$, $\varrho \geq 0$, $p_0, p_1 \in (1, \infty)$ we have*
$$ \left({}_0 H_{p_0}^{s'}(H_{p_1}^{r'}), \ {}_0 H_{p_0}^{s+s'}(H_{p_1}^{r'}) \cap {}_0 H_{p_0}^{s'}(H_{p_1}^{r+r'}) \right)_{\theta, p} $$
$$ = \left({}_0 H_{p_0}^{s'}(H_{p_1}^{r'}), \ {}_0 H_{p_0}^{s+s'}(H_{p_1}^{r'}) \right)_{\theta, p} \cap \left({}_0 H_{p_0}^{s'}(H_{p_1}^{r'}), \ {}_0 H_{p_0}^{s'}(H_{p_1}^{r+r'}) \right)_{\theta, p} $$
where $\theta \in (0, 1)$ and $p \in (1, \infty)$.

2.3. H^∞-calculus of N-parabolic mixed-order systems

Proof. For $s = 0$ or $r = 0$ the assertion is trivial. Let $s, r > 0$, $Z := {}_0H^{s'}_{p_0}(H^{r'}_{p_1})$, $Y := {}_0H^{s'+s}_{p_0}(H^{r'}_{p_1})$, and let $A := \Lambda_r(\mathcal{D}^Z_+)$ where $\mathcal{D}^Z_+$ stands for the Z-realization of ∇_+. According to Proposition 1.92 we have $D(A) = {}_0H^{s'}_{p_0}(H^{r'+r}_{p_1})$ with equivalent norms. In the following we show that we can apply Remark 2.73.

(i) There exist $\delta_0(r) > 0$ and $\tau(r) > 0$ such that for all $t < 0$, $\delta \leq \delta_0(r)$, and $z \in \Sigma^n_\delta$ we have $t - \Lambda_r(z) \in -(\tau(r) + S_{\pi/2})$. In particular, this yields $\varphi_t := (t - \Lambda_r)^{-1} \in H^\infty(\Sigma^n_\delta)$. With Theorem 1.26 we then obtain $t \in \rho(A)$ and $(t - A)^{-1} = \varphi_t(\mathcal{D}^Z_+)$ for all $t < 0$. Using a homogeneity argument we derive the boundedness of the function

$$\chi(t, z) := t\varphi_t(z), \quad (t, z) \in (-\infty, 0) \times \Sigma^n_\delta.$$

Therefore, there exists $C > 0$ such that $\|\chi(t, \cdot)\|_\infty \leq C$ for all $t < 0$. We get $\|\chi(t, \mathcal{D}^Z_+)\|_{L(Z)} \leq C$ for all $t < 0$. Altogether we obtain $\|t(t - A)^{-1}u\|_Z = \|\chi(t, \mathcal{D}^Z_+)u\|_Z \leq C\|u\|_Z$ for all $t < 0$ and $u \in Z$.

(ii) The assertion of (i) is also true for the operator $B := \Lambda_r(\mathcal{D}^Y_+)$ and due to the pointwise definition of the natural extension, cf. Definition and Lemma 1.85, we easily see that $(t - A)^{-1}|_Y = (t - B)^{-1} \in L(Y)$. Altogether we get $\|t(t - A)^{-1}u\|_Y \leq C\|u\|_Y$ for all $t < 0$ and $u \in Y$. □

Special cases of the next embedding result can be found in [EPS03, Remark 5.3] and [MS12, Proposition 3.2], for example. However, the authors of [MS12] consider spaces with a type of polynomial weight in the time domain. In [MS12, Proposition 3.2] the authors impose the additional restriction $s \in (0, 2)$, which appears because they consider powers of the time derivative $(1 - \partial_t)^s$. These powers cannot be controlled for large s. Some arguments of the proofs given there are related to arguments used here but we use the abstract interpolation result of Remark 2.73, respectively Lemma 2.74, instead of powers of the time derivative. With this approach we can drop the restriction on s.

Proposition 2.75 (Embedding result II). *For all $s' \geq 0$, $r' \in \mathbb{R}$, $s, r > 0$, $\varrho \geq 0$, $1 < p_0, p_1 < \infty$, and $\sigma \in (0, 1)$ we have the embedding*

$${}_0B^{s'+s}_{p_0}(H^{r'}_{p_1}) \cap {}_0H^{s'}_{p_0}(B^{r'+r}_{p_1 p_0}) \hookrightarrow {}_0B^{s'+\sigma s}_{p_0}(H^{r'+(1-\sigma)r}_{p_1}) \cap {}_0H^{s'+\sigma s}_{p_0}(B^{r'+(1-\sigma)r}_{p_1 p_0}).$$

Proof. Using Proposition 1.69 and Remark 1.67 (ii) we obtain

$$\left[{}_0H^{s'}_{p_0}(H^{r'}_{p_1}), {}_0H^{s'+s+\varepsilon}_{p_0}(H^{r'}_{p_1}) \cap {}_0H^{s'}_{p_0}(H^{r'+r+\delta}_{p_1})\right]_\theta$$
$$\hookrightarrow \left[{}_0H^{s'}_{p_0}(H^{r'}_{p_1}), {}_0H^{s'+s+\varepsilon}_{p_0}(H^{r'}_{p_1})\right]_\theta \cap \left[{}_0H^{s'}_{p_0}(H^{r'}_{p_1}), {}_0H^{s'}_{p_0}(H^{r'+r+\delta}_{p_1})\right]_\theta$$
$$= {}_0H^{s'+s-\varepsilon}_{p_0}(H^{r'}_{p_1}) \cap {}_0H^{s'}_{p_0}(H^{r'+r-\delta}_{p_1})$$

where $\varepsilon > 0$, $\delta := \varepsilon r/s$, and $\theta := (s-\varepsilon)/(s+\varepsilon)$. Due to the reiteration theorem (Theorem 1.40) we get

$$\left(\left[{}_0H_{p_0}^{s'}(H_{p_1}^{r'}), {}_0H_{p_0}^{s'+s+\varepsilon}(H_{p_1}^{r'}) \cap {}_0H_{p_0}^{s'}(H_{p_1}^{r'+r+\delta})\right]_\theta,\right.$$
$$\left.{}_0H_{p_0}^{s'+s+\varepsilon}(H_{p_1}^{r'}) \cap {}_0H_{p_0}^{s'}(H_{p_1}^{r'+r+\delta})\right)_{\frac{1}{2},p_0}$$
$$= \left({}_0H_{p_0}^{s'}(H_{p_1}^{r'}), {}_0H_{p_0}^{s'+s+\varepsilon}(H_{p_1}^{r'}) \cap {}_0H_{p_0}^{s'}(H_{p_1}^{r'+r+\delta})\right)_{\eta,p_0}$$

for $\eta := (1-\frac{1}{2})\theta + \frac{1}{2}\cdot 1 = s/(s+\varepsilon)$. Using Lemma 2.74, Remark 1.67 (ii), and Proposition 1.69 we get

$$\left({}_0H_{p_0}^{s'}(H_{p_1}^{r'}), {}_0H_{p_0}^{s'+s+\varepsilon}(H_{p_1}^{r'}) \cap {}_0H_{p_0}^{s'}(H_{p_1}^{r'+r+\delta})\right)_{\eta,p_0} = {}_0B_{p_0}^{s'+s}(H_{p_1}^{r'}) \cap {}_0H_{p_0}^{s'}(B_{p_1p_0}^{r'+r}).$$

Altogether we derive the embedding

$${}_0B_{p_0}^{s'+s}(H_{p_1}^{r'}) \cap {}_0H_{p_0}^{s'}(B_{p_1p_0}^{r'+r}) \tag{2.54}$$
$$\hookrightarrow \left({}_0H_{p_0}^{s'+s-\varepsilon}(H_{p_1}^{r'}) \cap {}_0H_{p_0}^{s'}(H_{p_1}^{r'+r-\delta}), {}_0H_{p_0}^{s'+s+\varepsilon}(H_{p_1}^{r'}) \cap {}_0H_{p_0}^{s'}(H_{p_1}^{r'+r+\delta})\right)_{\frac{1}{2},p_0}.$$

Lemma 2.61 yields the embeddings

$${}_0H_{p_0}^{s'+s-\varepsilon}(H_{p_1}^{r'}) \cap {}_0H_{p_0}^{s'}(H_{p_1}^{r'+r-\delta}) \hookrightarrow {}_0H_{p_0}^{s'+\sigma s-\varepsilon}(H_{p_1}^{r'+(1-\sigma)r}),$$
$${}_0H_{p_0}^{s'+s+\varepsilon}(H_{p_1}^{r'}) \cap {}_0H_{p_0}^{s'}(H_{p_1}^{r'+r+\delta}) \hookrightarrow {}_0H_{p_0}^{s'+\sigma s+\varepsilon}(H_{p_1}^{r'+(1-\sigma)r}).$$

With these embeddings and (2.54) we obtain

$${}_0B_{p_0}^{s'+s}(H_{p_1}^{r'}) \cap {}_0H_{p_0}^{s'}(B_{p_1p_0}^{r'+r})$$
$$\hookrightarrow \left({}_0H_{p_0}^{s'+\sigma s-\varepsilon}(H_{p_1}^{r'+(1-\sigma)r}), {}_0H_{p_0}^{s'+\sigma s+\varepsilon}(H_{p_1}^{r'+(1-\sigma)r})\right)_{\frac{1}{2},p_0}$$
$$= {}_0B_{p_0}^{s'+\sigma s}(H_{p_1}^{r'+(1-\sigma)r}),$$

which yields the first claimed embedding.

The second embedding can be obtained in the same way by using (2.54) and the embeddings

$${}_0H_{p_0}^{s'+s-\varepsilon}(H_{p_1}^{r'}) \cap {}_0H_{p_0}^{s'}(H_{p_1}^{r'+r-\delta}) \hookrightarrow {}_0H_{p_0}^{s'+\sigma s}(H_{p_1}^{r'+(1-\sigma)r-\delta}),$$
$${}_0H_{p_0}^{s'+s+\varepsilon}(H_{p_1}^{r'}) \cap {}_0H_{p_0}^{s'}(H_{p_1}^{r'+r+\delta}) \hookrightarrow {}_0H_{p_0}^{s'+\sigma s}(H_{p_1}^{r'+(1-\sigma)r+\delta}). \qquad \square$$

Note that we have a linking between the outer scale H_{p_0} and the inner scale $B_{p_1p_0}$ due to the occurrence of p_0 in $B_{p_1p_0}$. This coupling always appears if we apply Proposition 1.69.

2.3. H^∞-calculus of N-parabolic mixed-order systems

Proposition 2.76 (Embedding result III). *For all $s' > 0$, $r' \in \mathbb{R}$, $s, r \geq 0$, $\varrho \geq 0$, $1 < p_0, p_1 < \infty$, and $\sigma \in (0,1)$ we have the embeddings*

$$_0B_{p_0,\varrho}^{s'+s}(H_{p_1}^{r'}) \cap {}_0B_{p_0,\varrho}^{s'}(H_{p_1}^{r'+r}) \hookrightarrow {}_0H_{p_0,\varrho}^{s'+\sigma s}(B_{p_1 p_0}^{r'+(1-\sigma)r}), \qquad (2.55)$$

$$_0H_{p_0,\varrho}^{s'+s}(B_{p_1 p_0}^{r'}) \cap {}_0H_{p_0,\varrho}^{s'}(B_{p_1 p_0}^{r'+r}) \hookrightarrow {}_0B_{p_0,\varrho}^{s'+\sigma s}(H_{p_1}^{r'+(1-\sigma)r}). \qquad (2.56)$$

Proof. We only prove (2.55), the proof of (2.56) follows the same ideas. We also restrict ourselves to the case $\varrho = 0$. Let $\varepsilon > 0$ and $\delta := \varepsilon r/s$ and define the Newton polygon $N := N(\{(s,0),(0,r)\})$. For $\mathcal{V} := {}_0B_{p_0}^{s'}(H_{p_1}^{r'})$ and $\mathcal{W}_\pm := {}_0H_{p_0}^{s' \pm \varepsilon}(H_{p_1}^{r'})$ we use the results of Proposition 2.60 on the isomorphisms $\Phi_N(\mathcal{D}_+^{\mathcal{W}_\pm})$ where again $\mathcal{D}_+^{\mathcal{W}_\pm}$ stands for the $\mathcal{W}_\pm$-realization of ∇_+. Due to $\mathcal{W}_+, \mathcal{V} \hookrightarrow \mathcal{W}_-$ we have

$$\left[\Phi_N(\mathcal{D}_+^{\mathcal{W}_-})^{-1}\right]\Big|_{\mathcal{W}_+} = \Phi_N(\mathcal{D}_+^{\mathcal{W}_+})^{-1}, \quad \left[\Phi_N(\mathcal{D}_+^{\mathcal{W}_-})^{-1}\right]\Big|_{\mathcal{V}} = \Phi_N(\mathcal{D}_+^{\mathcal{V}})^{-1}. \quad (2.57)$$

By (2.57) and Lemma 1.52 we get

$$\left({}_0H_{p_0}^{s'+s-\varepsilon}(H_{p_1}^{r'}) \cap {}_0H_{p_0}^{s'-\varepsilon}(H_{p_1}^{r'+r}), \; {}_0H_{p_0}^{s'+s+\varepsilon}(H_{p_1}^{r'}) \cap {}_0H_{p_0}^{s'+\varepsilon}(H_{p_1}^{r'+r})\right)_{1/2, p_0}$$

$$= \left(\Phi_N(\mathcal{D}_+^{\mathcal{W}_-})^{-1}({}_0H_{p_0}^{s'-\varepsilon}(H_{p_1}^{r'})), \; \Phi_N(\mathcal{D}_+^{\mathcal{W}_+})^{-1}({}_0H_{p_0}^{s'+\varepsilon}(H_{p_1}^{r'}))\right)_{1/2, p_0}$$

$$= \left(\Phi_N(\mathcal{D}_+^{\mathcal{W}_-})^{-1}({}_0H_{p_0}^{s'-\varepsilon}(H_{p_1}^{r'})), \; \Phi_N(\mathcal{D}_+^{\mathcal{W}_-})^{-1}({}_0H_{p_0}^{s'+\varepsilon}(H_{p_1}^{r'}))\right)_{1/2, p_0}$$

$$= \Phi_N(\mathcal{D}_+^{\mathcal{W}_-})^{-1}\left(\left({}_0H_{p_0}^{s'-\varepsilon}(H_{p_1}^{r'}), \; {}_0H_{p_0}^{s'+\varepsilon}(H_{p_1}^{r'})\right)_{1/2, p_0}\right)$$

$$= \Phi_N(\mathcal{D}_+^{\mathcal{W}_-})^{-1}\left({}_0B_{p_0 p_0}^{s'}(H_{p_1}^{r'})\right) = \Phi_N(\mathcal{D}_+^{\mathcal{V}})^{-1}\left({}_0B_{p_0 p_0}^{s'}(H_{p_1}^{r'})\right)$$

$$= {}_0B_{p_0}^{s'+s}(H_{p_1}^{r'}) \cap {}_0B_{p_0}^{s'}(H_{p_1}^{r'+r}). \qquad (2.58)$$

Using (2.58), Proposition 1.69, and the embeddings

$$_0H_{p_0}^{s'+s-\varepsilon}(H_{p_1}^{r'}) \cap {}_0H_{p_0}^{s'-\varepsilon}(H_{p_1}^{r'+r}) \hookrightarrow {}_0H_{p_0}^{s'+\sigma s}(H_{p_1}^{r'+(1-\sigma)r-\delta}),$$

$$_0H_{p_0}^{s'+s+\varepsilon}(H_{p_1}^{r'}) \cap {}_0H_{p_0}^{s'+\varepsilon}(H_{p_1}^{r'+r}) \hookrightarrow {}_0H H_{p_0}^{s'+\sigma s}(H_{p_1}^{r'+(1-\sigma)r+\delta})$$

from Lemma 2.61 (i) we obtain

$$_0B_{p_0}^{s'+s}(H_p^{r'}) \cap {}_0B_{p_0}^{s'}(H_{p_1}^{r'+r})$$

$$\hookrightarrow \left({}_0H_{p_0}^{s'+\sigma s}(H_{p_1}^{r'+(1-\sigma)r-\delta}), \; {}_0H_{p_0}^{s'+\sigma s}(H_{p_1}^{r'+(1-\sigma)r+\delta})\right)_{1/2, p_0}$$

$$= {}_0H_{p_0}^{s'+\sigma s}(B_{p_1 p_0}^{r'+(1-\sigma)r}),$$

which proves (2.55). □

The embeddings which we have proven in this subsection are very useful for the application of the main result on mixed-order systems but they are also of interest in themselves. Therefore, we also want to state these results for spaces with more general domains in the space variable.

Remark 2.77 (Further embeddings). Let $\Omega \subseteq \mathbb{R}^n$ be sufficiently smooth (e.g., Ω satisfies a strong local Lipschitz condition) such that there exist bounded extension and restriction operators in the Bessel potential and Besov scale. All embeddings in Lemma 2.61, Proposition 2.75, and Proposition 2.76 remain valid if we replace $\mathbb{R}^n$ by Ω, i.e., we replace $_0\mathcal{F}_\varrho^\sigma(\mathbb{R}_+, \mathcal{K}^\eta(\mathbb{R}^n, X))$ by the space $_0\mathcal{F}_\varrho^\sigma(\mathbb{R}_+, \mathcal{K}^\eta(\Omega, X))$. This can be obtained by Lemma 1.53 and Lemma 1.50.

With the help of the above embedding results, we can give another sufficient condition for compatibility (see Definition 2.68) of the tuples $\mathbb{H}$ and $\mathbb{F}$.

Before we give a sufficient condition for the embeddings (2.51) and (2.52), we introduce a useful definition for this context.

Definition 2.78 (Admissible scale). Let μ_1 and μ_2 be convex increasing order functions such that $\mu_1 - \mu_2$ is an order function. Let the scale

$$(\mathcal{F}_\ell, \mathcal{K}_\ell) \in \{(H_{p_0}, B_{p_1 p_0}), (B_{p_0, p_0}, H_{p_1})\}, \quad \ell = 0, \ldots, M$$

be given such that there exist $\tau \in \{0, \ldots, M-1\}$ with

$$\begin{aligned}(\mathcal{F}_\ell, \mathcal{K}_\ell) &= (H_{p_0}, B_{p_1 p_0}), & \ell \in \{0, \ldots, \tau\}, \\ (\mathcal{F}_\ell, \mathcal{K}_\ell) &= (B_{p_0 p_0}, H_{p_1}), & \ell \in \{\tau+1, \ldots, M\}.\end{aligned}$$

The scale $(\mathcal{F}_\ell, \mathcal{K}_\ell)_{\ell=0,\ldots,M}$ is then called (μ_1, μ_2)-*admissible* if we have

$$\begin{aligned}(b_\tau(\mu_2), m_\tau(\mu_2)) &\neq (b_{\tau+1}(\mu_2), m_{\tau+1}(\mu_2)), & \text{if } \mu_1 - \mu_2 \text{ is convex,} \\ (b_\tau(\mu_1), m_\tau(\mu_1)) &\neq (b_{\tau+1}(\mu_1), m_{\tau+1}(\mu_1)), & \text{if } \mu_1 - \mu_2 \text{ is concave.}\end{aligned}$$

Note that this definition is also meaningful if $\mu_1 - \mu_2$ has trivial index, i.e., there exists $\alpha, \beta \in \mathbb{R}$ such that $(\mu_1 - \mu_2)(\gamma) = \alpha\gamma + \beta$ for all $\gamma \geq 0$ and therefore $\mu_1 - \mu_2$ is convex as well as concave.

Proposition 2.79 (Compatibility condition II). *Assume that in the situation of Definition 2.68, the tuples $\mathbb{H} = \prod_{i=1}^m \mathbb{H}_i$ and $\mathbb{F} = \prod_{j=1}^m \mathbb{F}_j$ satisfy conditions (i) and (ii). Let*

$$(\mathcal{F}_\ell, \mathcal{K}_\ell) \in \{(H_{p_0}, B_{p_1 p_0}), (B_{p_0 p_0}, H_{p_1})\}, \quad \ell = 0, \ldots, M.$$

If the scale $(\mathcal{F}_\ell, \mathcal{K}_\ell)_{\ell=0,\ldots,M}$ is $(\mu_{\mathbb{H}_i}, \mu_{\mathbb{F}_j})$-admissible for all $i, j = 1, \ldots, m$, then embedding conditions (2.51) and (2.52) are fulfilled, and $\mathbb{H}, \mathbb{F}$ is compatible with $\mathscr{L}$.

2.3. H^∞-calculus of N-parabolic mixed-order systems

Proof. Let $i, j \in \{1, \ldots, m\}$ and assume that $s_j + t_i$ is convex. As in the proof of Proposition 2.72 we set $\mu_1(\gamma) := \mu_{\mathbb{F}_i}(\gamma) + \beta_1$ and $\mu_2(\gamma) := (s_j + t_i)(\gamma) + \alpha\gamma + \beta_2$ where $\beta_1 := \beta_{\mu_{\mathbb{F}_i}}$, $\alpha := \alpha_{s_j+t_i}$, and $\beta_2 := \beta_{s_j+t_i}$. Note that we have $\mu_{\mathbb{H}_i}(\gamma) = \mu_{\mathbb{F}_j}(\gamma) + (s_j + t_i)(\gamma) = (\mu_1 + \mu_2)(\gamma) - \alpha\gamma - (\beta_1 + \beta_2)$. For fixed $\ell, \kappa \in \{0, \ldots, M\}$ we define

$$\sigma := \sigma_{ji}(\ell, \kappa, t) = m_\ell(\mu_{\mathbb{F}_j}) + m_\kappa(s_j + t_i),$$
$$\eta := \eta_{ji}(\ell, \kappa, t) = b_\ell(\mu_{\mathbb{F}_j}) + b_\kappa(s_j + t_i)$$

and get $\sigma + \alpha = m_\ell(\mu_1) + m_\kappa(\mu_2)$ and $\eta + \beta_1 + \beta_2 = b_\ell(\mu_1) + b_\kappa(\mu_2)$. Lemma 2.35 already yields that $(\eta + \beta_1 + \beta_2, \sigma + \alpha) \in N(\mu_1 + \mu_2)$. Compared to the proof of Proposition 2.72 we have to determine the position of the tuple $(\eta + \beta_1 + \beta_2, \sigma + \alpha)$ more precisely. Here we can benefit from the results of Lemma 2.38. Let $\Gamma = \Gamma(\mu_1 + \mu_2)$ have the same meaning as in Lemma 2.38 and define $N := N(\mu_1 + \mu_2)$.

(i) Let $(\eta + \beta_1 + \beta_2, \sigma + \alpha) \in N \setminus \Gamma$. This case follows by an application of Lemma 2.61 (i), Proposition 2.75, Proposition 2.76, and the trivial embeddings

$$_0H^s_{p_0}(\mathbb{R}_+, Y),\ _0B^s_{p_0}(\mathbb{R}_+, Y) \hookrightarrow\ _0H^t_{p_0}(\mathbb{R}_+, Y) \cap\ _0B^t_{p_0}(\mathbb{R}_+, Y),$$
$$H^s_{p_1}(\mathbb{R}^n, X),\ B^s_{p_1 p_0}(\mathbb{R}^n, X) \hookrightarrow H^t_{p_1}(\mathbb{R}^n, X) \cap B^t_{p_1 p_0}(\mathbb{R}^n, X)$$

for $s > t$.

(ii) Let $(\eta + \beta_1 + \beta_2, \sigma + \alpha) \in \Gamma \setminus N_V$ and $k \in \{0, \ldots, I(\mu_1)\}$ with

$$i_k(\mu_1) = i_k(\mu_{\mathbb{F}_j}) \leq \ell < i_{k+1}(\mu_1) = i_{k+1}(\mu_{\mathbb{F}_j}).$$

Due to Lemma 2.38 we may assume without loss of generality that

$$p_1 \neq 0,\quad i_{p_1}(\mu_2) = i_k(\mu_1) =: j,$$

and $i_{p_1 - 1}(\mu_2) \leq \kappa < i_{p_1}(\mu_2)$. Then $(b_j(\mu_1 + \mu_2), m_j(\mu_1 + \mu_2))$ and $(b_{j-1}(\mu_1 + \mu_2), m_{j-1}(\mu_1 + \mu_2))$ are the endpoints of the edge including $(\eta + \beta_1 + \beta_2, \sigma + \alpha)$. The assumption on the scales then yields

$$\bigcap_{r=0}^M {}_0\mathcal{F}^{m_r(\mu_1+\mu_2)}_r(\mathcal{K}^{b_r(\mu_1+\mu_2)}_r)$$
$$\hookrightarrow {}_0\mathcal{F}^{m_j(\mu_1+\mu_2)}_j(\mathcal{K}^{b_j(\mu_1+\mu_2)}_j) \cap {}_0\mathcal{F}^{m_{j-1}(\mu_1+\mu_2)}_{j-1}(\mathcal{K}^{b_{j-1}(\mu_1+\mu_2)}_{j-1})$$
$$\hookrightarrow {}_0\mathcal{F}^{\sigma+\alpha}_\ell(\mathcal{K}^{\eta+\beta_1+\beta_2}_\ell)$$

due to Lemma 2.61 (i), Proposition 2.75, Proposition 2.76, and $(\mathcal{F}_\ell, \mathcal{K}_\ell) = (\mathcal{F}_j, \mathcal{K}_j)$. Note that $\tau \notin (i_k(\mu_1), i_{k+1}(\mu_1))$ for τ as in Definition 2.78. Defining $N' := N(\{(\alpha, \beta_1 + \beta_2)\})$ the isomorphism $\Phi_{N'}(\mathcal{D}^{(\varrho)}_+)$ then yields $\mathbb{H}_i \hookrightarrow {}_0\mathcal{F}^\sigma_\ell(\mathcal{K}^\eta_\ell)$.

(iii) Let $(\eta + \beta_1 + \beta_2, \sigma + \alpha) \in N_V$ and $k \in \{0, \ldots, I(\mu_1)\}$ with
$$i_k(\mu_1) = i_k(\mu_{\mathbb{F}_j}) \leq \ell < i_{k+1}(\mu_1) = i_{k+1}(\mu_{\mathbb{F}_j}).$$

According to Lemma 2.38 we then have $i_{p_1}(\mu_2) \leq \kappa < i_{p_2}(\mu_2)$ where $i_{p_1}(\mu_2)$ and $i_{p_2}(\mu_2)$ are given as in Lemma 2.38. Due to Lemma 2.38 we derive
$$(\eta + \beta_1 + \beta_2, \sigma + \alpha) = (b_j(\mu_1 + \mu_2), m_j(\mu_1 + \mu_2)) \in N_V$$

where
$$j := \begin{cases} i_k(\mu_1), & \kappa < i_k(\mu_1), \\ \kappa, & i_k(\mu_1) \leq \kappa < i_{k+1}(\mu_1), \\ i_{k+1}(\mu_1) - 1, & \kappa \geq i_{k+1}(\mu_1). \end{cases}$$

Due to our assumptions we have $(\mathcal{F}_\ell, \mathcal{K}_\ell) = (\mathcal{F}_j, \mathcal{K}_j)$. As we have $\tau \notin (i_k(\mu_1), i_{k+1}(\mu_1))$, this yields
$$\bigcap_{r=0}^{M} {}_0\mathcal{F}_r^{m_r(\mu_1+\mu_2)}(\mathcal{K}_r^{b_r(\mu_1+\mu_2)}) \hookrightarrow {}_0\mathcal{F}_j^{m_j(\mu_1+\mu_2)}(\mathcal{K}_j^{b_j(\mu_1+\mu_2)})$$
$$= {}_0\mathcal{F}_\ell^{\sigma+\alpha}(\mathcal{K}_\ell^{\eta+\beta_1+\beta_2}).$$

As in part (ii) the mapping properties of the operator $\Phi_{N'}(\mathcal{D}_+)$ then show $\mathbb{H}_i \hookrightarrow {}_0\mathcal{F}^\sigma(\mathcal{K}^\eta)$.

The embeddings $\mathbb{F}_j \hookrightarrow \mathbb{F}_{ij}$ can be proved in exactly the same way if $s_j + t_i$ is concave. $\square$

Finally, we want to state a corollary in which we present a condensed version of Theorem 2.69 which is sufficient for many applications.

Corollary 2.80. *Let X be a Banach space of class $\mathcal{HT}$ with property (α). Let $\mathscr{L} \in [H_P(S_\theta \times \Sigma_\delta^n)]^{m \times m}$, $\theta > \pi/2$, be an N-parabolic mixed-order system such that for each $i, j = 1, \ldots, m$ the order function $s_j + t_i$ is convex and increasing or concave and decreasing. Let $\varrho \geq 0$, $s'_\ell \geq 0$, $r'_\ell \in \mathbb{R}$, $\ell = 0, \ldots, M$, such that*
$$\mu_{\mathbb{H}_i}(\gamma) := \max_\ell \{[s'_\ell + m_\ell(t_i)]\gamma + r'_\ell + b_\ell(t_i)\}, \quad \gamma \geq 0,$$
$$\mu_{\mathbb{F}_j}(\gamma) := \max_\ell \{[s'_\ell - m_\ell(s_j)]\gamma + r'_\ell - b_\ell(s_j)\}, \quad \gamma \geq 0, \quad i, j = 1, \ldots m$$

are convex increasing order functions. Furthermore, let the scale
$$(\mathcal{F}_\ell, \mathcal{K}_\ell) \in \{(B_{pp}, H_p), (H_p, B_{pp})\}, \quad 1 < p < \infty, \quad \ell = 0, \ldots, M$$

be $(\mu_{\mathbb{H}_i}, \mu_{\mathbb{F}_j})$-admissible for all $i, j = 1, \ldots m$ and let
$$s'_\ell > \max\{\max\{-m_\ell(t_i), m_\ell(s_i)\} : i = 1, \ldots, m\} \tag{2.59}$$

2.3. H^∞-calculus of N-parabolic mixed-order systems

for all $\ell \in \{0, \ldots, \tau\}$ where τ is taken from Definition 2.78. Using the same notation as in Definition 1.71 we define for $i, j = 1, \ldots m$ the spaces

$$\mathbb{H}_i := \bigcap_{\ell=0}^{M} {}_0\mathcal{F}_\ell^{s'_\ell + m_\ell(t_i)}(\mathcal{K}_\ell^{r'_\ell + b_\ell(t_i)}), \qquad \mathbb{F}_j := \bigcap_{\ell=0}^{M} {}_0\mathcal{F}_\ell^{s'_\ell - m_\ell(s_j)}(\mathcal{K}_\ell^{r'_\ell - b_\ell(s_j)}).$$

Then there exists $\sigma_0 > 0$ such that for all $\sigma \geq \sigma_0$,

$$[\mathscr{L}_\sigma(\mathcal{D}_+^{(\varrho)})]|_\mathbb{H} \in L_{\mathrm{Isom}}(\mathbb{H}, \mathbb{F}) \quad \text{and} \quad ([\mathscr{L}_\sigma(\mathcal{D}_+^{(\varrho)})]|_\mathbb{H})^{-1} = [\mathscr{L}_\sigma^{-1}(\mathcal{D}_+^{(\varrho)})]|_\mathbb{F}$$

where $\mathbb{H} := \prod_{i=1}^m \mathbb{H}_i$ and $\mathbb{F} := \prod_{i=1}^m \mathbb{F}_i$.

Proof. The order functions $s_j + t_i$ are convex and increasing or concave and decreasing, we have $\alpha_{s_j + t_i} = 0$ or $\alpha_{-s_j - t_i} = 0$, respectively. This yields $\delta_{i,1} = \delta_{j,2} = 0$ for all $i, j = 1, \ldots, m$. Due to Proposition 2.79, the tuples $\mathbb{H}$, $\mathbb{F}$ are compatible with $\mathscr{L}$. Assumption (2.59) ensures that the time-regularity is positive for the Besov scales. Hence the assertion follows from Theorem 2.69. □

Remark 2.81 (Possible extensions). It should be possible to extend the theory to general order functions such that we can drop the assumption that the functions $(s_j + t_i)_{ij}$ have to be convex or concave. In this case we also have to introduce more general weight functions, i.e., we have to allow arbitrary quotients of the form W_{ν_1}/W_{ν_2} in Definition 2.24. Up to now, from the view of applications, there is no need to generalize the theory in this direction.

Chapter 3

Triebel-Lizorkin spaces and the L_p-L_q-setting

In this chapter we want to generalize the results of Section 2.3 on mixed-order systems to the scale of Triebel-Lizorkin spaces. Triebel-Lizorkin spaces naturally appear in parabolic L_p-L_q-theory as the trace spaces of the solution. This fact was independently noted in [Ber87a], [Wei05], and [DHP07]. In Chapter 4 we will show that we are able to establish an L_p-L_q-theory for free boundary value problems by using mixed-order systems on Triebel-Lizorkin spaces. We explicitly consider the L_p-L_q two-phase Stefan problem with Gibbs-Thomson correction in Section 4.8.

Our philosophy in treating vector-valued Triebel-Lizorkin spaces is to present them as a complex interpolation space $F^s_{pq} = [H^s_{p_0}, B^s_{p_1 p_1}]_\theta$. This representation holds in the scalar-valued case but fails in general in the vector-valued case. Therefore we prove a related representation by ourselves in Corollary 3.11, which reads as follows

$$F^s_{pq}(\mathbb{R}^m, [H, X]_\theta) = [H^s_{p_0}(\mathbb{R}^m, H), B^s_{p_1 p_1}(\mathbb{R}^m, X)]_\theta \qquad (3.1)$$

for a Banach space X and a Hilbert space H. Using (3.1) we can transfer the results in Chapter 1 and Section 2.3 to vector-valued Triebel-Lizorkin spaces. So we explicitly derive that the realization of $\nabla_+ := (\partial_t, \nabla_x)$ admits a bounded joint H^∞-calculus on vector-valued Triebel-Lizorkin spaces. Moreover, we get the analog result of Theorem 2.69 for vector-valued Triebel-Lizorkin spaces.

To establish compatibility embeddings as in Proposition 2.75 we need the concepts of scalar-valued anisotropic Triebel-Lizorkin spaces with mixed norms. Literature concerning these spaces is rare and therefore we will restrict ourselves to scalar-valued mixed-order systems.

In the next sections we give the basic definition of several types of Triebel-Lizorkin spaces and state some results on them. The presented approach to Triebel-Lizorkin spaces on half-spaces is rather the same as in Section 1.2, where we have used the concept of retractions.

As mentioned, the consideration of Triebel-Lizorkin spaces is mainly motivated by trace results in the L_p-L_q-setting. For the investigation of boundary value problems in this setting, we include a section on singular integral operators in L_p-L_q-spaces which will be needed for the applications in Chapter 4.

3.1 Vector-valued Triebel-Lizorkin spaces and interpolation

Motivation. We start with definition and interpolation properties of vector-valued Triebel-Lizorkin spaces. Here we follow the work of H. Triebel [Tri97] and H.J. Schmeißer and W. Sickel [SS05], for definition of the spaces by the so-called decomposition method. This will give us the possibility to define vector-valued Triebel-Lizorkin spaces in the half-space by a retraction-coretraction argument and to obtain results on complex interpolation. In particular, we will consider Bessel space valued Triebel-Lizorkin spaces and complex interpolation results for them.

Definition 3.1 ([Tri97, Definition 15.4], [SS05, Definition 1]). Let $\psi \in \mathscr{D}(\mathbb{R}^m)$ with $\psi(x) \in [0,1]$, $x \in \mathbb{R}^m$, and $\psi(x) = 1$ if $|x| \leq A$, and $\psi(x) = 0$ if $|x| > B$ for $0 < A < B < \infty$. We define the smooth *dyadic decomposition* of unity by

$$\varphi_0 := \psi, \quad \varphi_1(x) := \varphi_0(x/2) - \varphi_0(x), \quad \varphi_j(x) := \varphi_1(2^{-j+1}x), \quad j \geq 2.$$

For $s \geq 0$, $1 < p, q < \infty$, and any Banach space X we define the X-valued Triebel-Lizorkin space $F_{pq}^s(\mathbb{R}^m, X)$ as the set of all $u \in \mathscr{S}'(\mathbb{R}^m, X)$ with

$$\|u\|_{F_{pq}^s(\mathbb{R}^m, X)} := \left\| \left\| (2^{sj} \mathscr{F}^{-1}[\varphi_j \mathscr{F} f])_{j \in \mathbb{N}_0} \right\|_{\ell_q(X)} \right\|_{L_p(\mathbb{R}^m)} < \infty.$$

Remark 3.2. In [RS96, Proposition 6] one can find a characterization of vector-valued Triebel-Lizorkin spaces by differences.

It is astonishing that the Banach space valued Triebel-Lizorkin scale does not cover the Bessel potential scale in general. This statement is based on some results in Littlewood-Paley theory and will be clarified with the next remark.

Remark 3.3. (i) Let X be a Banach space and $s \geq 0$, $1 < p < \infty$. We then have

$$H_p^s(\mathbb{R}^m, X) = F_{p2}^s(\mathbb{R}^m, X)$$

if and only if X can be renormed as a Hilbert space. This result can be found in [RdFT87, p. 283], [SS05, Remark 5], [Ama09, Remark 4.5.3], and the references given therein.

3.1. Vector-valued Triebel-Lizorkin spaces and interpolation

(ii) For all Banach spaces X, $s > 0$, and $1 < p < \infty$ we have
$$B_{pp}^s(\mathbb{R}^m, X) = F_{pp}^s(\mathbb{R}^m, X).$$
This observation is obvious due to the parallel definitions of Besov and Triebel-Lizorkin spaces (cf. Definition 1.55 (iii)) and Fubini's theorem.

In Section 1.2 we defined all Besov and Bessel potential spaces on half-spaces by retractions. In the next lines we want to generalize these concepts to vector-valued Triebel-Lizorkin spaces. As in Section 1.2 we use the mappings r^+, r_0^+, e^+, and e_0^+, which were introduced in Definitions 1.61 and 1.63.

Definition 3.4. Let X be a Banach space of class $\mathcal{HT}$ and $s \geq 0$, $1 < p, q < \infty$. Following the approach of [Ama09] for Bessel potential and Besov spaces, we define X-valued *Triebel-Lizorkin spaces* on the half-space by
$$F_{pq}^s(\mathbb{R}_+^m, X) := r^+ \left(F_{pq}^s(\mathbb{R}^m, X) \right),$$
$$_0F_{p,q}^s(\mathbb{R}_+^m, X) := r_0^+ \left(F_{pq}^s(\mathbb{R}^m, X) \right)$$
with the quotient norms
$$\|u\|_{F_{pq}^s(\mathbb{R}_+^m, X)} := \inf \left\{ \|f\|_{F_{pq}^s(\mathbb{R}^m, X)} : f \in F_{pq}^s(\mathbb{R}^m, X) \text{ with } r^+(f) = u \right\},$$
$$\|u\|_{_0F_{p,q}^s(\mathbb{R}_+^m, X)} := \inf \left\{ \|f\|_{F_{pq}^s(\mathbb{R}^m, X)} : f \in F_{pq}^s(\mathbb{R}^m, X) \text{ with } r_0^+(f) = u \right\}.$$
The operators r^+ and r_0^+ were introduced in Definitions 1.61 and 1.63.

Similar to Remark 1.70, we define Triebel-Lizorkin spaces with exponential weights.

Definition 3.5 (**Triebel-Lizorkin spaces with exponential weights**). Let X be a Banach space of class $\mathcal{HT}$ and $1 < p, q < \infty$. For $s > 0$ and exponential weight $\varrho \geq 0$ we define
$$_0F_{pq,\varrho}^s(\mathbb{R}_+, X) := \mathscr{M}_{-\varrho} \left({}_0F_{pq}^s(\mathbb{R}_+, X) \right)$$
with norm
$$\|f\|_{_0F_{pq,\varrho}^s(\mathbb{R}_+, X)} := \|\mathscr{M}_\varrho f\|_{_0F_{pq}^s(\mathbb{R}_+, X)}, \quad f \in {}_0F_{pq,\varrho}^s(\mathbb{R}_+, X).$$

Remark 3.6 (**Properties of Triebel-Lizorkin spaces with exponential weights**). Let X be a Banach space of class $\mathcal{HT}$, $s > 0$, $\varrho \geq 0$, and $1 < p, q < \infty$. From the definitions we already derive
$$[\mathscr{M}_\varrho]\big|_{_0F_{pq,\varrho}^s(\mathbb{R}_+, X)} \in L_{\text{Isom}} \left({}_0F_{pq,\varrho}^s(\mathbb{R}_+, X), {}_0F_{pq}^s(\mathbb{R}_+, X) \right).$$

Proposition 3.7. *For each Banach space X of class $\mathcal{HT}$ we have the embeddings*
$$F_{pq}^{s'+s}(\mathbb{R}^m, X) \hookrightarrow H_p^{s'}(\mathbb{R}^m, X),$$
$$_0F_{pq}^{s'+s}(\mathbb{R}_+^m, X) \hookrightarrow {}_0H_p^{s'}(\mathbb{R}_+^m, X),$$
$$_0F_{pq,\varrho}^{s'+s}(\mathbb{R}_+, X) \hookrightarrow {}_0H_{p,\varrho}^{s'}(\mathbb{R}_+, X)$$
for $s' \geq 0$, $s > 0$, $1 < p, q < \infty$, and $\varrho \geq 0$.

Proof. Using [SS01, Proposition 2 (ii)] we derive
$$F_{pq}^{s'+s}(\mathbb{R}^m, X) \hookrightarrow B_{p,\max\{p,q\}}^{s'+s}(\mathbb{R}^m, X).$$

Employing Theorem 1.56 (iii), there exists $\theta \in (0,1)$ such that
$$B_{p,\max\{p,q\}}^{s'+s}(\mathbb{R}^m, X) = \left(H_p^{s'}(\mathbb{R}^m, X), H_p^{s'+s+1}(\mathbb{R}^m, X)\right)_{\theta, \max\{p,q\}} \hookrightarrow H_p^{s'}(\mathbb{R}^m, X).$$

A usual retraction argument, cf. Lemma 1.53, and the exponential weight isomorphisms then also yield the other claimed embeddings. $\square$

Remark 3.8. (i) The vector-valued space $F_{pq}^s(\mathbb{R}^m, X)$ $(s > 0)$ is a retract of $L_p(\mathbb{R}^m, \ell_q^s(X))$ where
$$\ell_q^s(X) := \left\{(x_k)_{k \in \mathbb{N}_0} \subseteq X \colon \sum_{k=0}^{\infty} 2^{skq}\|x_k\|_X^q < \infty\right\},$$
$$\|(x_k)_{k \in \mathbb{N}_0}\|_{\ell_q^s(X)} := \left(\sum_{k=0}^{\infty} 2^{skq}\|x_k\|_X^q\right)^{1/q}.$$

This is well-known in the scalar case (cf. [Tri78, Theorem 2.4.2]) and the proof can be lifted to the vector-valued case (cf. [SS01, Proof of Proposition 12]). See also [Ama00, Theorem 3.1] for a similar proof and a corresponding result for vector-valued Besov spaces.

(ii) For an arbitrary interpolation couple $\{X_0, X_1\}$, $s_0, s_1 > 0$, $1 < p_0, p_1, q_0, q_1 < \infty$, and $\theta \in (0,1)$ we obtain
$$\left[F_{p_0 q_0}^{s_0}(\mathbb{R}^m, X_0), F_{p_1 q_1}^{s_1}(\mathbb{R}^m, X_1)\right]_\theta = F_{p_\theta q_\theta}^{s_\theta}(\mathbb{R}^m, [X_0, X_1]_\theta), \quad (3.2)$$
with $1/p_\theta = (1-\theta)/p_0 + \theta/p_1$, $1/q_\theta = (1-\theta)/q_0 + \theta/q_1$, and $s_\theta = (1-\theta)s_0 + \theta s_1$. The equality in (3.2) immediately follows from (i), Lemma 1.51, Theorem 1.41, and Theorem 1.42.

Remark 3.9. For $s > 0$, $1 < p, q < \infty$, and any Banach space X we have
$$\mathscr{D}(\mathbb{R}_+, X) \xhookrightarrow{d} {}_0F_{pq}^s(\mathbb{R}_+, X). \quad (3.3)$$

The embedding $\mathscr{S}(\mathbb{R}, X) \xhookrightarrow{d} F_{pq}^s(\mathbb{R}, X)$ can be proved by a retraction argument similar to the proof for Besov spaces in [Ama09, p. 52]. Another retraction argument then yields ${}_0\mathscr{S}(\mathbb{R}_+, X) \xhookrightarrow{d} {}_0F_{pq}^s(\mathbb{R}_+, X)$. Using $\mathscr{D}(\mathbb{R}, X) \xhookrightarrow{d} \mathscr{S}(\mathbb{R}, X)$ (cf. [Ama03, Lemma 4.1.4]) we obtain (3.3).

The next proposition is our starting-point for a transfer of the results in Section 2.3 to the Triebel-Lizorkin scale. More precisely, it allows us to transfer results by complex interpolation.

3.1. Vector-valued Triebel-Lizorkin spaces and interpolation

Proposition 3.10. *Let $\{X, H\}$ be an interpolation couple where X is a Banach space of class $\mathcal{HT}$ and H is a Hilbert space. Let $1 < p < \infty$, $s > 0$, $\varrho \geq 0$, and*

$$q \in \left(\frac{2p}{p+1}, 2p\right) \tag{3.4}$$

(cf. Figure 3.1). Then there exist $1 < p_0, p_1 < \infty$ and $\theta \in (0, 1)$ such that $1/p = (1-\theta)/p_0 + \theta/p_1$ and $1/q = (1-\theta)/2 + \theta/p_1$. With this we obtain

$$F_{pq}^s(\mathbb{R}^m, [H, X]_\theta) = \left[H_{p_0}^{s_0}(\mathbb{R}^m, H), B_{p_1 p_1}^{s_1}(\mathbb{R}^m, X)\right]_\theta, \tag{3.5}$$

$$F_{pq}^s(\mathbb{R}_+^m, [H, X]_\theta) = \left[H_{p_0}^{s_0}(\mathbb{R}_+^m, H), B_{p_1 p_1}^{s_1}(\mathbb{R}_+^m, X)\right]_\theta, \tag{3.6}$$

$$_0F_{pq}^s(\mathbb{R}_+^m, [H, X]_\theta) = \left[_0H_{p_0}^{s_0}(\mathbb{R}_+^m, H), _0B_{p_1 p_1}^{s_1}(\mathbb{R}_+^m, X)\right]_\theta, \tag{3.7}$$

$$_0F_{pq,\varrho}^s(\mathbb{R}_+, [H, X]_\theta) = \left[_0H_{p_0,\varrho}^{s_0}(\mathbb{R}_+, H), _0B_{p_1 p_1,\varrho}^{s_1}(\mathbb{R}_+, X)\right]_\theta \tag{3.8}$$

for all $s_0, s_1 > 0$ with $s = (1-\theta)s_0 + \theta s_1$.

Proof. (I) The space H is a Hilbert space and therefore we obtain for all $p_0, p_1 \in (1, \infty)$ and $s_0, s_1 > 0$,

$$H_{p_0}^{s_0}(\mathbb{R}^m, H) = F_{p_0 2}^{s_0}(\mathbb{R}^m, H), \quad B_{p_1 p_1}^{s_1}(\mathbb{R}^m, X) = F_{p_1 p_1}^{s_1}(\mathbb{R}^m, X)$$

due to Remark 3.3 (i) and (ii).

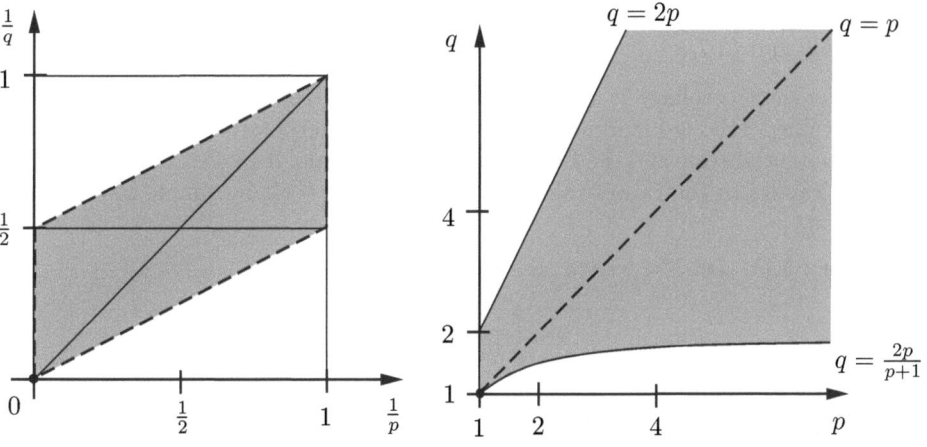

Figure 3.1: Illustration of the set of tuples (p, q) satisfying (3.4)

As one can easily see in Figure 3.1 the convex hull of the set $\{(a, b) \in (0, 1)^2 : b = 1/2 \text{ or } a = b\}$ is given by

$$Q := \{(a, b) \in (0, 1)^2 : a/2 < b < (1 + a)/2\}.$$

Due to assumption (3.4) we have $(1/p, 1/q) \in Q$ and therefore there exist $p_0, p_1 \in (1, \infty)$ and $\theta \in (0, 1)$ with

$$\begin{pmatrix} 1/p \\ 1/q \end{pmatrix} = (1 - \theta) \begin{pmatrix} 1/p_0 \\ 1/2 \end{pmatrix} + \theta \begin{pmatrix} 1/p_1 \\ 1/p_1 \end{pmatrix}.$$

Using Remark 3.8 (ii) we directly obtain

$$\left[H_{p_0}^{s_0}(\mathbb{R}^m, H), B_{p_1 p_1}^{s_1}(\mathbb{R}^m, X)\right]_\theta = \left[F_{p_0 2}^{s_0}(\mathbb{R}^m, H), F_{p_1 p_1}^{s_1}(\mathbb{R}^m, X)\right]_\theta$$
$$= F_{pq}^s(\mathbb{R}^m, [H, X]_\theta),$$

which proves the equality (3.5).

(II) According to Remark 1.67 (i) we have the retraction and coretraction

$$r^+ \in L(H_{p_0}^{s_0}(\mathbb{R}^m, H), H_{p_0}^{s_0}(\mathbb{R}_+^m, H)) \cap L(B_{p_1 p_1}^{s_1}(\mathbb{R}^m, X), B_{p_1 p_1}^{s_1}(\mathbb{R}_+^m, X)),$$

$$e^+ \in L(H_{p_0}^{s_0}(\mathbb{R}_+^m, H), H_{p_0}^{s_0}(\mathbb{R}^m, H)) \cap L(B_{p_1 p_1}^{s_1}(\mathbb{R}_+^m, X), B_{p_1 p_1}^{s_1}(\mathbb{R}^m, X)).$$

Thus, (3.5) and Lemma 1.51 yield

$$F_{pq}^s(\mathbb{R}_+^m, [H, X]_\theta) = r^+ F_{pq}^s(\mathbb{R}^m, [H, X]_\theta)$$
$$= r^+ \left[H_{p_0}^{s_0}(\mathbb{R}^m, H), B_{p_1 p_1}^{s_1}(\mathbb{R}^m, X)\right]_\theta$$
$$= \left[H_{p_0}^{s_0}(\mathbb{R}_+^m, H), B_{p_1 p_1}^{s_1}(\mathbb{R}_+^m, X)\right]_\theta,$$

which proves (3.6).

(III) Using r_0^+ and e_0^+ instead of r^+ and e^+ the proof of (3.7) is essentially the same as in (II). The result of (3.8) then follows from Proposition 3.6, Lemma 1.52, and (3.7). □

The next corollary is very useful because it gives the possibility to obtain Bessel-valued Triebel-Lizorkin spaces by complex interpolation of the "tame" pure Bessel potential scale and the Bessel-valued Besov scale. If (3.4) is fulfilled, then we can lift results to the Bessel-valued Triebel-Lizorkin scale by simple interpolation arguments.

Corollary 3.11. Let $1 < p < \infty$, $s > 0$, $r \in \mathbb{R}$, and $\varrho \geq 0$. Then for all

$$q \in \left(\frac{2p}{p+1}, 2p\right)$$

there exist $\theta \in (0, 1)$ and $1 < p_0, p_1 < \infty$ such that $1/p = (1 - \theta)/p_0 + \theta/p_1$ and $1/q = (1 - \theta)/2 + \theta/p_1$. With this we obtain

$$F_{pq}^s(\mathbb{R}^m, H_q^r(\mathbb{R}^n)) = \left[H_{p_0}^{s_0}(\mathbb{R}^m, H_2^r(\mathbb{R}^n)), B_{p_1 p_1}^{s_1}(\mathbb{R}^m, H_{p_1}^r(\mathbb{R}^n))\right]_\theta, \quad (3.9)$$

$$F_{pq}^s(\mathbb{R}_+^m, H_q^r(\mathbb{R}^n)) = \left[H_{p_0}^{s_0}(\mathbb{R}_+^m, H_2^r(\mathbb{R}^n)), B_{p_1 p_1}^{s_1}(\mathbb{R}_+^m, H_{p_1}^r(\mathbb{R}^n))\right]_\theta, \quad (3.10)$$

$${}_0 F_{pq}^s(\mathbb{R}_+^m, H_q^r(\mathbb{R}^n)) = \left[{}_0 H_{p_0}^{s_0}(\mathbb{R}_+^m, H_2^r(\mathbb{R}^n)), {}_0 B_{p_1 p_1}^{s_1}(\mathbb{R}_+^m, H_{p_1}^r(\mathbb{R}^n))\right]_\theta, \quad (3.11)$$

$${}_0 F_{pq,\varrho}^s(\mathbb{R}_+, H_q^r(\mathbb{R}^n)) = \left[{}_0 H_{p_0,\varrho}^{s_0}(\mathbb{R}_+, H_2^r(\mathbb{R}^n)), {}_0 B_{p_1 p_1,\varrho}^{s_1}(\mathbb{R}_+, H_{p_1}^r(\mathbb{R}^n))\right]_\theta \quad (3.12)$$

for all $s_0, s_1 > 0$ with $s = (1 - \theta)s_0 + \theta s_1$.

3.1. Vector-valued Triebel-Lizorkin spaces and interpolation

Proof. Let p_0, p_1, and θ be as in Proposition 3.10. Defining $H := H_2^s(\mathbb{R}^n)$ and $X := H_{p_1}^r(\mathbb{R}^n)$ we get $[H, X]_\theta = H_q^r(\mathbb{R}^n)$. Hence Proposition 3.10 yields the claim. □

Corollary 3.12. *Let $1 < p < \infty$, $s > 0$, $r \in \mathbb{R}$, and $q \in \left(\frac{2p}{p+1}, 2p\right)$. Then we have*

$$e^+ \in L(F_{pq}^s(\mathbb{R}_+^m, H_q^r(\mathbb{R}^n)), F_{pq}^s(\mathbb{R}^m, H_q^r(\mathbb{R}^n))),$$
$$r^+ \in L(F_{pq}^s(\mathbb{R}^m, H_q^r(\mathbb{R}^n)), F_{pq}^s(\mathbb{R}_+^m, H_q^r(\mathbb{R}^n))),$$
$$e_0^+ \in L({}_0F_{pq}^s(\mathbb{R}_+^m, H_q^r(\mathbb{R}^n)), F_{pq}^s(\mathbb{R}^m, H_q^r(\mathbb{R}^n))),$$
$$r_0^+ \in L(F_{pq}^s(\mathbb{R}^m, H_q^r(\mathbb{R}^n)), {}_0F_{pq}^s(\mathbb{R}_+^m, H_q^r(\mathbb{R}^n))).$$

Hence r^+, r_0^+, e^+, and e_0^+, which are introduced in Definitions 1.61 and 1.63, are retractions and coretractions on the space $F_{pq}^s(\mathbb{R}^m, H_q^r(\mathbb{R}^n))$.

Proof. This follows immediately by an interpolation argument from Remark 1.67 (i) and Corollary 3.11 with $s_0 = s_1 = s$. □

Corollary 3.11 shows that we can reach all Bessel potential space valued isotropic Triebel-Lizorkin spaces $F_{pq}^s(\mathbb{R}^m, H_q^r(\mathbb{R}^n))$ by interpolation of the scales $H_p(H_2)$ and $B_{qq}(H_q)$ if p and q fulfill (3.4). This representation is used frequently to derive results for the Bessel-valued Triebel-Lizorkin scale. For example we can show that the definition of ${}_0F_{pq}^s(\mathbb{R}_+^m, H_q^r(\mathbb{R}^n))$ is equivalent to the definition by vanishing traces. In order to do this we need the following abstract interpolation result, which can be found in [LM68] or [KMM07, Theorem 7.10].

Remark 3.13. Let $\{X_0, X_1\}$, $\{Z_0, Z_1\}$, and $\{Y_0, Y_1\}$ be interpolation couples with

$$X_0 \cap X_1 \stackrel{d}{\hookrightarrow} X_i, \quad Z_0 \cap Z_1 \stackrel{d}{\hookrightarrow} Z_i, \quad Y_i \hookrightarrow Z_i, \quad i = 0, 1$$

and suppose that there exists an operator $D: X_0 + X_1 \to Z_0 + Z_1$ such that $D|_{X_i} \in L(X_i, Z_i)$ for $i = 0, 1$. For $i = 0, 1$ define the spaces

$$X_i(D) := \{u \in X_i : Du \in Y_i\}, \quad \|u\|_{X_i(D)} := \|u\|_{X_i} + \|Du\|_{Y_i}, \quad u \in X_i(D).$$

If there exist operators $G: Z_0 + Z_1 \to X_0 + X_1$ and $K: Z_0 + Z_1 \to Y_0 + Y_1$ with $G|_{Z_i} \in L(Z_i, X_i)$, $K|_{Z_i} \in L(Z_i, Y_i)$ for $i = 0, 1$ and

$$DGu = u + Ku, \quad u \in Z_i, \quad i = 0, 1,$$

then we derive

$$[X_0(D), X_1(D)]_\theta = \{u \in [X_0, X_1]_\theta : Du \in [Y_0, Y_1]_\theta\}, \quad \theta \in (0, 1).$$

Proposition 3.14. *Let $s > 0$, $r \in \mathbb{R}$, and $k \in \mathbb{N}_0$ such that $k + 1/p < s < k + 1 + 1/p$. Then we have*

$${}_0F_{pq}^s(\mathbb{R}_+, H_q^r(\mathbb{R}^n)) = \left\{u \in F_{pq}^s(\mathbb{R}_+, H_q^r(\mathbb{R}^n)) : u^{(j)}(0) = 0, \text{ for all } 0 \leq j \leq k\right\}$$

for all $p \in (1, \infty)$ and $q \in (2p/(1+p), 2p)$. Due to Proposition 3.7 the traces are classical.

Proof. According to Corollary 3.11 there exist constants $p_0, p_1 \in (1, \infty)$ and $\theta \in (0,1)$ such that we have $1/p = (1-\theta)/p_0 + \theta/p_1$ and $1/q = (1-\theta)/2 + \theta/p_1$. For $s_j := s + 1/p_j - 1/p$, $j = 0, 1$, we derive

$$s_0 \in (k + 1/p_0, k + 1 + 1/p_0), \quad s_1 \in (k + 1/p_1, k + 1 + 1/p_1),$$

and $s = (1-\theta)s_0 + \theta s_1$. Corollary 3.11 then yields

$$_0F^s_{pq}(\mathbb{R}_+, H^r_q(\mathbb{R}^n)) = \left[_0H^{s_0}_{p_0}(\mathbb{R}_+, H^r_2(\mathbb{R}^n)), _0B^{s_1}_{p_1 p_1}(\mathbb{R}_+, H^r_{p_1}(\mathbb{R}^n))\right]_\theta.$$

Here it is crucial that we can choose s_0 and s_1 such that the Bessel potential space as well as the Besov space possess the same number of vanishing traces. This is necessary to apply Remark 3.13.

In order to apply Remark 3.13 we define

$$X_0 := H^{s_0}_{p_0}(\mathbb{R}_+, H^r_2(\mathbb{R}^n)), \qquad Z_0 := \prod_{j=0}^{k} H^r_2(\mathbb{R}^n),$$

$$X_1 := B^{s_1}_{p_1 p_1}(\mathbb{R}_+, H^r_{p_1}(\mathbb{R}^n)), \qquad Z_1 := \prod_{j=0}^{k} H^r_{p_1}(\mathbb{R}^n),$$

$Y_0 := Y_1 := \{0\}$, and the trace operator $Du := (u^{(j)}(0))_{j=0,\ldots,k}$ with the corresponding extension operator G. These operators are well defined and possess the requested boundedness according to [Ama09, Theorem 4.5.4, Theorem 4.6.3]. Additionally, we get $DGu = u$ for all $u \in X_0 + X_1$ so that we can define $K := 0$. The density assumptions of Remark 3.13 are obvious. We then have

$$X_0(D) = \{u \in H^{s_0}_{p_0}(\mathbb{R}_+, H^r_2(\mathbb{R}^n)) : Du = 0\}$$
$$= {_0H^{s_0}_{p_0}}(\mathbb{R}_+, H^r_2(\mathbb{R}^n)), \quad \|u\|_{X_0(D)} = \|u\|_{_0H^{s_0}_{p_0}(\mathbb{R}_+, H^r_2(\mathbb{R}^n))},$$
$$X_1(D) = \{u \in B^{s_1}_{p_1 p_1}(\mathbb{R}_+, H^r_{p_1}(\mathbb{R}^n)) : Du = 0\}$$
$$= {_0B^{s_1}_{p_1 p_1}}(\mathbb{R}_+, H^r_{p_1}(\mathbb{R}^n)), \quad \|u\|_{X_1(D)} = \|u\|_{_0B^{s_1}_{p_1 p_1}(\mathbb{R}_+, H^r_{p_1}(\mathbb{R}^n))}$$

by Remark 1.67 (iii). Hence we obtain

$$_0F^s_{pq}(\mathbb{R}_+, H^r_q(\mathbb{R}^n)) = \left[_0H^{s_0}_{p_0}(\mathbb{R}_+, H^r_2(\mathbb{R}^n)), _0B^{s_1}_{p_1 p_1}(\mathbb{R}_+, H^r_{p_1}(\mathbb{R}^n))\right]_\theta$$
$$= \{u \in [X_0, X_1]_\theta : Du = 0\}$$
$$= \left\{u \in F^s_{pq}(\mathbb{R}_+, H^r_q(\mathbb{R}^n)) : (u^{(j)}(0))_{j=0,\ldots,k} = 0\right\}$$

by (3.10) and (3.11). □

3.2 Anisotropic Triebel-Lizorkin spaces and representation by intersections

> *Motivation.* In Section 2.3 we noticed the importance of compatibility embeddings as in Proposition 2.75 and Proposition 2.76. Thus, it is worthwhile to consider such compatibility embeddings for Triebel-Lizorkin spaces. As in Section 2.3 we are faced with the interpolation of intersections of spaces and therefore it is helpful to have a characterization of these intersections as anisotropic Triebel-Lizorkin spaces. Hence, we provide such a characterization as well as the behavior under interpolation in this section.

Lebesgue, Bessel potential, and Besov spaces with mixed norms are well-known in the literature, see for example in [BP61], [Bug71], [BIN78], and [BIN79]. However, there is not much literature concerning Triebel-Lizorkin spaces with mixed norms. In [JS08] one can find general results on the trace problem of scalar-valued anisotropic Triebel-Lizorkin spaces with mixed norms. There, the authors show that the trace spaces are also of this type but there is no representation of the anisotropic Triebel-Lizorkin spaces with mixed norms by an intersection of spaces as in [Ber87a], [Wei05], and [DHP07]. In the following we want to show that the anisotropic Triebel-Lizorkin spaces with mixed norms used in [JS08] can be represented as an intersection of two spaces in some special cases. First, we state some results of M.Z. Berkolaiko which are of interest for our purpose.

For anisotropic Besov and Bessel potential spaces a representation by intersections can be found in [Ama09, Theorem 3.6.3, Theorem 3.7.2] for the non-mixed norm case. These proofs, however, cannot be carried over directly to the mixed norm case because we cannot change the order of variables.

We want to state the definitions of the anisotropic Triebel-Lizorkin spaces with mixed norms given in [Ber85] and [JS08].

Definition 3.15 (Anisotropic Triebel-Lizorkin space with mixed norms). (i) Let X be an arbitrary Banach space and $\vec{p} \in (1, \infty)^n$. Then we define $L_{\vec{p}}(\mathbb{R}^n, X)$ as the space of all measurable functions $f: \mathbb{R}^n \to X$ such that

$$\|f\|_{L_{\vec{p}}(\mathbb{R}^n, X)} := \left(\int_{\mathbb{R}} \left(\cdots \left(\int_{\mathbb{R}} \|f(x_1, \ldots, x_n)\|_X^{p_1} dx_1 \right)^{p_2/p_1} \cdots \right)^{p_n/p_{n-1}} dx_n \right)^{1/p_n}$$
$$< \infty.$$

(ii) ([Ber85]): Let $1 < c_1 < c_2 < \infty$ and $\tau \in C^\infty(\mathbb{R})$ with $\tau(\zeta) \in [0,1]$ such that $\tau(\zeta) = 1$ for $|\zeta| \leq c_1$ and $\tau(\zeta) = 0$ for $|\zeta| > c_2$. Let $\vec{r} \in (0, \infty)^n$ and for all $\xi \in \mathbb{R}^n$ let $\mu_0(\xi) := \prod_{j=1}^n \tau(\xi_j)$ and $\mu_k(\xi) := \mu_0(2^{-k/r_1}\xi_1, \ldots, 2^{-k/r_n}\xi_n)$. Then we define the partition of unity $\sigma_0 := \mu_0$ and $\sigma_k := \mu_k - \mu_{k-1}$ for

$k \in \mathbb{N}$. For $\vec{p} \in (1,\infty)^n$ and $q \in (1,\infty)$ we define $L_{\vec{p},q}^{\vec{r}}(\mathbb{R}^n)$ to be the space of all measurable functions $f \colon \mathbb{R}^n \to \mathbb{C}$ such that

$$\|f\|_{L_{\vec{p},q}^{\vec{r}}(\mathbb{R}^n)} := \|f\|_{L_{\vec{p}}(\mathbb{R}^n)} + \left\|(2^j\mathscr{F}^{-1}[\sigma_j \mathscr{F} f])_{j \in \mathbb{N}_0}\right\|_{L_{\vec{p}}(\mathbb{R}^n, \ell_q)} < \infty.$$

Here the anisotropy is given by $\vec{r}$.

(iii) ([JS08]): For $\vec{a} \in (0,\infty)^n$ and $\xi \in \mathbb{R}^n$ we define the *anisotropic distance function* $|\xi|_{\vec{a}}$ as the unique $t > 0$ such that

$$\frac{\xi_1^2}{t^{2a_1}} + \cdots + \frac{\xi_n^2}{t^{2a_n}} = 1.$$

Let $\psi \in C^\infty(\mathbb{R})$ with $\psi(t) \in [0,1]$ and $\psi(t) = 1$ if $t \leq 11/10$ and $\psi(t) = 0$ if $t > 13/10$. Additionally, we define $\Psi_j(\xi) := \psi(2^{-j}|\xi|_{\vec{a}})$ for $j \in \mathbb{N}_0$ and the partition $\Phi_0 := 0$ and $\Phi_j := \Psi_j - \Psi_{j-1}$ for $j \in \mathbb{N}$. For $\vec{p} \in (1,\infty)^n$, $q \in (1,\infty)$, and $s \geq 0$ we define $F_{\vec{p},q}^{s,\vec{a}}(\mathbb{R}^n)$ to be the space of all $u \in \mathscr{S}'(\mathbb{R}^n)$ such that

$$\|u\|_{F_{\vec{p},q}^{s,\vec{a}}(\mathbb{R}^n)} := \left\|(2^{sj}\mathscr{F}^{-1}[\Phi_j \mathscr{F} f])_{j \in \mathbb{N}_0}\right\|_{L_{\vec{p}}(\mathbb{R}^n, \ell_q)} < \infty.$$

Here the anisotropy is given by $\vec{a}$.

Remark 3.16. To avoid confusion we want to mention that, in the works of M.Z. Berkolaiko, J. Johnsen and W. Sickel the last variable x_n is associated with the time variable and $(x_1, \ldots, x_{n-1})$ are associated with the space variables. The definitions of $L_{\vec{p},q}^{\vec{r}}(\mathbb{R}^n)$ and $F_{\vec{p},q}^{s,\vec{a}}(\mathbb{R}^n)$ differ in their decomposition and partition of unity, respectively. In Definition 3.15 (ii) a partition based on a cubic structure is used whereas in Definition 3.15 (iii) the partition is based on an ellipsoid structure. It is conjectured that both definitions are equivalent but this is not necessary for our purposes. We only need the relation given in Remark 3.19 below.

Definition 3.17 (Anisotropic Bessel potential space with mixed norms)**.** For $\vec{p} \in (1,\infty)^n$, $\vec{a} \in (0,\infty)^n$, and $s \geq 0$ we define $H_{\vec{p}}^{s,\vec{a}}(\mathbb{R}^n)$ as the set of all $u \in L_{\vec{p}}(\mathbb{R}^n)$ such that

$$\mathscr{F}^{-1}(1 + |\xi|_{\vec{a}}^2)^{s/2}\mathscr{F} u \in L_{\vec{p}}(\mathbb{R}^n).$$

The norm is then given by

$$\|u\|_{H_{\vec{p}}^{s,\vec{a}}(\mathbb{R}^n)} := \|\mathscr{F}^{-1}(1+|\xi|_{\vec{a}}^2)^{s/2}\mathscr{F} u\|_{L_{\vec{p}}(\mathbb{R}^n)}, \quad u \in H_{\vec{p}}^{s,\vec{a}}(\mathbb{R}^n).$$

Remark 3.18. Let $\vec{p} \in (1,\infty)^n$, $\vec{a} \in (0,\infty)^n$, and $s \geq 0$. If $\vec{m} := (m_1, \ldots, m_n) := (s/a_1, \ldots, s/a_n) \in \mathbb{N}_0^n$, then we have

$$H_{\vec{p}}^{s,\vec{a}}(\mathbb{R}^n) = W_{\vec{p}}^{\vec{m}}(\mathbb{R}^n)$$

3.2. Anisotropic Triebel-Lizorkin spaces and representation by intersections

where the anisotropic Sobolev space $W_{\vec{p}}^{\vec{m}}(\mathbb{R}^n)$ consists of all $u \in L_{\vec{p}}(\mathbb{R}^n)$ such that

$$\partial^{m_k} u \in L_{\vec{p}}(\mathbb{R}^n), \quad k = 1, \ldots, n,$$

cf. [JS08, Proposition 2.10 (ii)]. The canonical norm is given by

$$\|u\|_{W_{\vec{p}}^{\vec{m}}(\mathbb{R}^n)} := \|u\|_{L_{\vec{p}}(\mathbb{R}^n)} + \sum_{k=1}^{n} \|\partial_k^{m_k} u\|_{L_{\vec{p}}(\mathbb{R}^n)}.$$

Remark 3.19 ([Ber87a, Theorem 4 B], [JS08, Proposition 2.10 (i)]). For all $\vec{p} \in (1, \infty)^n$, $\vec{r} \in (0, \infty)^n$, and $s \geq 0$ we have

$$H_{\vec{p}}^{s,\vec{a}}(\mathbb{R}^n) = L_{\vec{p},2}^{\vec{r}}(\mathbb{R}^n) = F_{\vec{p},2}^{s,\vec{a}}(\mathbb{R}^n), \quad \vec{a} := (s/r_1, \ldots, s/r_n)$$

where $H_{\vec{p}}^{s,\vec{a}}(\mathbb{R}^n)$ denotes the anisotropic Bessel potential spaces with mixed norms. In Soviet mathematical literature the Bessel potential spaces are usually called Liouville spaces and are denoted by "L_p^s".

Remark 3.20 ([JS08, Proposition 3.15]). Let $\vec{a} \in (0, \infty)^n$, $\vec{p} \in (1, \infty)^n$, $q \in (1, \infty)$, and $s, r \geq 0$ with $s - r \geq 0$. Then we have the isomorphism

$$\Lambda_{r,\vec{a}} := \mathscr{F}^{-1}(1 + |\xi|_{\vec{a}}^2)^{r/2}\mathscr{F} \in L_{\text{Isom}}\left(F_{\vec{p},q}^{s,\vec{a}}(\mathbb{R}^n), F_{\vec{p},q}^{s-r,\vec{a}}(\mathbb{R}^n)\right).$$

Remark 3.21 (Trace result of M.Z. Berkolaiko).

(i) In [Ber85, Theorem 2], respectively [Ber87b, Corollary 1], M.Z. Berkolaiko proved the following trace result for the semi-isotropic case: If $\kappa := 1 - (p_\nu r_\nu)^{-1} > 0$, $\nu \in \{1, \ldots, n\}$, and if there exists $r > 0$ such that $r_{\nu+1} = \cdots = r_n = r$, then

$$\gamma_{x_\nu=0}\left(L_{\vec{p},2}^{\vec{r}}(\mathbb{R}^n)\right) = L_{\vec{p}_z}\left(\mathbb{R}^{n-\nu}, B_{\vec{p}_w, p_\nu}^{\vec{\rho}_w}(\mathbb{R}^{\nu-1})\right) \cap L_{\vec{p}_z, p_\nu}^{\rho}\left(\mathbb{R}^{n-\nu}, L_{\vec{p}_w}(\mathbb{R}^{\nu-1})\right)$$

where $\vec{\rho}_w := (\kappa r_1, \ldots, \kappa r_{\nu-1})$, $\rho := \kappa r$, $\vec{p}_w := (p_1, \ldots, p_{\nu-1})$, and $\vec{p}_z := (p_{\nu+1}, \ldots, p_n)$ (i.e., the restriction operator is continuous and there exists a continuous extension operator). For the definition of the anisotropic Besov space and the vector-valued Triebel-Lizorkin space by differences we refer to [Ber87b]. In the following we only need the isotropic and non-mixed norm versions of these spaces.

(ii) Note that for the trace in the "space" variable $\gamma_{x_{n-1}=0}$, i.e., $\nu = n - 1$ (see Remark 3.16) the assumption $r_{\nu+1} = \cdots = r_n = r$ is always fulfilled.

(iii) Let $\nu = n - 1$, $\vec{r} := (l, \ldots, l, t) \in (0, \infty)^n$, and $\vec{p} := (q, \ldots, q, p) \in (1, \infty)^n$ be such that we have $\kappa = 1 - (lq)^{-1} > 0$. Then we get

$$\vec{p}_w = (q, \ldots, q) \in (1, \infty)^{n-2}, \quad p_{n-1} = q, \quad \vec{p}_z = p,$$
$$r = t, \quad \vec{\rho}_w := (\kappa l, \ldots, \kappa l), \quad \rho = \kappa t.$$

Hence (i) reads as follows:

$$\gamma_{x_{n-1}=0}\left(L_{\vec{p},2}^{\vec{r}}(\mathbb{R}^n)\right) = L_p(\mathbb{R}, B_{qq}^{\kappa l}(\mathbb{R}^{n-2})) \cap L_{p,q}^{\kappa t}(\mathbb{R}, L_q(\mathbb{R}^{n-2})). \qquad (3.13)$$

At this point we want to emphasize that on the right-hand side of (3.13) only isotropic vector-valued spaces appear. In this semi-isotropic and semi-mixed norm situation the variables x_1 and x_{n-1} can be interchanged without consequences. Hence, we also get

$$\gamma_{x_1=0}\left(L_{\vec{p},2}^{\vec{r}}(\mathbb{R}^n)\right) = L_p(\mathbb{R}, B_{qq}^{\kappa l}(\mathbb{R}^{n-2})) \cap L_{p,q}^{\kappa t}(\mathbb{R}, L_q(\mathbb{R}^{n-2})). \qquad (3.14)$$

(iv) In the works of M.Z. Berkolaiko the vector-valued Triebel-Lizorkin spaces on the right-hand side of (3.13) and (3.14) are defined by differences. Due to Remark 3.2 this yields the same space as in Definition 3.1. Thus, we finally obtain

$$\gamma_{x_1=0}\left(L_{\vec{p},2}^{\vec{r}}(\mathbb{R}^n)\right) = L_p(\mathbb{R}, B_{qq}^{\kappa l}(\mathbb{R}^{n-2})) \cap F_{pq}^{\kappa t}(\mathbb{R}, L_q(\mathbb{R}^{n-2})).$$

Remark 3.22 (Trace result of J. Johnsen and W. Sickel). In [JS08, Theorem 2.2] one can find the following results:

(i) Let $\vec{p} \in (1,\infty)^n$, $\vec{a} \in (0,\infty)^n$, and $s > a_1/p_1$. Then we have

$$\gamma_{x_1=0}\left(F_{\vec{p},2}^{s,\vec{a}}(\mathbb{R}^n)\right) = F_{\vec{p}'',p_1}^{s-a_1/p_1,\vec{a}''}(\mathbb{R}^{n-1})$$

where $\vec{a}'' := (a_2,\ldots,a_n)$ and $\vec{p}'' := (p_2,\ldots,p_n)$ (i.e., the restriction operator is continuous and there exists a continuous extension operator).

(ii) Let $\vec{p} = (q,\ldots,q,p) \in (1,\infty)^n$ and $\vec{a} = (l^{-1},\ldots,l^{-1},t^{-1}) \in (0,\infty)^n$ such that $s > 1/(lq)$. Then (i) reads as follows:

$$\gamma_{x_1=0}\left(F_{\vec{p},2}^{s,\vec{a}}(\mathbb{R}^n)\right) = F_{\vec{p}'',q}^{s-1/(lq),\vec{a}''}(\mathbb{R}^{n-1})$$

where $\vec{a}'' = (l^{-1},\ldots,l^{-1},t^{-1})$ and $\vec{p}'' = (q,\ldots,q,p)$.

Proposition 3.23 (Representation of anisotropic Triebel-Lizorkin spaces by intersections). Let $s > 0$, $\vec{a} = (l^{-1},\ldots,l^{-1},t^{-1}) \in (0,\infty)^n$, $p,q \in (1,\infty)$, and $\vec{p} := (q,\ldots,q,p) \in (1,\infty)^n$. Then we obtain the following representation of anisotropic spaces with mixed norms:

$$F_{\vec{p},q}^{s,\vec{a}}(\mathbb{R}^n) = F_{pq}^{st}(\mathbb{R}, L_q(\mathbb{R}^{n-1})) \cap L_p(\mathbb{R}, B_{qq}^{sl}(\mathbb{R}^{n-1})). \qquad (3.15)$$

Proof. Let $s > 0$, $\vec{a} = (l^{-1},\ldots,l^{-1},t^{-1}) \in (0,\infty)^n$, $p,q \in (1,\infty)$, and $\vec{p} := (q,\ldots,q,p) \in (1,\infty)^n$. Then we define $s' := s + 1/(lq) > 1/(lq)$ and get

$$F_{\vec{p},q}^{s,\vec{a}}(\mathbb{R}^n) = F_{\vec{p},q}^{s'-1/(lq),\vec{a}}(\mathbb{R}^n).$$

3.2. Anisotropic Triebel-Lizorkin spaces and representation by intersections

Due to Remark 3.22 (ii) we have

$$\gamma_{x_1=0}\left(F^{s',\vec{a}_+}_{\vec{p}_+,2}(\mathbb{R}^{n+1})\right) = F^{s'-1/(lq),\vec{a}}_{\vec{p},q}(\mathbb{R}^n)$$

where $\vec{a}_+ := (l^{-1},\ldots,l^{-1},t^{-1}) \in (0,\infty)^{n+1}$ and $\vec{p}_+ := (q,\ldots,q,p) \in (1,\infty)^{n+1}$. According to Remark 3.19 we have $F^{s',\vec{a}_+}_{\vec{p}_+,2}(\mathbb{R}^{n+1}) = L^{\vec{r}_+}_{\vec{p}_+,2}(\mathbb{R}^{n+1})$ where $\vec{r}_+ := (s'l,\ldots,s'l,s't) \in (0,\infty)^{n+1}$. In this situation Remark 3.21 (iv) states that

$$\gamma_{x_1=0}\left(L^{\vec{r}_+}_{\vec{p}_+,2}(\mathbb{R}^{n+1})\right) = L_p(\mathbb{R}, B^{\kappa s'l}_{qq}(\mathbb{R}^{n-1})) \cap F^{\kappa s't}_{pq}(\mathbb{R}, L_q(\mathbb{R}^{n-1}))$$

$$= L_p(\mathbb{R}, B^{s'l-1/q}_{qq}(\mathbb{R}^{n-1})) \cap F^{s't-t/(lq)}_{pq}(\mathbb{R}, L_q(\mathbb{R}^{n-1}))$$

$$= L_p(\mathbb{R}, B^{sl}_{qq}(\mathbb{R}^{n-1})) \cap F^{st}_{pq}(\mathbb{R}, L_q(\mathbb{R}^{n-1}))$$

with $\kappa = 1 - 1/(s'lq)$. Next, we want to apply Lemma 1.54 with

$$X := H^{s',\vec{a}_+}_{\vec{p}_+}(\mathbb{R}^{n+1}), \quad Y_0 := F^{st}_{pq}(\mathbb{R}, L_q(\mathbb{R}^{n-1})) \cap L_p(\mathbb{R}, B^{sl}_{qq}(\mathbb{R}^{n-1})),$$

$$Y_1 := F^{s,\vec{a}}_{\vec{p},q}(\mathbb{R}^n).$$

Remarks 3.21 and 3.22 then yield that there exist trace operators

$$R_0 := \gamma^{\mathrm{B}}_{x_1=0} \in L(X, Y_0), \quad R_1 := \gamma^{\mathrm{JS}}_{x_1=0} \in L(X, Y_1)$$

with corresponding extension operators

$$E_0 := \mathrm{ext}^{\mathrm{B}}_{x_1=0} \in L(Y_0, X), \quad E_1 := \mathrm{ext}^{\mathrm{JS}}_{x_1=0} \in L(Y_1, X).$$

We have $D := \mathscr{S}(\mathbb{R}^{n+1}) \overset{d}{\hookrightarrow} X$ and trivially $R_0 f = R_1 f$ for all $f \in D$. Lemma 1.54 then yields $Y_0 = Y_1$ with equivalence of norms. □

The next aim is to show that anisotropic Triebel-Lizorkin spaces with mixed norms are an interpolation scale. This can be proved by a common retraction argument for which we need some results on Fourier-multipliers on mixed L_p-spaces.

Remark 3.24 (Fourier multipliers on spaces with mixed norms, cf. [Hyt05, Theorem 2.2]**).** Let X be a Banach space of class $\mathcal{HT}$ with property (α) and $\vec{p} \in (1,\infty)^n$. If

$$m \in C^n(\mathbb{R}^n \setminus \{0\}, L(X))$$

is given such that $\{\xi^\alpha D^\alpha m(\xi) \colon \xi \in \mathbb{R}^n \setminus \{0\}, \alpha \in \{0,1\}^n\} \subseteq L(X)$ is $\mathcal{R}$-bounded, then m is an $L_{\vec{p}}$-Fourier multiplier, i.e., we have

$$\mathrm{op}[m] \in L(L_{\vec{p}}(\mathbb{R}^n, X)).$$

Lemma 3.25. *Using the functions $(\Phi_k)_{k \in \mathbb{N}_0}$ introduced in Definition 3.15 (iii) we define $\widetilde{\Phi}_0 := \Phi_0 + \Phi_1$ and $\widetilde{\Phi}_k := \Phi_{k-1} + \Phi_k + \Phi_{k+1}$ for $k \in \mathbb{N}$.*

(i) For all $\alpha \in \mathbb{N}_0^n$ there exists $C(\alpha, \vec{a}) > 0$ such that

$$\|\xi^\alpha D^\alpha \Phi_k\|_\infty \leq C(\alpha, \vec{a}), \quad k \in \mathbb{N}_0, \qquad (3.16)$$
$$\|\xi^\alpha D^\alpha \widetilde{\Phi}_k\|_\infty \leq 3C(\alpha, \vec{a}), \quad k \in \mathbb{N}_0. \qquad (3.17)$$

(ii) Let $\vec{p} \in (1, \infty)^n$, $q \in (1, \infty)$, and $g = (g_k)_{k \in \mathbb{N}_0} \in L_{\vec{p}}(\mathbb{R}^n, \ell_q)$. Then the series $\sum_{k=0}^\infty \mathscr{F}^{-1} \widetilde{\Phi}_k \mathscr{F} g_k$ is convergent in $\mathscr{S}'(\mathbb{R}^n)$.

Proof. (i) For $\vec{a} \in (0, \infty)^n$ we introduce the notation

$$t^{\vec{a}} x := (t^{a_1} x_1, \ldots, t^{a_n} x_n), \quad t^{s\vec{a}} x := (t^s)^{\vec{a}} x$$

where $t \geq 0$, $x \in \mathbb{R}^n$, and $s \in \mathbb{R}$. Due to $t\eta(\xi) = \eta(t^{\vec{a}}\xi)$ $(t > 0)$ for $\eta(\xi) := |\xi|_{\vec{a}}$ we derive

$$(\partial_j \eta)(\xi) = (\partial_j \eta)(t^{\vec{a}} \xi) \cdot t^{a_j - 1}, \quad t > 0.$$

Setting $t := |\xi|_{\vec{a}}^{-1}$ we get

$$|\partial_j \eta(\xi)| = |\partial_j \eta(|\xi|_{\vec{a}}^{-\vec{a}} \xi)| \cdot |\xi|_{\vec{a}}^{1-a_j} \leq C(\vec{a}) |\xi|_{\vec{a}}^{1-a_j}, \quad \xi \in \mathbb{R}^n \setminus \{0\}$$

due to the compactness of $\{\zeta \in \mathbb{R}^n \setminus \{0\} : |\zeta|_{\vec{a}} = 1\}$ and $\left| |\xi|_{\vec{a}}^{-\vec{a}} \xi \right|_{\vec{a}} = |\xi|_{\vec{a}}^{-1} |\xi|_{\vec{a}} = 1$. With this we get for $k \geq 1$

$$|\xi_j| |\partial_j \Phi_k(\xi)| = |\xi_j| |2^{-k} (\partial_j \psi)(2^{-k} |\xi|_{\vec{a}}) - 2^{-(k-1)} (\partial_j \psi)(2^{-(k-1)} |\xi|_{\vec{a}})| |\partial_j \eta(\xi)|$$
$$\leq 3 \cdot 2^{-k} \|\partial_j \psi\|_\infty \cdot \chi_{\operatorname{supp} \Phi_k}(\xi) |\xi_j| |\partial_j \eta(\xi)|$$
$$\leq 3 \cdot 2^{-k} C(\vec{a}) \|\partial_j \psi\|_\infty \cdot \chi_{\operatorname{supp} \Phi_k}(\xi) |\xi_j| |\xi|_{\vec{a}}^{1-a_j}.$$

For $\xi \in \operatorname{supp} \Phi_k$ we have $|\xi_j| \leq 11/10 \cdot 2^{ka_j}$ and $|\xi|_{\vec{a}} \leq 11/10 \cdot 2^k$, which yield

$$|\xi_j| |\partial_j \Phi_k(\xi)| \leq C \|\partial_j \psi\|_\infty, \quad \xi \in \mathbb{R}^n.$$

Iterating these arguments we obtain (3.16). The second assertion (3.17) then follows from (3.16) and the definition of $\widetilde{\Phi}_k$.

(ii) The Fourier transform is continuous in $\mathscr{S}'(\mathbb{R}^n)$, and therefore it suffices to show that the scalar sequence $\left(\sum_{k=0}^N [\widetilde{\Phi}_k \mathscr{F} g_k](f) \right)_{N \in \mathbb{N}_0}$ converges for all $f \in$

3.2. Anisotropic Triebel-Lizorkin spaces and representation by intersections

$\mathscr{S}(\mathbb{R}^n)$. For $f \in \mathscr{S}(\mathbb{R}^n)$ we get

$$\sum_{k=0}^{N} \left|[\widetilde{\Phi}_k \mathscr{F} g_k](f)\right| = \sum_{k=0}^{N} \left|g_k(\mathscr{F}(\widetilde{\Phi}_k f))\right|$$

$$\leq \int_{\mathbb{R}^n} \sum_{k=0}^{N} |g_k(x)| \cdot |(\mathscr{F}(\widetilde{\Phi}_k f))(x)| dx$$

$$\leq \int_{\mathbb{R}^n} \Big(\sum_{k=0}^{N} |g_k(x)|^q\Big)^{1/q} \cdot \Big(\sum_{k=0}^{N} |(\mathscr{F}(\widetilde{\Phi}_k f))(x)|^{q'}\Big)^{1/q'} dx$$

$$\leq \|(g_k)_k\|_{L_{\vec{p}}(\mathbb{R}^n, \ell_q(\{1,\ldots,N\}))} \cdot \|(\mathscr{F}(\widetilde{\Phi}_k f))_k\|_{L_{\vec{p}'}(\mathbb{R}^n, \ell_{q'}(\{0,\ldots,N\}))}$$

$$\leq \|(g_k)_k\|_{L_{\vec{p}}(\mathbb{R}^n, \ell_q(\mathbb{N}_0))} \cdot \|(\mathscr{F}(\widetilde{\Phi}_k f))_k\|_{L_{\vec{p}'}(\mathbb{R}^n, \ell_{q'}(\{0,\ldots,N\}))} \quad (3.18)$$

by Hölder's inequality in $\ell_1(\{0,\ldots,N\})$ and $L_1(\mathbb{R}^n)$ with $1/q + 1/q' = 1$, $1/p_k + 1/p'_k = 1$ ($k = 1,\ldots,n$), and $\vec{p}' := (p'_1,\ldots,p'_n)$. Note that Hölder's inequality for mixed norms follows by an iteration of the one-dimensional Hölder inequality, see for example [BIN78, Ch. 1, 2.4] or [AF03, 2.49]. For all $x \in \mathbb{R}^n$ and $\alpha, \gamma \in \mathbb{N}_0^n$ we get

$$\|x^\gamma D^\alpha(\widetilde{\Phi}_k f)(x)\|_{\ell_{q'}(\mathbb{N}_0)} \leq \Big(\sum_{k=0}^{\infty} \Big(\sum_{\beta \leq \alpha} \binom{\alpha}{\beta} |D^{\alpha-\beta}\widetilde{\Phi}_k(x)| |x^\gamma D^\beta f(x)|\Big)^{q'}\Big)^{1/q'}$$

$$\leq \sum_{\beta \leq \alpha} \binom{\alpha}{\beta} \Big(\sum_{k=0}^{\infty} \Big(|D^{\alpha-\beta}\widetilde{\Phi}_k(x)| |x^\gamma D^\beta f(x)|\Big)^{q'}\Big)^{1/q'}$$

$$\leq \sum_{\beta \leq \alpha} \binom{\alpha}{\beta} \|x^\gamma D^\beta f\|_\infty \Big(\sum_{k=0}^{\infty} \|D^{\alpha-\beta}\widetilde{\Phi}_k\|_\infty^{q'} \chi_{\text{supp }\widetilde{\Phi}_k}(x)\Big)^{1/q'}.$$

With the same arguments as in the proof of (i) we can show that there exists $C(\alpha, \gamma) > 0$ such that $\|D^{\alpha-\beta}\widetilde{\Phi}_k\|_\infty \leq C(\alpha, \gamma)$ for all $k \in \mathbb{N}_0$. It is obvious that we have $\#\{k \in \mathbb{N}_0 : x \in \text{supp }\widetilde{\Phi}_k\} \leq 4$, so we can show that

$$\sup_{x \in \mathbb{R}^n} \|x^\gamma D^\alpha(\widetilde{\Phi}_k f)(x)\|_{\ell_{q'}(\mathbb{N}_0)} \leq C(\alpha, \gamma, f), \quad k \in \mathbb{N}_0.$$

This proves $F := (\widetilde{\Phi}_k f)_{k \in \mathbb{N}_0} \in \mathscr{S}(\mathbb{R}^n, \ell_{q'}(\mathbb{N}_0))$ and therefore

$$\mathscr{F} F = (\mathscr{F}(\widetilde{\Phi}_k f))_{k \in \mathbb{N}_0} \in \mathscr{S}(\mathbb{R}^n, \ell_{q'}(\mathbb{N}_0)) \hookrightarrow L_{\vec{p}'}(\mathbb{R}^n, \ell_{q'}(\mathbb{N}_0)).$$

From (3.18) we then derive

$$\sum_{k=0}^{N} \left|[\widetilde{\Phi}_k \mathscr{F} g_k](f)\right| \leq \|(g_k)_k\|_{L_{\vec{p}}(\mathbb{R}^n, \ell_q(\mathbb{N}_0))} \cdot \|(\mathscr{F}(\widetilde{\Phi}_k f))_k\|_{L_{\vec{p}'}(\mathbb{R}^n, \ell_{q'}(\mathbb{N}_0))}$$

for $N \in \mathbb{N}_0$ which yields the convergence. $\square$

Proposition 3.26. Let $\vec{p}_0, \vec{p}_1 \in (1, \infty)^n$, $1 < q_0, q_1 < \infty$, $\vec{a} \in (0, \infty)^n$, and $s > 0$. Then we have

$$\left[F^{s,\vec{a}}_{\vec{p}_0,q_0}(\mathbb{R}^n), F^{s,\vec{a}}_{\vec{p}_1,q_1}(\mathbb{R}^n)\right]_\theta = F^{s,\vec{a}}_{\vec{p}_\theta,q_\theta}(\mathbb{R}^n), \quad \theta \in (0,1)$$

where $\vec{p}_\theta := (p_{\theta,1}, \ldots, p_{\theta,n})$ and q_θ are given by

$$\frac{1}{q_\theta} = \frac{1-\theta}{q_0} + \frac{\theta}{q_1}, \quad \frac{1}{p_{\theta,k}} = \frac{1-\theta}{p_{0,k}} + \frac{\theta}{p_{1,k}}, \quad k = 1, \ldots, n.$$

Proof. We use the standard method of retractions and coretractions together with Lemma 1.51. First, we consider the case $s = 0$ and define for $\vec{p} \in (1, \infty)^n$ and $q \in (1, \infty)$,

$$S \colon F^{0,\vec{a}}_{\vec{p},q}(\mathbb{R}^n) \to L_{\vec{p}}(\mathbb{R}^n, \ell_q),$$
$$f \mapsto \left(\mathscr{F}^{-1}(\Phi_j \mathscr{F} f)\right)_{j \in \mathbb{N}_0}.$$

The definition of $\|\cdot\|_{F^{0,\vec{a}}_{\vec{p},q}(\mathbb{R}^n)}$ trivially yields that S is an isometry and hence

$$S \in L\left(F^{0,\vec{a}}_{\vec{p},q}(\mathbb{R}^n), L_{\vec{p}}(\mathbb{R}^n, \ell_q)\right).$$

To obtain a corresponding retraction we define

$$R \colon L_{\vec{p}}(\mathbb{R}^n, \ell_q) \to \mathscr{S}'(\mathbb{R}^n),$$
$$(g_j)_{j \in \mathbb{N}_0} \mapsto \sum_{k=0}^\infty \mathscr{F}^{-1}(\widetilde{\Phi}_k \mathscr{F} g_k),$$

where the series converges in $\mathscr{S}'(\mathbb{R}^n)$ according to Lemma 3.25 (ii). Considering the supports of Φ_j and $\widetilde{\Phi}_j$ we get

$$\sum_{k=0}^\infty \Phi_j \widetilde{\Phi}_k \mathscr{F} g_k = \sum_{k=\max\{j-2,0\}}^{j+2} \Phi_j \widetilde{\Phi}_k \mathscr{F} g_k.$$

With this we easily see that

$$\|R(g_j)_{j\in\mathbb{N}_0}\|_{F^{0,\vec{a}}_{\vec{p},q}(\mathbb{R}^n)} = \left\|\left(\mathscr{F}^{-1}\left[\Phi_j \mathscr{F} \sum_{k=0}^\infty \mathscr{F}^{-1}(\widetilde{\Phi}_k \mathscr{F} g_k)\right]\right)_{j\in\mathbb{N}_0}\right\|_{L_{\vec{p}}(\mathbb{R}^n,\ell_q)}$$
$$= \left\|\left(\mathscr{F}^{-1}\left[\sum_{k=0}^\infty m_{jk} \mathscr{F} g_k\right]\right)_{j\in\mathbb{N}_0}\right\|_{L_{\vec{p}}(\mathbb{R}^n,\ell_q)}$$

for $(g_j)_{j \in \mathbb{N}_0} \in L_{\vec{p}}(\mathbb{R}^n, \ell_q)$ where $m_{jk} := \Phi_j \widetilde{\Phi}_k$ for $j, k \in \mathbb{N}_0$.

3.2. Anisotropic Triebel-Lizorkin spaces and representation by intersections 159

In contrast to other proofs on this topic we use operator-valued Fourier multipliers associated with $\mathcal{R}$-boundedness. In the following we want to apply Remark 3.24, where the necessary $\mathcal{R}$-boundedness can be proved by the square function estimate (Theorem 1.9). Note that ℓ_q is of class $\mathcal{HT}$ and has property (α) due to Remarks 1.13 and 1.15. We define the symbol

$$m\colon \mathbb{R}^n \setminus \{0\} \to L(\ell_q(\mathbb{N}_0)),$$
$$\xi \mapsto (m_{jk}(\xi))_{j,k \in \mathbb{N}_0}$$

and get, for all $(x_k)_{k \in \mathbb{N}_0} \in \ell_q$ and $\xi \in \mathbb{R}^n \setminus \{0\}$,

$$\|[m(\xi)]((x_k)_k)\|_{\ell_q} = \left\|\left(\sum_{k=0}^{\infty} m_{jk}(\xi) x_k\right)_{j \in \mathbb{N}_0}\right\|_{\ell_q}$$

$$= \left\|\left(\sum_{k=\max\{j-2,0\}}^{j+2} m_{jk}(\xi) x_k\right)_{j \in \mathbb{N}_0}\right\|_{\ell_q}$$

$$\leq \left(\sum_{j=0}^{\infty}\left(\sum_{k=\max\{j-2,0\}}^{j+2} |x_k|\right)^q\right)^{1/q} \leq 5\|(x_j)_j\|_{\ell_q}$$

due to $|m_{jk}(\xi)| \leq 1$. Hence, m is well-defined and we trivially have $m \in C^{\infty}(\mathbb{R}^n \setminus \{0\}, L(\ell_q(\mathbb{N}_0)))$. According to Lemma 3.25 we can show that there exists $C > 0$ such that

$$|\xi^{\alpha} D^{\alpha} m_{jk}(\xi)| \leq C, \quad \xi \in \mathbb{R}^n \setminus \{0\}, \, \alpha \in \{0,1\}^n, \, j,k \in \mathbb{N}_0.$$

Let $N \in \mathbb{N}$, $T_1, \ldots, T_N \in \{\xi^{\alpha} D^{\alpha} m(\xi) \colon \xi \in \mathbb{R}^n \setminus \{0\}, \alpha \in \{0,1\}^n\}$, $f_1, \ldots, f_N \in \ell^q$. Then we get

$$\left\|\left(\sum_{i=1}^{N} |T_i f_i|^2\right)^{1/2}\right\|_{\ell_q} = \left(\sum_{j=0}^{\infty}\left(\sum_{i=1}^{N} |(T_i f_i)_j|^2\right)^{q/2}\right)^{1/q}$$

$$\leq C\left(\sum_{j=0}^{\infty}\left(\sum_{i=1}^{N}\left(\sum_{k=\max\{j-2,0\}}^{j+2} |(f_i)_k|\right)^2\right)^{q/2}\right)^{1/q} \quad (3.19)$$

due to

$$|(T_i f_i)_j| \leq C \sum_{k=\max\{j-2,0\}}^{j+2} |(f_i)_k|.$$

With the triangle inequality for the Euclidean norm and for $\|\cdot\|_{\ell_q}$ we then derive

$$\left\|\left(\sum_{i=1}^{N} |T_i f_i|^2\right)^{1/2}\right\|_{\ell_q} \leq 5C \left\|\left(\sum_{i=1}^{N} |f_i|^2\right)^{1/2}\right\|_{\ell_q}.$$

from (3.19). Hence this yields

$$R \in L\left(L_{\vec{p}}(\mathbb{R}^n, \ell_q), F_{\vec{p},q}^{0,\vec{a}}(\mathbb{R}^n)\right)$$

by Remark 3.24 and Theorem 1.9.

For $f \in F_{\vec{p},q}^{0,\vec{a}}(\mathbb{R}^n)$ we deduce

$$RSf = \sum_{k=0}^{\infty} \mathscr{F}^{-1}(\widetilde{\Phi}_k \Phi_k \mathscr{F} f) = \sum_{k=0}^{\infty} \mathscr{F}^{-1}(\Phi_k \mathscr{F} f) = \mathscr{F}^{-1}\left(\sum_{k=0}^{\infty} \Phi_k \mathscr{F} f\right) = f$$

due to $\widetilde{\Phi}_k(\xi) = 1$ for all $\xi \in \operatorname{supp} \Phi_k$ ($k \in \mathbb{N}_0$) and $\sum_{k=0}^{\infty} \Phi_k = 1$. So we have proved that R is a retraction with corresponding coretraction S. This implies that

$$\left[F_{\vec{p}_0,q_0}^{0,\vec{a}}(\mathbb{R}^n), F_{\vec{p}_1,q_1}^{0,\vec{a}}(\mathbb{R}^n)\right]_\theta = F_{\vec{p}_\theta,q_\theta}^{0,\vec{a}}(\mathbb{R}^n), \quad \theta \in (0,1)$$

by Lemma 1.51, Theorem 1.41, and Theorem 1.42. The general case then follows by Remark 3.20 and the compatibility of interpolation and isomorphisms stated in Lemma 1.52. □

3.3 Auxiliary results on Bessel-valued Triebel-Lizorkin spaces

Motivation. Based on the interpolation results of the previous sections, we can now study the joint time-space H^∞-calculus in Bessel-valued Triebel-Lizorkin spaces. We will do this in analogy to Chapter 1 and Section 2.3 and obtain similar results on the realization of N-parabolic scalar symbols in the setting of Bessel-valued Triebel-Lizorkin spaces.

Similar to Definition 1.71 we define an abbreviation for Bessel-valued Triebel-Lizorkin spaces.

Definition 3.27 (Spaces of mixed scales). For $\varrho \geq 0$, $1 < p < \infty$, and $q \in (2p/(1+p), 2p)$, $s > 0$, $r \in \mathbb{R}$, and

$$\mathcal{F} = F_{pq}, \quad \mathcal{K} = H_q$$

we define

$$_0\mathcal{F}_\varrho^s(\mathcal{K}^r) := {}_0F_{pq,\varrho}^s(\mathbb{R}_+, H_q^r(\mathbb{R}^n)).$$

Remark 3.28. The restriction on q always comes from the fact that we obtain Bessel-valued Triebel-Lizorkin spaces by interpolation of a pure Bessel potential scale and the Bessel-valued Besov scale, cf. Corollary 3.11. The Besov spaces can

3.3. Auxiliary results on Bessel-valued Triebel-Lizorkin spaces

also be obtained by interpolation of Bessel spaces of course. In this sense our approach to Triebel-Lizorkin spaces is purely Bessel oriented.

It might be possible to drop the restriction on q by an approach which follows the intrinsic structure of the Bessel-valued Triebel-Lizorkin spaces. The results of Section 2.3 mainly depend on Fourier multiplier theorems. Versions of Fourier multiplier theorems on Triebel-Lizorkin spaces can be found in [BK05] and [BK09], for example.

a) The joint time-space H^∞-calculus on Bessel-valued Triebel-Lizorkin spaces

As in Section 1.3 b) we define the time derivative operator. Let $1 < p < \infty$, $2p/(1+p) < q < 2p$, $\varrho \geq 0$, $s > 0$, and $r \in \mathbb{R}$. Then we define the Bessel-valued time-derivative operator on the space $\mathcal{M} := {}_0F^s_{pq,\varrho}(\mathbb{R}_+, H^r_q(\mathbb{R}^n))$ by

$$\mathcal{D}_t\colon {}_0F^s_{pq,\varrho}(\mathbb{R}_+, H^r_q(\mathbb{R}^n)) \supseteq D(\mathcal{D}_t) \to {}_0F^s_{pq,\varrho}(\mathbb{R}_+, H^r_q(\mathbb{R}^n)),$$
$$u \mapsto \partial_t u$$

where $D(\mathcal{D}_t) := {}_0F^{s+1}_{pq,\varrho}(\mathbb{R}_+, H^r_q(\mathbb{R}^n))$. If the dependence on the space $\mathcal{M}$ is of importance, we will also write $\mathcal{D}_t^{\mathcal{M}}$ instead of $\mathcal{D}_t$. This operator inherits all properties from the time derivative operators considered in Section 1.3 b).

As in Section 1.3 c) we define the natural extension of the $H^r_q(\mathbb{R}^n)$-realization $\mathcal{D}$ of ∇ to the Bessel-valued Triebel-Lizorkin space ${}_0F^s_{pq,\varrho}(\mathbb{R}_+, H^r_q(\mathbb{R}^n))$.

Definition and Lemma 3.29 (Natural extension on Triebel-Lizorkin spaces)**.** *Let X, Y be Banach spaces of class $\mathcal{HT}$. For a densely defined closed linear operator $A\colon X \supseteq D(A) \to Y$ the natural extension of A to ${}_0F^s_{pq,\varrho}(\mathbb{R}_+, X)$ given by*

$$A^+\colon {}_0F^s_{pq,\varrho}(\mathbb{R}_+, X) \supseteq D(A^+) \to {}_0F^s_{pq,\varrho}(\mathbb{R}_+, Y),$$
$$u \mapsto A \circ u$$

is a well-defined operator with dense domain $D(A^+) := {}_0F^s_{pq,\varrho}(\mathbb{R}_+, D(A))$.

Proof. We can directly show

$$A^+ \in L({}_0F^s_{pq,\varrho}(\mathbb{R}_+, D(A)), {}_0F^s_{pq,\varrho}(\mathbb{R}_+, Y))$$

by using the definition of the norms. Hence A^+ is well defined. As in the proof of Definition and Lemma 1.85 we have

$$\mathscr{D}(\mathbb{R}_+,) \otimes D(A) \stackrel{d}{\hookrightarrow} \mathscr{D}(\mathbb{R}_+, X) \stackrel{d}{\hookrightarrow} {}_0F^s_{pq,\varrho}(\mathbb{R}_+, X)$$

according to Remark 3.9. Due to $\mathscr{D}(\mathbb{R}_+) \otimes D(A) \subseteq {}_0F^s_{pq,\varrho}(\mathbb{R}_+, D(A))$ the operator A^+ is densely defined. $\square$

As before, we will use the same notation $\mathcal{D}$ instead of $\mathcal{D}_+$ for the natural extension of the $H_p^r(\mathbb{R}^n)$-realization of ∇. The result of Lemma 1.86 also carries over to the setting of Triebel-Lizorkin spaces.

Lemma 3.30. *In the same situation as in Definition and Lemma 3.29 we have*

$$A^+ \in L({}_0F_{pq,\varrho}^s(\mathbb{R}_+, X), {}_0F_{pq,\varrho}^s(\mathbb{R}_+, Y))$$

if $A \in L(X,Y)$. In particular, we have

$$A^+ \in L_{\mathrm{Isom}}({}_0F_{pq,\varrho}^s(\mathbb{R}_+, X), {}_0F_{pq,\varrho}^s(\mathbb{R}_+, Y))$$

if and only if $A \in L_{\mathrm{Isom}}(X,Y)$. In both cases we have

$$\|A^+\|_{L({}_0F_{pq,\varrho}^s(\mathbb{R}_+,X),{}_0F_{pq,\varrho}^s(\mathbb{R}_+,Y))} \leq \|A\|_{L(X,Y)}.$$

To obtain the bounded joint time-space H^∞-calculus on Triebel-Lizorkin spaces there are two strategies. The first one is to adapt the proof in Section 1.3 c), where the result is derived by an iterative approach. The second one is a direct interpolation argument by Theorem 1.45, which we will use here. So the analog to Theorem 1.89 reads as follows:

Theorem 3.31 (Joint time-space H^∞-calculus on Triebel-Lizorkin spaces). *Let $1 < p < \infty$, $2p/(1+p) < q < 2p$, $r \in \mathbb{R}$, $s > 0$, $\varrho \geq 0$, and*

$$\mathcal{W} := {}_0\mathcal{F}_\varrho^s(\mathcal{K}^r) = {}_0F_{pq,\varrho}^s(\mathbb{R}_+, H_q^r(\mathbb{R}^n)), \quad (\mathcal{F}, \mathcal{K}) := (F_{pq}, H_q).$$

Let $\mathcal{N} := H_q^r(\mathbb{R}^n)$. We denote the $\mathcal{N}$-realization of ∇ by $\mathcal{D}^\mathcal{N}$ and the $\mathcal{W}$-realization of $(\sigma + \partial_t, \nabla)$ by

$$\mathcal{D}_{+,\sigma}^\mathcal{W} := (\sigma + \mathcal{D}_t^\mathcal{W}, (\mathcal{D}^\mathcal{N})^+), \quad \sigma \geq 0.$$

Then for $\Omega := S_\theta \times \Sigma_\delta^n$, $\theta > \pi/2$, we get the following results.

(i) *The tuple $\mathcal{D}_{+,\sigma}^\mathcal{W}$ has a bounded joint $H^\infty(\Omega)$-calculus on $\mathcal{W}$.*

(ii) *The H^∞-calculus in the space $\mathcal{W}$ is compatible with the H^∞-calculus in the space $\mathcal{W}' := L_{p,\varrho}(\mathbb{R}_+, H_q^r(\mathbb{R}^n))$, i.e.,*

$$f(\mathcal{D}_{+,\sigma}^\mathcal{W}) \subseteq f(\mathcal{D}_{+,\sigma}^{\mathcal{W}'})$$

holds for all $f \in H_P(\Omega)$.

(iii) *We have $f(\mathcal{D}_{+,\sigma}^\mathcal{W}) = f_\sigma(\mathcal{D}_+^\mathcal{W})$ for all $f \in H_P(\Omega)$.*

(iv) *If we define $\mathcal{V} := {}_0F_{pq,\varrho+\sigma}^s(\mathbb{R}_+, H_q^r(\mathbb{R}^n))$, then we have*

$$f(\mathcal{D}_+^\mathcal{V})u = \mathcal{M}_\sigma^{-1} f(\mathcal{D}_{+,\sigma}^\mathcal{W}) \mathcal{M}_\sigma u$$

for all $f \in H_P(\Omega)$ and $u \in \mathcal{V}$.

3.3. Auxiliary results on Bessel-valued Triebel-Lizorkin spaces

Proof. (i) Let $p_0, p_1, q_0, q_1 \in (1, \infty)$, and $\theta \in (0,1)$ be given as in Corollary 3.11. Considering the operator tuples

$$\mathcal{D}_{+,\sigma}^{\mathcal{W}_0} := (\sigma + \mathcal{D}_t^{\mathcal{W}_0}, (\mathcal{D}^{\mathcal{N}_0})^+) \quad \text{and} \quad \mathcal{D}_{+,\sigma}^{\mathcal{W}_1} := (\sigma + \mathcal{D}_t^{\mathcal{W}_1}, (\mathcal{D}^{\mathcal{N}_1})^+)$$

with

$$\mathcal{W}_0 := {}_0H_{p_0,\varrho}^s(\mathbb{R}_+, H_2^r(\mathbb{R}^n)), \qquad \mathcal{N}_0 := H_2^r(\mathbb{R}^n),$$
$$\mathcal{W}_1 := {}_0B_{p_1p_1,\varrho}^s(\mathbb{R}_+, H_{p_1}^r(\mathbb{R}^n)), \qquad \mathcal{N}_1 := H_{p_1}^r(\mathbb{R}^n),$$

we see that $\mathcal{D}_{+,\sigma}^{\mathcal{W}_0}$ and $\mathcal{D}_{+,\sigma}^{\mathcal{W}_0}$ fulfill the assumptions of Theorem 1.45. The compatibility condition for the resolvents (1.9) can be obtained by Lemma 1.76 (ii) and the representation by Fourier multipliers in Theorem 1.81 and Theorem 1.84. According to Corollary 3.11 we have

$$[\mathcal{W}_0, \mathcal{W}_1]_\theta = \mathcal{W}, \quad [D(\mathcal{D}_t^{\mathcal{W}_0}), D(\mathcal{D}_t^{\mathcal{W}_1})]_\theta = D(\mathcal{D}_t^{\mathcal{W}}).$$

As in the proof of Theorem 1.81 (i) we have to show an interpolation result for the domains $D(\mathcal{D}_j^{\mathcal{N}_0})$ and $D(\mathcal{D}_j^{\mathcal{N}_1})$ but the situation here is somewhat different because the spaces $\mathcal{N}_0$ and $\mathcal{N}_1$ are not contained in each other. Due to Theorem 1.81 (i) we have $1 \in \rho(\mathcal{D}_j^{\mathcal{N}_0}) \cap \rho(\mathcal{D}_j^{\mathcal{N}_1})$. With this we define the operator

$$A\colon \mathcal{N}_0 + \mathcal{N}_1 \to D(\mathcal{D}_j^{\mathcal{N}_0}) + D(\mathcal{D}_j^{\mathcal{N}_1}),$$
$$u = u_0 + u_1 \mapsto (1 - \mathcal{D}_j^{\mathcal{N}_0})^{-1} u_0 + (1 - \mathcal{D}_j^{\mathcal{N}_k})^{-1} u_1$$

and obtain $A|_{\mathcal{N}} \in L_{\text{Isom}}(\mathcal{N}, X)$ with $X := [D(\mathcal{D}_j^{\mathcal{N}_0}), D(\mathcal{D}_j^{\mathcal{N}_1})]_\theta$. The representation of the resolvents by Fourier multipliers and Lemma 1.76 (ii) then yields the representation $A|_{\mathcal{N}} = [\text{op}[m]^{(q)}]|_{\mathcal{N}}$ by the L_q-Fourier multiplier $m(\xi) := (1 - i\xi_j)^{-1}$. Note that we already have $1 \in \rho(\mathcal{D}_j^{\mathcal{N}})$ and

$$[\text{op}[m]^{(q)}]|_{\mathcal{N}} = (1 - \mathcal{D}_j^{\mathcal{N}})^{-1} \in L_{\text{Isom}}(\mathcal{N}, D(\mathcal{D}_j^{\mathcal{N}}))$$

according to Theorem 1.81. Thus, we obtain

$$\left[D(\mathcal{D}_j^{\mathcal{N}_0}), D(\mathcal{D}_j^{\mathcal{N}_1}) \right]_\theta = D(\mathcal{D}_j^{\mathcal{N}}). \tag{3.20}$$

It is obvious that $D(\mathcal{D}_j^{\mathcal{N}_0})$ is a Hilbert space with the graph norm $\|u\| := (\|u\|_{\mathcal{N}_0}^2 + \|\mathcal{D}_j^{\mathcal{N}_0} u\|_{\mathcal{N}_0}^2)^{1/2}$, $u \in D(\mathcal{D}_j^{\mathcal{N}_0})$. Proposition 3.10 and (3.20) then yield

$$\left[D((\mathcal{D}_j^{\mathcal{N}_0})^+), D((\mathcal{D}_j^{\mathcal{N}_1})^+) \right]_\theta = \left[{}_0H_{p_0,\varrho}^s(\mathbb{R}_+, D(\mathcal{D}_j^{\mathcal{N}_0})), {}_0B_{p_1p_1,\varrho}^s(\mathbb{R}_+, D(\mathcal{D}_j^{\mathcal{N}_1})) \right]_\theta$$
$$= {}_0F_{pq,\varrho}^s \left(\mathbb{R}_+, \left[D(\mathcal{D}_j^{\mathcal{N}_0}), D(\mathcal{D}_j^{\mathcal{N}_1}) \right]_\theta \right)$$
$$= {}_0F_{pq,\varrho}^s (\mathbb{R}_+, D(\mathcal{D}_j^{\mathcal{N}}))$$
$$= D((\mathcal{D}_j^{\mathcal{N}})^+).$$

Altogether, we obtain that $\mathcal{D}^{\mathcal{W}}_{+,\sigma}$ is the interpolated tuple of $\mathcal{D}^{\mathcal{W}_0}_{+,\sigma}$ and $\mathcal{D}^{\mathcal{W}_1}_{+,\sigma}$ in Theorem 1.45. Theorem 1.45 then yields the sectoriality and bisectoriality of $\sigma + \mathcal{D}^{\mathcal{W}}_t$ and $(\mathcal{D}^{\mathcal{N}}_j)^+$, $j = 1, \ldots, n$, respectively.

(ii) Due to Proposition 3.7 the representation of $f(\mathcal{D}^{\mathcal{W}}_{+,\sigma})$ can be derived by the compatibility result in Proposition 1.35.

(iii) This follows from Lemma 1.34.

(iv) Using Proposition 1.32 this similarly follows as in Theorem 1.89. □

Remark 3.32 (Properties of $\mathrm{op}_+[\Psi_s]$ and $\mathrm{op}[\Lambda_r]$ on Triebel-Lizorkin spaces). For $s \geq 0$, $r, r' \in \mathbb{R}$, $\varrho \geq 0$, $1 < p < \infty$, $s' \geq 0$, and $q \in (2p/(1+p), 2p)$, we have

$$\left[\mathrm{op}[\Lambda_r]|_{H_q^{r'+r}(\mathbb{R}^n)}\right]^+ \in L_{\mathrm{Isom}}({}_0F^{s'}_{pq,\varrho}(\mathbb{R}_+, H_q^{r'+r}(\mathbb{R}^n)), {}_0F^{s'}_{pq,\varrho}(\mathbb{R}_+, H_q^{r'}(\mathbb{R}^n))),$$

$$\mathrm{op}_+[\Psi_s]|_{{}_0F^{s'+s}_{pq,\varrho}(\mathbb{R}_+, H_q^{r'}(\mathbb{R}^n))} \in L_{\mathrm{Isom}}({}_0F^{s'+s}_{pq,\varrho}(\mathbb{R}_+, H_q^{r'}(\mathbb{R}^n)), {}_0F^{s'}_{pq,\varrho}(\mathbb{R}_+, H_q^{r'}(\mathbb{R}^n))),$$

where $\mathrm{op}_+[\Psi_s] := \mathscr{M}^{-1}_\varrho r_0^+ \mathrm{op}[\Psi_s(\cdot + \varrho)]e_0^+ \mathscr{M}_\varrho$. This result follows directly from (3.12), Lemma 3.29, Lemma 1.58, and Proposition 1.79.

Proposition 3.33 (Shifts on Triebel-Lizorkin spaces). Let $1 < p < \infty$, $2p/(1+p) < q < 2p$, $r' \in \mathbb{R}$, $s' > 0$, $s, r \geq 0$, $\varrho \geq 0$, and $\mathcal{W} := {}_0F^{s'}_{pq,\varrho}(H_q^{r'})$. Then the symbols Λ_r and Ψ_s give rise to the isomorphisms

$$\Lambda_r(\mathcal{D}^{\mathcal{W}}_+) = \left[\mathrm{op}[\Lambda_r]|_{H_q^{r'+r}(\mathbb{R}^n)}\right]^+ \in L_{\mathrm{Isom}}\left({}_0F^{s'}_{pq,\varrho}(H_q^{r'+r}), {}_0F^{s'}_{pq,\varrho}(H_q^{r'})\right),$$

$$\Psi_s(\mathcal{D}^{\mathcal{W}}_+) = \mathrm{op}_+[\Psi_s]|_{{}_0F^{s'+s}_{pq,\varrho}(H_q^{r'})} \in L_{\mathrm{Isom}}\left({}_0F^{s'+s}_{pq,\varrho}(H_q^{r'}), {}_0F^{s'}_{pq,\varrho}(H_q^{r'})\right),$$

and we have $D(\Lambda_r(\mathcal{D}^{\mathcal{W}}_+)) = {}_0F^{s'}_{pq,\varrho}(H_q^{r'+r})$, $D(\Psi_s(\mathcal{D}^{\mathcal{W}}_+)) = {}_0F^{s'+s}_{pq,\varrho}(H_q^{r'})$.

Proof. It is obvious that $\Lambda_r^{-1} \in H^\infty(\Omega)$ and $\Lambda_r \in H_P(\Omega)$ with $\Omega := S_\theta \times \Sigma^n_\delta$, $\theta > \pi/2$. From Theorem 3.31 (ii), Proposition 1.92, and Remark 3.32 we infer

$$(\Lambda_r^{-1})(\mathcal{D}^{\mathcal{W}}_+) = (\Lambda_r^{-1})(\nabla^{\mathcal{W}'}_+)|_{\mathcal{W}}, \quad \mathcal{W}' := L_{p,\varrho}(\mathbb{R}_+, H_q^{r'}(\mathbb{R}^n)),$$

$$= \left[\mathrm{op}[\Lambda_{-r}]|_{H_q^{r'}(\mathbb{R}^n)}\right]^+ \in L_{\mathrm{Isom}}(\mathcal{W}, F^{s'}_{pq,\varrho}(H_q^{r'+r})).$$

The statement on the domains follows as in the proof of Proposition 1.92 by Theorem 1.26.

The results on $\Psi_s(\mathcal{D}^{\mathcal{W}}_+)$ can be proved in the same way. □

b) H^∞-calculus of N-parabolic symbols on Bessel-valued Triebel-Lizorkin spaces

The shift operators in Proposition 2.60 were an essential ingredient for the proof of the main result on mixed-order systems in Section 2.3. Thus, we now have to provide a version for Triebel-Lizorkin spaces to derive an analogon.

3.3. Auxiliary results on Bessel-valued Triebel-Lizorkin spaces

Proposition 3.34. *Let $J \in \mathbb{N}_0$, $N_V := (r_i, s_i)_{i=0,\ldots,J+1} \subseteq [0, \infty)^2$ be the vertices (starting at the origin and being indexed in the counter-clockwise direction) of a Newton polygon N, $1 < p < \infty$, $2p/(1+p) < q < 2p$, $s' > 0$, and $r' \in \mathbb{R}$. If Φ_N is defined as in Proposition 2.60 and $\mathcal{W} := F^{s'}_{pq,\varrho}(H^{r'}_q)$, then for all $\sigma \geq 0$ the operator $\Phi_N(\mathcal{D}^\mathcal{W}_{+,\sigma})$ is invertible with*

$$D(\Phi_N(\mathcal{D}^\mathcal{W}_{+,\sigma})) = \bigcap_{(r,s) \in N_V} F^{s'+s}_{pq,\varrho}(H^{r'+r}_q),$$

$$\Phi_N(\mathcal{D}^\mathcal{W}_{+,\sigma}) \in L_{\text{Isom}}\left(\bigcap_{(r,s) \in N_V} F^{s'+s}_{pq,\varrho}(H^{r'+r}_q), \mathcal{W}\right).$$

Proof. The proof of Proposition 2.60 can be literally transferred using Theorem 1.26, Proposition 3.33, and (2.39). □

The next theorem is the analog of Theorem 2.62 and Theorem 2.63 for Triebel-Lizorkin spaces.

Theorem 3.35 (H^∞-**calculus for symbols in H_P on Triebel-Lizorkin spaces**). *Let $P \in H_P(\Omega)$, $\Omega := S_\theta \times \Sigma^n_\delta$, $\theta > \pi/2$, $1 < p < \infty$, $2p/(1+p) < q < 2p$, $s' > 0$, $r' \in \mathbb{R}$, and $\mathcal{W} := F^{s'}_{pq,\varrho}(H^{r'}_q)$. Then we deduce the following assertions.*

(i) *Let μ be an upper convex order function of P with $\alpha_\mu < s'$ and define the strictly positive order function $\mu_+(\gamma) := \mu(\gamma) + \alpha_\mu \gamma + \beta_\mu$. If we define $\mathcal{W}_- := {}_0 F^{s''}_{pq,\varrho}(H^{r''}_q)$ with $s'' := s' - \alpha_\mu > 0$ and $r'' := r' - \beta_\mu \in \mathbb{R}$, then we get, for all $\sigma \geq \sigma_0(P, \mu)$,*

$$D(P_\sigma(\mathcal{D}^{\mathcal{W}_-}_+)) \supseteq \bigcap_{\ell=0}^{M} {}_0 F^{s'+m_\ell(\mu)}_{pq,\varrho}(H^{r'+b_\ell(\mu)}_q)$$

$$= \bigcap_{(r,s) \in N_V(\mu_+)} {}_0 F^{s''+s}_{pq,\varrho}(H^{r''+r}_q) =: \mathcal{V}$$

and the restriction of the maximal realization to $\mathcal{V}$ yields the bounded operator $P_\sigma(\mathcal{D}_+)|_\mathcal{V} \in L(\mathcal{V}, \mathcal{W})$.

(ii) *Let μ be an upper concave order function of P with $\alpha_{-\mu} < s'$ and define the strictly positive order function $\mu_+(\gamma) := -\mu(\gamma) + \alpha_{-\mu}\gamma + \beta_{-\mu}$. If we define $\mathcal{W}_- := {}_0 F^{s''}_{pq,\varrho}(H^{r''}_q)$ with $s'' := s' - \alpha_{-\mu} > 0$ and $r'' := r' - \beta_{-\mu} \in \mathbb{R}$, then we get, for all $\sigma \geq \lambda_0(P, \mu)$,*

$$D(P_\sigma(\mathcal{D}^{\mathcal{W}_-}_+)) \supseteq \mathcal{W} \qquad (3.21)$$

and the restriction of the maximal realization to $\mathcal{W}$ yields the bounded operator $P_\sigma(\mathcal{D}_+)|_\mathcal{W} \in L(\mathcal{W}, \mathcal{V})$ with

$$\mathcal{V} := \bigcap_{\ell=0}^{M} {}_0 F^{s'-m_\ell(\mu)}_{pq,\varrho}(H^{r'-b_\ell(\mu)}_q) = \bigcap_{(r,s) \in N_V(\mu_+)} {}_0 F^{s''+s}_{pq,\varrho}(H^{r''+r}_q).$$

Proof. Using Proposition 3.34 instead of Proposition 2.60 the proofs are literally the same as in Theorem 2.62 and Theorem 2.63. □

Corollary 3.36. *Let $1 < p < \infty$, $2p/(1+p) < q < 2p$, $\varrho \geq 0$, $s' > 0$ and $r' \in \mathbb{R}$. Further let $\Omega := S_\theta \times \Sigma_\delta^n$, $\theta > \pi/2$. If $P \in S_N(\overline{\Omega})$ is an N-parabolic symbol, then there exists $\sigma > 0$ such that*

$$P_\sigma(\mathcal{D}_+) \in L_{\text{Isom}}\left(\bigcap_{(r,s) \in N_V(P)} {}_0F^{s'+s}_{pq,\varrho}(H_q^{r'+r}), {}_0F^{s'}_{pq,\varrho}(H_q^{r'})\right).$$

Proof. This proof can also be carried over literally from Corollary 2.65 by using Theorem 3.35 instead of Theorems 2.62 and 2.63. □

3.4 Mixed-order systems on Triebel-Lizorkin spaces

> *Motivation.* In order to formulate an analog version of Theorem 2.69 for Triebel-Lizorkin spaces we have to provide the same types of tools as in Section 2.3. Therefore we have to prove compatibility embeddings for the Bessel-valued Triebel-Lizorkin spaces. The representation of anisotropic Triebel-Lizorkin spaces with mixed norms by intersections in Proposition 3.23 crystallizes to be very helpful to derive these embeddings.
> Using the above results for the scales (F_{pq}, H_q) and (H_p, B_{pp}), we can prove the analog of the main result Theorem 2.69 for Triebel-Lizorkin spaces.

Proposition 3.37. *For all $s' \geq 0$, $r' \in \mathbb{R}$, $s, r \geq 0$, $\sigma \in (0,1)$, $\varrho \geq 0$, $p < \infty$, and $q \in (2p/(p+1), 2p)$ we have the embeddings*

$$_0F^{s'+s}_{pq,\varrho}(\mathbb{R}_+, H_q^{r'}(\mathbb{R}^n)) \cap {}_0F^{s'}_{pq,\varrho}(\mathbb{R}_+, H_q^{r'+r}(\mathbb{R}^n)) \hookrightarrow {}_0F^{s'+\sigma s}_{pq,\varrho}(\mathbb{R}_+, H_q^{r'+(1-\sigma)r}(\mathbb{R}^n)), \tag{3.22}$$

$$_0H^{s'+s}_{p,\varrho}(\mathbb{R}_+, B_{qq}^{r'}(\mathbb{R}^n)) \cap {}_0H^{s'}_{p,\varrho}(\mathbb{R}_+, B_{qq}^{r'+r}(\mathbb{R}^n)) \hookrightarrow {}_0H^{s'+\sigma s}_{p,\varrho}(\mathbb{R}_+, B_{qq}^{r'+(1-\sigma)r}(\mathbb{R}^n)). \tag{3.23}$$

Proof. The embedding (3.23) is already included in Lemma 2.61. Using Proposition 3.34 instead of Proposition 2.60 we can prove (3.22) in exactly the same way as Lemma 2.61. □

Proposition 3.38. *For all $s' \geq 0$, $r' \in \mathbb{R}$, $s, r > 0$, $\sigma \in (0,1)$, $\varrho \geq 0$, $1 < p < \infty$, and $q \in (2p/(1+p), 2p)$ we have the embedding*

$$_0F^{s'+s}_{pq,\varrho}(\mathbb{R}_+, H_q^{r'}(\mathbb{R}^n)) \cap {}_0H^{s'}_{p,\varrho}(\mathbb{R}_+, B_{qq}^{r'+r}(\mathbb{R}^n))$$
$$\hookrightarrow {}_0F^{s'+\sigma s}_{pq,\varrho}(\mathbb{R}_+, H_q^{r'+(1-\sigma)r}(\mathbb{R}^n)) \cap {}_0H^{s'+\sigma s}_{p,\varrho}(\mathbb{R}_+, B_{qq}^{r'+(1-\sigma)r}(\mathbb{R}^n)).$$

3.4. Mixed-order systems on Triebel-Lizorkin spaces

Proof. To prove this assertion we use the representation of anisotropic Triebel-Lizorkin spaces with mixed norms given in Proposition 3.23. For p_0, p_1, and θ as in Corollary 3.11 we define

$$\vec{a} := (r^{-1}, \ldots, r^{-1}, s^{-1}) \in (0, \infty)^{n+1}, \qquad \vec{p} := (q, \ldots, q, p) \in (1, \infty)^{n+1},$$
$$\vec{p_0} := (2, \ldots, 2, p_0) \in (1, \infty)^{n+1}, \qquad \vec{p_1} := (p_1, \ldots, p_1) \in (1, \infty)^{n+1}$$

and derive

$$F_{pq}^s(\mathbb{R}, L_q(\mathbb{R}^n)) \cap L_p(\mathbb{R}, B_{qq}^r(\mathbb{R}^n)) = F_{\vec{p},q}^{1,\vec{a}}(\mathbb{R}^{n+1}) = \left[F_{\vec{p_0},2}^{1,\vec{a}}(\mathbb{R}^{n+1}), F_{\vec{p_1},p_1}^{1,\vec{a}}(\mathbb{R}^{n+1})\right]_\theta$$
$$= \left[H_{p_0}^s(\mathbb{R}, L_2(\mathbb{R}^n)) \cap L_{p_0}(\mathbb{R}, H_2^r(\mathbb{R}^n)), B_{p_1 p_1}^s(\mathbb{R}, L_{p_1}(\mathbb{R}^n)) \cap L_{p_1}(\mathbb{R}, B_{p_1 p_1}^r(\mathbb{R}^n))\right]_\theta$$
(3.24)

by Propositions 3.23 and 3.26. Employing Lemmas 1.50 and 1.51 we can apply r_0^+ to equation (3.24) and obtain

$$_0 F_{pq}^s(\mathbb{R}_+, L_q(\mathbb{R}^n)) \cap L_p(\mathbb{R}_+, B_{qq}^r(\mathbb{R}^n))$$
$$= \left[_0 H_{p_0}^s(\mathbb{R}_+, L_2(\mathbb{R}^n)) \cap L_{p_0}(\mathbb{R}_+, H_2^r(\mathbb{R}^n)),\right.$$
$$\left._0 B_{p_1 p_1}^s(\mathbb{R}_+, L_{p_1}(\mathbb{R}^n)) \cap L_{p_1}(\mathbb{R}_+, B_{p_1 p_1}^r(\mathbb{R}^n))\right]_\theta.$$

Let $N := N(\{(r', s')\})$ and let $\Phi = \Phi_N(\mathcal{D}_+)$ be the associated shift operator (cf. Propositions 2.60 and 3.34). Define

$$X := {}_0 F_{pq}^{s'+s}(\mathbb{R}_+, H_q^{r'}(\mathbb{R}^n)) \cap {}_0 H_p^{s'}(\mathbb{R}_+, B_{qq}^{r'+r}(\mathbb{R}^n)).$$

Using Lemma 1.52 we get

$$X = \left[\Phi^{-1}\left({}_0 H_{p_0}^s(\mathbb{R}_+, L_2(\mathbb{R}^n)) \cap L_{p_0}(\mathbb{R}_+, H_2^r(\mathbb{R}^n))\right),\right.$$
$$\left.\Phi^{-1}\left({}_0 B_{p_1 p_1}^s(\mathbb{R}_+, L_{p_1}(\mathbb{R}^n)) \cap L_{p_1}(\mathbb{R}_+, B_{p_1 p_1}^r(\mathbb{R}^n))\right)\right]_\theta$$
$$= \left[{}_0 H_{p_0}^{s'+s}(\mathbb{R}_+, H_2^{r'}(\mathbb{R}^n)) \cap {}_0 H_{p_0}^{s'}(\mathbb{R}_+, H_2^{r'+r}(\mathbb{R}^n)),\right.$$
$$\left.{}_0 B_{p_1 p_1}^{s'+s}(\mathbb{R}_+, H_{p_1}^{r'}(\mathbb{R}^n)) \cap {}_0 H_{p_1}^{s'}(\mathbb{R}_+, B_{p_1 p_1}^{r'+r}(\mathbb{R}^n))\right]_\theta$$
$$\hookrightarrow \left[{}_0 H_{p_0}^{s'+\sigma s}(\mathbb{R}_+, H_2^{r'+(1-\sigma)r}(\mathbb{R}^n)), {}_0 B_{p_1 p_1}^{s'+\sigma s}(\mathbb{R}_+, H_{p_1}^{r'+(1-\sigma)r}(\mathbb{R}^n))\right]_\theta$$

by the embeddings of Lemma 2.61 and Proposition 2.75. Corollary 3.11 then yields

$$X \hookrightarrow {}_0 F_{pq}^{s'+\sigma s}(\mathbb{R}_+, H_q^{r'+(1-\sigma)r}(\mathbb{R}^n)).$$

Using $H_2^{r'+(1-\sigma)r}(\mathbb{R}^n) = B_{22}^{r'+(1-\sigma)r}(\mathbb{R}^n)$, Proposition 1.69, and the other embeddings in Proposition 2.75 we also derive

$$X \hookrightarrow \left[{}_0 H_{p_0}^{s'+\sigma s}(\mathbb{R}_+, B_{22}^{r'+(1-\sigma)r}(\mathbb{R}^n)), {}_0 H_{p_1}^{s'+\sigma s}(\mathbb{R}_+, B_{p_1 p_1}^{r'+(1-\sigma)r}(\mathbb{R}^n))\right]_\theta$$
$$= {}_0 H_p^{s'+\sigma s}(\mathbb{R}_+, B_{qq}^{r'+(1-\sigma)r}(\mathbb{R}^n))$$

in the same way. As before, the exponential weights can be treated by Remark 3.6. □

Proposition 3.39. *For all $s' > 0$, $r' \in \mathbb{R}$, $s, r \geq 0$, $\varrho \geq 0$, $\sigma \in (0,1)$, $1 < p < \infty$, and $q \in (2p/(p+1), 2p)$ we have the embeddings*

(i) $_0F_{pq,\varrho}^{s'+s}(\mathbb{R}_+, H_q^{r'}(\mathbb{R}^n)) \cap {_0F_{pq,\varrho}^{s'}}(\mathbb{R}_+, H_q^{r'+r}(\mathbb{R}^n))$
$$\hookrightarrow {_0H_{p,\varrho}^{s'+\sigma s}}(\mathbb{R}_+, B_{qq}^{r'+(1-\sigma)r}(\mathbb{R}^n)),$$

(ii) $_0H_{p,\varrho}^{s'+s}(\mathbb{R}_+, B_{qq}^{r'}(\mathbb{R}^n)) \cap {_0H_{p,\varrho}^{s'}}(\mathbb{R}_+, B_{qq}^{r'+r}(\mathbb{R}^n))$
$$\hookrightarrow {_0F_{pq,\varrho}^{s'+\sigma s}}(\mathbb{R}_+, H_q^{r'+(1-\sigma)r}(\mathbb{R}^n)).$$

Proof. (i) Let $N := N(\{(r,0),(0,s)\})$ and let $\Phi = \Phi_N(\mathcal{D}_+)$ be the associated shift operator (cf. Propositions 2.60 and 3.34). We then obtain

$$X := {_0F_{pq}^{s'+s}}(\mathbb{R}_+, H_q^{r'}(\mathbb{R}^n)) \cap {_0F_{pq}^{s'}}(\mathbb{R}_+, H_q^{r'+r}(\mathbb{R}^n))$$
$$= \Phi^{-1}({_0F_{pq}^{s'}}(\mathbb{R}_+, H_q^{r'}(\mathbb{R}^n))).$$

Due to (3.11) there exist $p_0, p_1 \in (1, \infty)$ and $\theta \in (0,1)$ such that

$$\Phi^{-1}({_0F_{pq}^{s'}}(\mathbb{R}_+, H_q^{r'}(\mathbb{R}^n)))$$
$$= \Phi^{-1}([{_0H_{p_0}^{s'}}(\mathbb{R}_+, H_2^{r'}(\mathbb{R}^n)), {_0B_{p_1 p_1}^{s'}}(\mathbb{R}_+, H_{p_1}^{r'}(\mathbb{R}^n))]_\theta)$$
$$= \left[\Phi^{-1}({_0H_{p_0}^{s'}}(\mathbb{R}_+, H_2^{r'}(\mathbb{R}^n))), \Phi^{-1}({_0B_{p_1 p_1}^{s'}}(\mathbb{R}_+, H_{p_1}^{r'}(\mathbb{R}^n)))\right]_\theta.$$

The established embeddings in Lemma 2.61 then yield

$$\Phi^{-1}({_0H_{p_0}^{s'}}(\mathbb{R}_+, H_2^{r'}(\mathbb{R}^n)))$$
$$= {_0H_{p_0}^{s'+s}}(\mathbb{R}_+, H_2^{r'}(\mathbb{R}^n)) \cap {_0H_{p_0}^{s'}}(\mathbb{R}_+, H_2^{r'+r}(\mathbb{R}^n))$$
$$\hookrightarrow {_0H_{p_0}^{s'+\sigma s}}(\mathbb{R}_+, H_2^{r'+(1-\sigma)r}(\mathbb{R}^n)), \tag{3.25}$$

$$\Phi^{-1}({_0B_{p_1 p_1}^{s'}}(\mathbb{R}_+, H_{p_1}^{r'}(\mathbb{R}^n)))$$
$$= {_0B_{p_1 p_1}^{s'+s}}(\mathbb{R}_+, H_{p_1}^{r'}(\mathbb{R}^n)) \cap {_0B_{p_1 p_1}^{s'}}(\mathbb{R}_+, H_{p_1}^{r'+r}(\mathbb{R}^n))$$
$$\hookrightarrow {_0H_{p_1}^{s'+\sigma s}}(\mathbb{R}_+, B_{p_1 p_1}^{r'+(1-\sigma)r}(\mathbb{R}^n)). \tag{3.26}$$

Altogether, we obtain with the two embeddings (3.25) and (3.26)

$$X \hookrightarrow \left[{_0H_{p_0}^{s'+\sigma s}}(\mathbb{R}_+, H_2^{r'+(1-\sigma)r}(\mathbb{R}^n)), {_0H_{p_1}^{s'+\sigma s}}(\mathbb{R}_+, B_{p_1 p_1}^{r'+(1-\sigma)r}(\mathbb{R}^n))\right]_\theta$$
$$= {_0H_p^{s'+\sigma s}}\left(\mathbb{R}_+, \left[H_2^{r'+(1-\sigma)r}(\mathbb{R}^n), B_{p_1 p_1}^{r'+(1-\sigma)r}(\mathbb{R}^n)\right]_\theta\right)$$
$$= {_0H_p^{s'+\sigma s}}(\mathbb{R}_+, B_{qq}^{r'+(1-\sigma)r}(\mathbb{R}^n)),$$

where we used Proposition 1.69. Note that we also used

$$H_2^{r'+(1-\sigma)r}(\mathbb{R}^n) = B_{22}^{r'+(1-\sigma)r}(\mathbb{R}^n)$$

in the last step.

3.4. Mixed-order systems on Triebel-Lizorkin spaces

(ii) Once more let $N := N(\{(r,0),(0,s)\})$ and let $\Phi = \Phi_N(\mathcal{D}_+)$ be the associated shift operator. The conditions on p and q then yield that there exist $p_0, p_1 \in (1, \infty)$ and $\theta \in (0,1)$ such that

$$_0F_{pq}^{s'+\sigma s}(\mathbb{R}_+, H_q^{r'+(1-\sigma r)}(\mathbb{R}^n))$$
$$= \left[_0H_{p_0}^{s'+\sigma s}(\mathbb{R}_+, H_2^{r'+(1-\sigma)r}(\mathbb{R}^n)), _0B_{p_1p_1}^{s'+\sigma s}(\mathbb{R}_+, H_{p_1}^{r'+(1-\sigma)r}(\mathbb{R}^n))\right]_\theta$$

(cf. Corollary 3.11). Using this and Proposition 1.69 we get

$$_0H_p^{s'+s}(\mathbb{R}_+, B_{qq}^{r'}(\mathbb{R}^n)) \cap _0H_p^{s'}(\mathbb{R}_+, B_{qq}^{r'+r}(\mathbb{R}^n))$$
$$= \Phi^{-1}\left(_0H_p^{s'}(\mathbb{R}_+, B_{qq}^{r'}(\mathbb{R}^n))\right)$$
$$= \Phi^{-1}\left(_0H_p^{s'}\left(\mathbb{R}_+, \left[H_2^{r'}(\mathbb{R}^n), B_{p_1p_1}^{r'}(\mathbb{R}^n)\right]_\theta\right)\right)$$
$$= \Phi^{-1}\left(\left[_0H_{p_0}^{s'}(\mathbb{R}_+, H_2^{r'}(\mathbb{R}^n)), _0H_{p_1}^{s'}(\mathbb{R}_+, B_{p_1p_1}^{r'}(\mathbb{R}^n))\right]_\theta\right)$$
$$= \left[\Phi^{-1}\left(_0H_{p_0}^{s'}(\mathbb{R}_+, H_2^{r'}(\mathbb{R}^n))\right), \Phi^{-1}\left(_0H_{p_1}^{s'}(\mathbb{R}_+, B_{p_1p_1}^{r'}(\mathbb{R}^n))\right)\right]_\theta \quad (3.27)$$

with $H_2^{r'}(\mathbb{R}^n) = B_{22}^{r'}(\mathbb{R}^n)$. As in (i) we use the established embeddings in Lemma 2.61 to derive

$$\Phi^{-1}\left(_0H_{p_0}^{s'}(\mathbb{R}_+, H_2^{r'}(\mathbb{R}^n))\right)$$
$$= _0H_{p_0}^{s'+s}(\mathbb{R}_+, H_2^{r'}(\mathbb{R}^n)) \cap _0H_{p_0}^{s'}(\mathbb{R}_+, H_2^{r'+r}(\mathbb{R}^n))$$
$$\hookrightarrow _0H_{p_0}^{s'+\sigma s}(\mathbb{R}_+, H_2^{r'+(1-\sigma)r}(\mathbb{R}^n)), \quad (3.28)$$

$$\Phi^{-1}\left(_0H_{p_1}^{s'}(\mathbb{R}_+, B_{p_1p_1}^{r'}(\mathbb{R}^n))\right)$$
$$= _0H_{p_1}^{s'+s}(\mathbb{R}_+, B_{p_1p_1}^{r'}(\mathbb{R}^n)) \cap _0H_{p_1}^{s'}(\mathbb{R}_+, B_{p_1p_1}^{r'+r}(\mathbb{R}^n))$$
$$\hookrightarrow _0H_{p_1}^{s'+\sigma s}(\mathbb{R}_+, B_{p_1p_1}^{r'+(1-\sigma)r}(\mathbb{R}^n)). \quad (3.29)$$

Therefore, with (3.27) and the embeddings (3.28) and (3.29) we get

$$_0H_p^{s'+s}(\mathbb{R}_+, B_{qq}^{r'}(\mathbb{R}^n)) \cap _0H_p^{s'}(\mathbb{R}_+, B_{qq}^{r'+r}(\mathbb{R}^n))$$
$$\hookrightarrow \left[_0H_{p_0}^{s'+\sigma s}(\mathbb{R}_+, H_2^{r'+(1-\sigma)r}(\mathbb{R}^n)), _0H_{p_1}^{s'+\sigma s}(\mathbb{R}_+, B_{p_1p_1}^{r'+(1-\sigma)r}(\mathbb{R}^n))\right]_\theta$$
$$= _0F_{pq}^{s'+\sigma s}(\mathbb{R}_+, H_q^{r'+(1-\sigma)r}(\mathbb{R}^n)). \qquad \square$$

Now we want to provide an analog of the main result on mixed-order systems, Theorem 2.69, for the scale of Triebel-Lizorkin spaces. In Section 3.3 we have derived all necessary statements for the scales (F_{pq}, H_q) and (H_p, B_{qq}) to formulate this analog result. To clarify the situation we state this explicitly in the next definition and theorem.

Definition 3.40 (Compatible tuple of spaces). Let $\mathscr{L} \in [H_P(S_\theta \times \Sigma_\delta^n)]^{m \times m}$, $\theta > \pi/2$, $\delta > 0$, be an N-parabolic mixed-order system with order functions s_j, t_k, $j, k = 1, \ldots, m$. Then in the Triebel-Lizorkin setting, a tuple of spaces $\mathbb{H} = \prod_{i=1}^{m} \mathbb{H}_i$ and $\mathbb{F} = \prod_{j=1}^{m} \mathbb{F}_j$ is called *compatible with* $\mathscr{L}$ if the following conditions are satisfied:

(i) For $i, j = 1, \ldots, m$, the spaces $\mathbb{H}_i, \mathbb{F}_j$ are of the form

$$\mathbb{H}_i := \bigcap_{\ell=0}^{M} {}_0\mathcal{F}_{\ell,\varrho}^{s'_\ell + m_\ell(t_i)}(\mathcal{K}_\ell^{r'_\ell + b_\ell(t_i)}), \quad \mathbb{F}_j := \bigcap_{\ell=0}^{M} {}_0\mathcal{F}_{\ell,\varrho}^{s'_\ell - m_\ell(s_j)}(\mathcal{K}_\ell^{r'_\ell - b_\ell(s_j)})$$

with $(\mathcal{F}_\ell, \mathcal{K}_\ell) \in \{(F_{pq}, H_q), (H_p, B_{qq})\}$, $1 < p, q < \infty$, $\varrho \geq 0$, $s'_\ell \geq 0$, and $r'_\ell \in \mathbb{R}$ ($\ell = 0, \ldots, M$).

(ii) For each $i, j = 1, \ldots, m$, the functions

$$\mu_{\mathbb{H}_i}(\gamma) := \max_\ell \{[s'_\ell + m_\ell(t_i)]\gamma + r'_\ell + b_\ell(t_i)\}, \quad \gamma \geq 0,$$

$$\mu_{\mathbb{F}_j}(\gamma) := \max_\ell \{[s'_\ell - m_\ell(s_j)]\gamma + r'_\ell - b_\ell(s_j)\}, \quad \gamma \geq 0,$$

are convex increasing order functions and

$$s'_\ell \geq \max\left\{\max_{i=1,\ldots,m}\{(\delta_{i,1} - m_\ell(t_i)), \max_{j=1,\ldots,m}(\delta_{j,2} + m_\ell(s_j))\}\right\} \quad (3.30)$$

for all $\ell = 0, \ldots, M$ where

$$\delta_{i,1} := \begin{cases} 0, & \text{if } s_j + t_i \text{ is not concave for all } j \in \{0, \ldots, m\}, \\ \max\{\alpha_{-s_j - t_i} : j \in \{0, \ldots, m\} \text{ such that } s_j + t_i \text{ is concave}\}, & \text{else,} \end{cases}$$

$$\delta_{j,2} := \begin{cases} 0, & \text{if } s_j + t_i \text{ is not convex for all } i \in \{0, \ldots, m\}, \\ \max\{\alpha_{s_j + t_i} : i \in \{0, \ldots, m\} \text{ such that } s_j + t_i \text{ is convex}\}, & \text{else.} \end{cases}$$

Note that for each $\ell \in \{0, \ldots, M\}$ we have to choose $(\mathcal{F}_\ell, \mathcal{K}_\ell) = (H_p, B_{qq})$ if equality holds in (3.30).

(iii) Embedding conditions: For $i, j = 1, \ldots, m$ we define

$$\mathbb{H}_{ij} := \bigcap_{\substack{\ell, \kappa = 0, \\ \ell \neq \kappa}}^{M} {}_0\mathcal{F}_{\ell,\varrho}^{\sigma_{ji}(\ell,\kappa,t)}(\mathcal{K}_\ell^{\eta_{ji}(\ell,\kappa,t)}), \quad \mathbb{F}_{ij} := \bigcap_{\substack{\ell, \kappa = 0, \\ \ell \neq \kappa}}^{M} {}_0\mathcal{F}_{\ell,\varrho}^{\sigma_{ji}(\ell,\kappa,s)}(\mathcal{K}_\ell^{\eta_{ji}(\ell,\kappa,s)})$$

where

$$\sigma_{ji}(\ell, \kappa, t) := s'_\ell - m_\ell(s_j) + m_\kappa(s_j) + m_\kappa(t_i),$$
$$\sigma_{ji}(\ell, \kappa, s) := s'_\ell + m_\ell(t_i) - [m_\kappa(s_j) + m_\kappa(t_i)],$$
$$\eta_{ji}(\ell, \kappa, t) := r'_\ell - b_\ell(s_j) + b_\kappa(s_j) + b_\kappa(t_i),$$
$$\eta_{ji}(\ell, \kappa, s) := r'_\ell + b_\ell(t_i) - [b_\kappa(s_j) + b_\kappa(t_i)].$$

3.4. Mixed-order systems on Triebel-Lizorkin spaces

Then the embeddings

$$\mathbb{H}_i \hookrightarrow \mathbb{H}_{ij} \quad \text{if } s_j + t_i \text{ is convex,} \tag{3.31}$$

$$\mathbb{F}_j \hookrightarrow \mathbb{F}_{ij} \quad \text{if } s_j + t_i \text{ is concave} \tag{3.32}$$

hold for $i, j = 1, \ldots m$.

Theorem 3.41 (Main Theorem on N-parabolic mixed-order systems on Triebel-Lizorkin spaces). *Let $\mathscr{L} \in [H_P(S_\theta \times \Sigma_\delta^n)]^{m \times m}$, $\theta > \pi/2$, $\delta > 0$, be an N-parabolic mixed-order system. Let the tuples $\mathbb{H} = \prod_{i=1}^m \mathbb{H}_i$ and $\mathbb{F} = \prod_{j=1}^m \mathbb{F}_j$ be compatible in the Triebel-Lizorkin setting with $\mathscr{L}$ as formulated in Definition 3.40. Then there exists $\sigma_0 > 0$ such that, for all $\sigma \geq \sigma_0$,*

$$[\mathscr{L}_\sigma(\mathcal{D}_+)]|_{\mathbb{H}} \in L_{\mathrm{Isom}}(\mathbb{H}, \mathbb{F}) \quad \text{and} \quad ([\mathscr{L}_\sigma(\mathcal{D}_+)]|_{\mathbb{H}})^{-1} = [\mathscr{L}_\sigma^{-1}(\mathcal{D}_+)]|_{\mathbb{F}}.$$

Proof. Using Theorem 3.35 instead of Theorem 2.62 and Theorem 2.63 the proof of Theorem 2.69 can be carried over literally. □

Similarly to Definition 2.78, we introduce an admissible Triebel-Lizorkin scale.

Definition 3.42 (Admissible Triebel-Lizorkin scale). Let μ_1 and μ_2 be convex increasing order functions such that $\mu_1 - \mu_2$ is an order function and let $1 < p < \infty$, $2p/(1+p) < q < 2p$. Let the scale

$$(\mathcal{F}_\ell, \mathcal{K}_\ell) \in \{(H_p, B_{qq}), (F_{pq}, H_q)\}, \quad \ell = 0, \ldots, M$$

be given such that there exists $\tau \in \{0, \ldots, M-1\}$ with

$$(\mathcal{F}_\ell, \mathcal{K}_\ell) = (H_p, B_{qq}), \quad \ell \in \{0, \ldots, \tau\},$$
$$(\mathcal{F}_\ell, \mathcal{K}_\ell) = (F_{pq}, H_q), \quad \ell \in \{\tau+1, \ldots, M\}.$$

The scale $(\mathcal{F}_\ell, \mathcal{K}_\ell)_{\ell=0,\ldots,M}$ is then called (μ_1, μ_2)-admissible if we have

$$(b_\tau(\mu_2), m_\tau(\mu_2)) \neq (b_{\tau+1}(\mu_2), m_{\tau+1}(\mu_2)), \quad \text{if } \mu_1 - \mu_2 \text{ is convex,}$$
$$(b_\tau(\mu_1), m_\tau(\mu_1)) \neq (b_{\tau+1}(\mu_1), m_{\tau+1}(\mu_1)), \quad \text{if } \mu_1 - \mu_2 \text{ is concave.}$$

Note that this definition is also meaningful if $\mu_1 - \mu_2$ has trivial index, i.e., if there exists $\alpha, \beta \in \mathbb{R}$ such that $(\mu_1 - \mu_2)(\gamma) = \alpha\gamma + \beta$ for all $\gamma > 0$ and therefore $\mu_1 - \mu_2$ is convex as well as concave.

Proposition 3.43 (Compatibility condition III). *Let us consider the situation of Theorem 3.41 with $1 < p < \infty$, $2p/(1+p) < q < 2p$, and*

$$(\mathcal{F}_\ell, \mathcal{K}_\ell) \in \{(H_p, B_{qq}), (F_{pq}, H_q)\}, \quad \ell = 0, \ldots, M.$$

If the scale $(\mathcal{F}_\ell, \mathcal{K}_\ell)_{\ell=0,\ldots,M}$ is $(\mu_{\mathbb{H}_i}, \mu_{\mathbb{F}_j})$-admissible for all $i, j = 0, \ldots, M$, then the embedding conditions (3.31) and (3.32) are satisfied, and the tuples $\mathbb{H}$, $\mathbb{F}$ are compatible with $\mathscr{L}$.

Proof. The proof is exactly the same as the proof of Proposition 2.79. Instead of the embeddings in Lemma 2.61 and Propositions 2.75 and 2.76 we here have to apply Propositions 3.37, 3.38, and 3.39. □

As in Section 2.3 we state a condensed version of Theorem 3.41.

Corollary 3.44. *Let $\mathscr{L} \in [H_P(S_\theta \times \Sigma_\delta^n)]^{m \times m}$, $\theta > \pi/2$, be an N-parabolic mixed-order system such that for each $i,j = 1, \ldots, m$ the order function $s_j + t_i$ is convex and increasing or concave and decreasing. Let $\varrho \geq 0$, $s'_\ell \geq 0$, $r'_\ell \in \mathbb{R}$ ($\ell = 0, \ldots, M$) such that*

$$\mu_{\mathbb{H}_i}(\gamma) := \max_\ell \{[s'_\ell + m_\ell(t_i)]\gamma + r'_\ell + b_\ell(t_i)\}, \quad \gamma \geq 0,$$

$$\mu_{\mathbb{F}_j}(\gamma) := \max_\ell \{[s'_\ell - m_\ell(s_j)]\gamma + r'_\ell - b_\ell(s_j)\}, \quad \gamma \geq 0, \quad i,j = 1, \ldots m$$

are convex increasing order functions. Furthermore, let the scale

$$(\mathcal{F}_\ell, \mathcal{K}_\ell) \in \{(F_{pq}, H_q), (H_p, B_{qq})\}, \quad 1 < p < \infty, \ 2p/(1+p) < q < 2p,$$

$\ell = 0, \ldots, M$, *be $(\mu_{\mathbb{H}_i}, \mu_{\mathbb{F}_j})$-admissible for all $i, j = 1, \ldots m$ and let*

$$s'_\ell > \max\{\max\{-m_\ell(t_i), m_\ell(s_i)\} : i = 1, \ldots, m\}$$

for all $\ell \in \{0, \ldots, \tau\}$ where τ is taken from Definition 3.42. Using the same notation as in Definition 3.27 we define for $i, j = 1, \ldots m$ the spaces

$$\mathbb{H}_i := \bigcap_{\ell=0}^{M} {}_0\mathcal{F}_{\ell,\varrho}^{s'_\ell + m_\ell(t_i)}(\mathcal{K}_\ell^{r'_\ell + b_\ell(t_i)}), \quad \mathbb{F}_j := \bigcap_{\ell=0}^{M} {}_0\mathcal{F}_{\ell,\varrho}^{s'_\ell - m_\ell(s_j)}(\mathcal{K}_\ell^{r'_\ell - b_\ell(s_j)}).$$

Then there exists $\sigma_0 > 0$ such that, for all $\sigma \geq \sigma_0$,

$$[\mathscr{L}_\sigma(\mathcal{D}_+)]|_{\mathbb{H}} \in L_{\text{Isom}}(\mathbb{H}, \mathbb{F}) \quad \text{and} \quad ([\mathscr{L}_\sigma(D_+)]|_{\mathbb{H}})^{-1} = [\mathscr{L}_\mu^{-1}(D_+)]|_{\mathbb{F}}$$

where $\mathbb{H} := \prod_{i=1}^m \mathbb{H}_i$ and $\mathbb{F} := \prod_{i=1}^m \mathbb{F}_i$.

Proof. As the order functions $s_j + t_i$ are convex and increasing or concave and decreasing, we have $\alpha_{s_j+t_i} = 0$ or $\alpha_{-s_j-t_i} = 0$, respectively. This yields $\delta_{i,1} = \delta_{j,2} = 0$ for all $i, j = 1, \ldots, m$. Due to Proposition 3.43, the tuples $\mathbb{H}$ and $\mathbb{F}$ are compatible with $\mathscr{L}$. Hence, the assertion follows from Theorem 3.41. □

3.5 Singular integral operators on L_p-L_q

Motivation. In most cases we can treat a boundary value problem on the half-space by partial Fourier and Laplace transform followed by a reduction to the boundary. Roughly speaking, we often derive a formula for the solution of a boundary value problem which is formally given by

$$u(t, x', x_n) = [h(\partial_t, \mathcal{D}_1, \ldots, \mathcal{D}_{n-1}, x_n) g](t, x') \qquad (3.33)$$

for $t > 0$ and $x = (x', x_n) \in \mathbb{R}^n_+$ where h is a suitable scalar function and g a function on the boundary, i.e., in the trace space of the canonical solution space for u. For the heat equation, for instance, the solution is given by $u(t, x', x_n) = (h(\partial_t, \mathcal{D}_1, \ldots, \mathcal{D}_{n-1}, x_n) g)(t, x')$ where g is the trace of u on the boundary and

$$h(\lambda, z, x_n) := \exp\left(-\sqrt{\lambda + |z|^2_-} \cdot x_n\right) \quad \text{with } |z|_- := \left(-\sum_{k=1}^{n-1} z_k^2\right)^{1/2}$$

for $(\lambda, z) \in S_\theta \times \Sigma_\delta^{n-1}$.

For this reason we have to concentrate on the parameter-dependent H^∞-calculus and the regularities in these parameters. Due to the fact that the operator $h(\partial_t, \mathcal{D}_1, \ldots, \mathcal{D}_{n-1}, x_n)$ in (3.33) only makes sense for fixed $x_n > 0$ we have to rearrange the order of arguments. In L_p-L_p-theory there is no trouble by this rearrangement but in L_p-L_q-theory some problems arise from the fact that we cannot apply Fubini's theorem for the time- and x_n-variable. To handle this problem we have to interpret the operator $h(\partial_t, \mathcal{D}_1, \ldots, \mathcal{D}_{n-1}, x_n)$ by a two step functional calculus. We define such an operator by singular integral operators in Definition 3.60. This definition is related to the so-called "Volevich trick", cf. [Vol65].

Singular integral operators on L_p-L_q can be found in [Fer87], for example. There the author considers operator-valued kernels of product type but this is not sufficient for our applications. To derive the desired regularities, associated with an order function, and the connection to the joint H^∞-calculus of $\mathcal{D}_+$ we have to develop own results on this topic.

a) Singular integral operators

In the sequel we define the class of integral kernels which are bounded from above by the kernel of the one-sided Hilbert transform and a weight function given by an order function. A special case of the kernels in the next definition can be found in [DHP03, Section 7.1], for instance.

Here and in the following we always assume $1 < p, q < \infty$. As singular integral operators appear typically in connection with boundary value problems, we consider the tuple $\nabla' := (\partial_1, \ldots, \partial_{n-1})$ instead of ∇. We will denote the $\mathcal{N}$-realization of ∇' by $(\mathcal{D}')^{\mathcal{N}}$, and the $\mathcal{N}$-realization of (∂_t, ∇') by $(\mathcal{D}'_+)^{\mathcal{N}}$. We will write $\mathcal{D}'$ and $\mathcal{D}'_+$ if the underlying space is clear. We also set $z' := (z_1, \ldots, z_{n-1})$.

Definition 3.45. Let μ be a strictly positive order function and $K \in \mathbb{N}_0$. Then we define
$$\mathcal{K}_K(S_\theta \times \Sigma_\delta^{n-1}, \mu), \quad \theta > \pi/2, \quad \delta > 0$$
as the set of all kernel functions $k \colon S_\theta \times \Sigma_\delta^{n-1} \times \mathbb{R}_+ \times \mathbb{R}_+ \to \mathbb{C}$ with
$$k(\cdot, \cdot, x_n, y_n) \in H^\infty(S_\theta \times \Sigma_\delta^{n-1}), \quad x_n, y_n > 0$$
and $k(\lambda, z', \cdot, y_n) \in C^K(\mathbb{R}_+, \mathbb{C})$ for all $(\lambda, z') \in S_\theta \times \Sigma_\delta^{n-1}$, $y_n > 0$ such that
$$|\partial_{x_n}^j k(\lambda, z', x_n, y_n)| \leq C \cdot \frac{(W_\mu(\lambda, z))^j}{x_n + y_n}$$
for all $(\lambda, z') \in S_\theta \times \Sigma_\delta^{n-1}$, $x_n, y_n > 0$, and $j = 0, \ldots, K$.

Remark 3.46 (One-sided Hilbert transform). For $u \in L_p(\mathbb{R}_+)$ we define the one-sided Hilbert transform by
$$(H_+ u)(x) := \int_0^\infty \frac{u(y)}{x+y} dy, \quad x \in \mathbb{R}_+.$$
Then we have $H_+ \in L(L_p(\mathbb{R}_+))$. This can be easily seen by a reflection argument and the boundedness of the Hilbert transform on $L_p(\mathbb{R})$.

In the following, let X be a Banach space of class $\mathcal{HT}$ with property (α). We fix the ground space $\mathcal{N} := H_q^r(\mathbb{R}^{n-1}, X)$ with $r \in \mathbb{R}$ and consider the $\mathcal{N}$-realization $\mathcal{D}' = (\mathcal{D}')^{\mathcal{N}}$ of ∇'.

Definition 3.47. For fixed $\lambda \in S_\theta$ and kernel $k \in \mathcal{K}_K(S_\theta \times \Sigma_\delta^{n-1}, \mu)$ we define the parameter-dependent *singular integral operator*
$$(G^{\mathcal{N}}[\lambda, k]g)(x_n) := \int_0^\infty [k(\lambda, \mathcal{D}', x_n, y_n)g(y_n)] dy_n, \quad x_n > 0$$
for $g \in L_q(\mathbb{R}_+, H_q^r(\mathbb{R}^{n-1}, X))$ and $r \in \mathbb{R}$. For fixed x_n and λ the integral exists due to Remark 3.46 and $\|k(\lambda, \mathcal{D}', x_n, y_n)\|_{L(H_q^r(\mathbb{R}^{n-1}, X))} \leq C(x_n + y_n)^{-1}$ for all $y_n > 0$. In particular, this yields $G^{\mathcal{N}}[\lambda, k] \in L(L_q(\mathbb{R}_+, H_q^r(\mathbb{R}^{n-1}, X)))$.

As an application of Proposition 1.10 we derive the next result.

Proposition 3.48. *Let $k \in \mathcal{K}_K(S_\theta \times \Sigma_\delta^{n-1}, \mu)$. Then the following assertions hold:*

(i) *The set*
$$\{k(\lambda, \mathcal{D}', x_n, y_n) \colon \lambda \in S_\theta\} \subseteq L(H_q^r(\mathbb{R}^{n-1}, X)), \quad x_n, y_n > 0$$

3.5. Singular integral operators on L_p-L_q

is $\mathcal{R}$-bounded with

$$\mathcal{R}_p(\{k(\lambda, \mathcal{D}', x_n, y_n) \colon \lambda \in S_\theta\}) \leq C \cdot \frac{1}{x_n + y_n}, \quad x_n, y_n > 0 \qquad (3.34)$$

for some constant $C > 0$.

(ii) *The set $\{G^{\mathcal{N}}[\lambda, k] \colon \lambda \in S_\theta\} \subseteq L(L_q(\mathbb{R}_+, H_q^r(\mathbb{R}^{n-1}, X)))$ is $\mathcal{R}$-bounded.*

Proof. (i) Due to the definition of $\mathcal{K}_M(S_\theta \times \Sigma_\delta^{n-1}, \mu)$ there exists $C > 0$ such that

$$(x_n + y_n) k(\cdot, \cdot, x_n, y_n) \in H^\infty(S_\theta \times \Sigma_\delta^{n-1}), \quad \|k(\cdot, \cdot, x_n, y_n)\|_\infty \leq C(x_n + y_n)^{-1}$$

for all $x_n, y_n > 0$. Theorem 1.81 and Lemma 1.23 directly yield the $\mathcal{R}$-boundedness of

$$\{f(\mathcal{D}') \colon f \in H^\infty(S_\theta \times \Sigma_\delta^{n-1}), \|f\|_\infty \leq C\} \subseteq L(H_q^r(\mathbb{R}^{n-1}, X)).$$

Thus, we derive (i) from Remark 1.7 (vi).

(ii) To apply Proposition 1.10 we define $\mathcal{K} := \{k(\lambda, \mathcal{D}', x_n, y_n) \colon \lambda \in S_\theta\}$ and

$$k_0(x_n, y_n) := (x_n + y_n)^{-1}, \quad x_n, y_n > 0.$$

The assumptions of Proposition 1.10 are fulfilled due to (3.34) and Remark 3.46. So we deduce the $\mathcal{R}$-boundedness of $\{G^{\mathcal{N}}[\lambda, k] \colon \lambda \in S_\theta\} \subseteq L(L_q(\mathbb{R}_+, H_q^r(\mathbb{R}^{n-1}, X)))$. $\square$

Definition 3.49. Let X be a Banach space of class $\mathcal{HT}$. We define the *rearranging operator* by

$$U \colon L_q(\mathbb{R}^{n-1}, L_q(\mathbb{R}_+, X)) \to L_q(\mathbb{R}_+, L_q(\mathbb{R}^{n-1}, X))$$

where $[(Uf)(x_n)](x') := [f(x')](x_n)$ for $f \in L_q(\mathbb{R}^{n-1}, L_q(\mathbb{R}_+, X))$, $x' \in \mathbb{R}^{n-1}$, $x_n > 0$.

Lemma 3.50. *We get*

$$U \in L_{\mathrm{Isom}}\left(H_q^r(\mathbb{R}^{n-1}, H_q^\ell(\mathbb{R}_+, X)), H_q^\ell(\mathbb{R}_+, H_q^r(\mathbb{R}^{n-1}, X))\right) \qquad (3.35)$$

for all $\ell, r \geq 0$.

Proof. The case $\ell, r = 0$ is based on Fubini's theorem and can be found in [AE09, III X 6.22], for example. The general case can be shown by easy calculations. $\square$

In the next proposition we introduce singular integral operators on L_p-L_q given by kernels in the class $\mathcal{K}_M(S_\theta \times \Sigma_\delta^n, \mu)$.

Definition and Proposition 3.51 (Singular integral operator on L_p-L_q). Let $k \in \mathscr{K}_K(S_\theta \times \Sigma_\delta^{n-1}, \mu)$ and $\mathcal{W} := {}_0H_{p,\varrho}^s(\mathbb{R}_+, L_q(\mathbb{R}_+, H_q^r(\mathbb{R}^{n-1}, X)))$, $s, r \geq 0, \varrho \geq 0$, $1 < p, q < \infty$. Then we define the singular integral operator realized on $\mathcal{W}$ by

$$G^{\mathcal{W}}[k] := T(\mathcal{D}_t^{\mathcal{W}}) \in L(\mathcal{W}) \tag{3.36}$$

where

$$T(\lambda) := (G^{\mathcal{N}}[\lambda, k])^+ \in L(\mathcal{W}), \quad \lambda \in S_\theta.$$

Proof. Due to Proposition 3.48 (ii) and Lemma 1.86 (ii) we have

$$[\lambda \in S_\theta \mapsto (G^{\mathcal{N}}[\lambda, k])^+] \in H_{\mathcal{R}}^\infty(S_\theta, \mathcal{B}_{\mathcal{D}_t^{\mathcal{M}}}))$$

where $\mathcal{B}_{\mathcal{D}_t^{\mathcal{M}}}$ is the commutator of $\mathcal{D}_t^{\mathcal{M}}$. An application of Theorem 1.28 and Theorem 1.84 then shows the asserted boundedness of $G^{\mathcal{W}}[k]$. □

To derive more information on the regularity in the x_n-variable of $G^{\mathcal{W}}[k]g$ we have to consider the behavior of the parameter-dependent H^∞-calculus under derivatives with respect to this parameter.

Lemma 3.52. *Let $s \geq 0$ and $m \in \mathbb{N}_0$. Then*

$${}_0H_{p,\varrho}^s(\mathbb{R}_+^{(t)}, H_q^m(\mathbb{R}_+^{(x)}, X))$$

is the set of all functions $f \in {}_0H_{p,\varrho}^s(\mathbb{R}_+^{(t)}, L_q(\mathbb{R}_+^{(x)}, X))$ such that

$$(\partial_x^+)^\ell f \in {}_0H_{p,\varrho}^s(\mathbb{R}_+^{(t)}, L_q(\mathbb{R}_+^{(x)}, X)) \tag{3.37}$$

for all $\ell \in \{0, \ldots, m\}$.

Proof. Let $f \in {}_0H_{p,\varrho}^s(\mathbb{R}_+^{(t)}, H_q^m(\mathbb{R}_+^{(x)}, X))$. Then it is obvious that (3.37) holds for all $\ell \leq m$ by the definition of the canonical extension.

Let $f \in {}_0H_{p,\varrho}^s(\mathbb{R}_+^{(t)}, L_q(\mathbb{R}_+^{(x)}, X))$ be such that (3.37) holds. We then have $\partial_x^\ell(f(t)) \in L_q(\mathbb{R}_+^{(x)}, X)$ for almost all $t \in \mathbb{R}_+$ and $\ell \leq m$. This obviously yields $f(t) \in H_q^m(\mathbb{R}_+^{(x)}, X)$ for almost all $t \in \mathbb{R}_+$. Using the definition of the norm in ${}_0H_{p,\varrho}^s(\mathbb{R}_+^{(t)}, H_q^m(\mathbb{R}_+^{(x)}, X))$ we obtain $f \in {}_0H_{p,\varrho}^s(\mathbb{R}_+^{(t)}, H_q^m(\mathbb{R}_+^{(x)}, X))$. □

In the next result we describe the derivatives of the H^∞-calculus depending on a parameter.

Lemma 3.53. *Let $I \subseteq \mathbb{R}$ be an interval, $K \in \mathbb{N}_0$, and let $f: \Sigma_\delta^{n-1} \times I \to \mathbb{C}$ be a function such that*

$$f(z', \cdot) \in C^K(I, \mathbb{C}), \quad z' \in \Sigma_\delta^{n-1},$$
$$(\partial_x^j f)(\cdot, x) \in H^\infty(\Sigma_\delta^{n-1}), \quad x \in I, \, j = 0, \ldots, K.$$

3.5. Singular integral operators on L_p-L_q

Let $\mathcal{N} = H_q^r(\mathbb{R}^{n-1}, X)$, and let $\mathcal{D}' = (\mathcal{D}')^{\mathcal{N}}$ be the $\mathcal{N}$-realization of ∇'. For $x \in I$, set $T(x) := f(\mathcal{D}', x)$. Then for every $g \in \mathcal{N}$ we have $T(\cdot)g \in C^K(I, \mathcal{N})$ and
$$\partial_x^j(T(x)g) = [(\partial_x^j f)(\mathcal{D}', x)]g, \quad x \in I, \quad j = 0, \ldots, K.$$

Proof. It is obvious that it is sufficient to consider the case $K = 1$. Let $g \in H_q^r(\mathbb{R}^{n-1}, X)$ and $(g_j)_{j \in \mathbb{N}} \subseteq \mathscr{S}(\mathbb{R}^n, X)$ with $g_j \to g$ in $H_q^r(\mathbb{R}^{n-1}, X)$. Then we have
$$\partial_x(T(x)g_j) = \partial_x(\mathscr{F}^{-1}f(i\xi', x)\mathscr{F}g_j) = \mathscr{F}^{-1}(\partial_x f)(i\xi', x)\mathscr{F}g_j$$
$$= (\partial_x f)(\mathcal{D}', x)g_j$$
where $\xi' := (\xi_1, \ldots, \xi_{n-1})$. Now it is easy to show $\partial_x(T(x)g) = (\partial_x f)(\mathcal{D}', x)g$ for all $x \in I$. □

In the following, let μ be a strictly positive order function, let $M \in \mathbb{N}$, $m_\ell(\mu) \geq 0$, $b_\ell(\mu) \geq 0$, $\ell = 0, \ldots, M$ be given as in Definition 2.21, and let X be a Banach space of class $\mathcal{HT}$ with property (α).

For the treatment of singular integral operators on Bessel potential spaces with mixed regularity we introduce an abbreviation for those spaces. We define
$$X_j(\mu, \kappa) := \bigcap_{\ell=0}^{M(\mu)} {}_0 H_{p,\varrho}^{j \cdot m_\ell(\mu)}(\mathbb{R}_+, H_q^\kappa(\mathbb{R}_+, H_q^{j \cdot b_\ell(\mu)}(\mathbb{R}^{n-1}, X))),$$
for $\varrho \geq 0$, $j \in \mathbb{N}_0$, and $\kappa \geq 0$. With the help of these spaces we can describe the gained regularity in the x_n-variable by a singular integral operator defined by (3.36).

Lemma 3.54. *Let $f \in H^\infty(S_\theta \times \Sigma_\delta^{n-1})$, $\theta > \pi/2$. Then we have*
$$(f(\lambda, (\mathcal{D}')^{\mathcal{N}}))^{+_n} = Uf(\lambda, (\mathcal{D}')^{\mathcal{M}})U^{-1}$$
where $\mathcal{N} := L_q(\mathbb{R}^{n-1}, X)$ and $\mathcal{M} := L_q(\mathbb{R}^{n-1}, L_q(\mathbb{R}_+, X))$. For the sake of clarity $+_n$ denotes the natural extension to $L_q(\mathbb{R}_+, L_q(\mathbb{R}^{n-1}, X))$.

Proof. We have $(f(\lambda, (\mathcal{D}')^{\mathcal{N}}))^{+_n} = f(\lambda, ((\mathcal{D}')^{\mathcal{N}})^{+_n})$ due to Lemma 1.87 (iv). With Proposition 1.32 and $U^{-1}((\mathcal{D}')^{\mathcal{N}})^{+_n}U = (\mathcal{D}')^{\mathcal{M}}$ we then get $f(\lambda, ((\mathcal{D}')^{\mathcal{N}})^{+_n}) = Uf(\lambda, (\mathcal{D}')^{\mathcal{M}})U^{-1}$, which yields the assertion. □

Lemma 3.55. *Let X be a Banach space of class $\mathcal{HT}$ with property (α). Let $k \in \mathscr{K}_K(S_\theta \times \Sigma_\delta^{n-1}, \mu)$,*
$$\mathcal{W} := L_{p,\varrho}(\mathbb{R}_+, L_q(\mathbb{R}_+, L_q(\mathbb{R}^{n-1}, X))),$$
$j = 0, \ldots, K$, and $f \in X_j(\mu, 0)$. Then we have
$$(\partial_n^j k)\Phi_{N(\mu)}^{-j} \in \mathscr{K}_0(S_\theta \times \Sigma_\delta^{n-1}, \mu),$$
$$\partial_n^j(G^{\mathcal{W}}[k]f) = G^{\mathcal{W}}[(\partial_n^j k)\Phi_{N(\mu)}^{-j}](U\Phi_{N(\mu)}^j(\mathcal{D}'_+)^{\mathcal{W}'}U^{-1}f)$$
where $\mathcal{W}' := L_{p,\varrho}(\mathbb{R}_+, L_q(\mathbb{R}^{n-1}, L_q(\mathbb{R}_+, X)))$.

Proof. For $g := U\Phi_{N(\mu)}^{j}(\mathcal{D}'_+)^{\mathcal{W}'} U^{-1} f \in \mathcal{W}$ we get
$$G^{\mathcal{W}}[k]f = G^{\mathcal{W}}[k]\left(U\Phi_{N(\mu)}^{-j}(\mathcal{D}'_+)^{\mathcal{W}'} U^{-1} g\right).$$

Defining $T(\lambda) := (G^{\mathcal{N}}[\lambda, k])^+$, $S(\lambda) := \left(U\Phi_{N(\mu)}^{-j}(\lambda, (\mathcal{D}')^{\mathcal{N}})U^{-1}\right)^+$, and $\mathcal{N} := L_q(\mathbb{R}^{n-1}, L_q(\mathbb{R}_+, X))$ we get
$$G^{\mathcal{W}}[k]f = T(\mathcal{D}_t^{\mathcal{W}})S(\mathcal{D}_t^{\mathcal{W}})g = (TS)(\mathcal{D}_t^{\mathcal{W}})g$$

due to Lemma 1.29 and Theorem 1.26 (ii). Lemma 3.54 then yields $S(\lambda) = [\Phi_{N(\mu)}^{-j}(\lambda, (\mathcal{D}'_+)^{\mathcal{N}'})]^{+n}$ with $\mathcal{N}' := L_q(\mathbb{R}^{n-1}, X)$. For all $h \in L_q(\mathbb{R}_+, L_q(\mathbb{R}^{n-1}, X))$ and $\lambda \in S_\theta$ we derive
$$\begin{aligned}(T(\lambda)S(\lambda)h)(x_n) &= \int_0^\infty k(\lambda, (\mathcal{D}')^{\mathcal{N}'}, x_n, y_n)(S(\lambda)h)(y_n) dy_n \\ &= \int_0^\infty \Phi_{N(\mu)}^{-j}(\lambda, (\mathcal{D}')^{\mathcal{N}'}) k(\lambda, (\mathcal{D}')^{\mathcal{N}'}, x_n, y_n) h(y_n) dy_n \\ &= (G^{\mathcal{W}}[\lambda, \Phi_{N(\mu)}^{-j} k]h)(x_n).\end{aligned}$$

Altogether, we obtain $G^{\mathcal{W}}[k]f = G^{\mathcal{W}}[\Phi_{N(\mu)}^{-j} k]g$. Now the assertions follow from Lemma 3.53. $\square$

Proposition 3.56. Let $k \in \mathscr{K}_K(S_\theta \times \Sigma_\delta^{n-1}, \mu)$ and
$$\mathcal{W} := L_{p,\varrho}(\mathbb{R}_+, L_q(\mathbb{R}_+, L_q(\mathbb{R}^{n-1}, X))).$$

Then we get
$$(G^{\mathcal{W}}[k])|_{X_K(\mu, 0)} \in L\Big(X_M(\mu, 0), \bigcap_{j=0}^K X_{K-j}(\mu, j)\Big).$$

Proof. For $K = 0$ the assertion follows from (3.36). Let $\Phi := \Phi_{N(\mu)}$, $K \geq 1$, $j \in \{0, \ldots, K\}$, and $m \in \{0, \ldots, j\}$. Let $f \in X_K(\mu, 0)$. We define $\mathcal{N} := L_q(\mathbb{R}^{n-1}, X)$, $\mathcal{M} := L_q(\mathbb{R}^{n-1}, L_q(\mathbb{R}_+, X))$ and derive
$$\partial_n^m(G^{\mathcal{W}}[k]f) = G^{\mathcal{W}}[\Phi^{-m}\partial_n^m k](U\Phi^m((\mathcal{D}'_+)^{\mathcal{W}'})U^{-1}f)$$

due to Lemma 3.55. It is easy to show that
$$U\Phi^m((\mathcal{D}'_+)^{\mathcal{W}})U^{-1} \in L(X_K(\mu, 0), X_{K-m}(\mu, 0))$$

and
$$\Phi^{-m}\partial_n^m k \in \mathscr{K}_0(S_\theta \times \Sigma_\delta^{n-1}, \mu).$$

We get $G^{\mathcal{W}}[\Phi^{-m}(\partial_n^m k)] \in L(X_{K-m}(\mu, 0), X_{K-j}(\mu, 0))$ due to (3.36) and the embedding $X_{K-m}(\mu, 0) \hookrightarrow X_{K-j}(\mu, 0)$. Altogether, this yields
$$\partial_n^m G^{\mathcal{W}}[k] \in L(X_K(\mu, 0), X_{K-j}(\mu, 0)), \text{ for all } m \in \{0, \ldots, j\}.$$

So we obtain $G^{\mathcal{W}}[k] \in L(X_K(\mu, 0), X_{K-j}(\mu, j))$ by Lemma 3.52. $\square$

b) Extension symbols

Definition 3.57 (Extension symbols). Let μ be a strictly positive order function and $K \in \mathbb{N}_0$. Then we define

$$\mathscr{E}_K(S_\theta \times \Sigma_\delta^{n-1}, \mu), \quad \theta > \pi/2$$

as the set of all functions $h\colon S_\theta \times \Sigma_\delta^{n-1} \times [0, \infty) \to \mathbb{C}$ with

$$h(\cdot, \cdot, x_n) \in H^\infty(S_\theta \times \Sigma_\delta^{n-1}), \quad x_n \geq 0$$

and $h(\lambda, z', \cdot) \in C^{K+1}([0, \infty), \mathbb{C})$ for all $(\lambda, z') \in S_\theta \times \Sigma_\delta^{n-1}$ such that

$$|\partial_n^j h(\lambda, z', x_n)| \leq C \cdot \frac{(W_\mu(\lambda, z))^{j-1}}{x_n}$$

for all $(\lambda, z') \in S_\theta \times \Sigma_\delta^{n-1}$, $x_n > 0$, and $j = 0, \ldots, K$.

Example 3.58. Let $\omega(\lambda, z') := (\rho\lambda + \mu|z'|_-^2)^{1/2}$, $|z'|_- := \left(-\sum_{k=1}^{n-1} z_k^2\right)^{1/2}$ with $\rho, \mu > 0$, and let

$$h(\lambda, z', x_n) := \exp(-\omega(\lambda, z')x_n), \quad (\lambda, z') \in S_\theta \times \Sigma_\delta^{n-1}, \; \theta > \pi/2, \; x_n > 0.$$

Then we have $h \in \mathscr{E}_2(S_\theta \times \Sigma_\delta^{n-1}, \mu)$ where $\mu(\gamma) := \max\{1, \gamma/2\}$, $\gamma > 0$. This typical symbol occurs in the solution formula of problems which are related to the heat equation.

The symbol class $\mathscr{E}_K(S_\theta \times \Sigma_\delta^{n-1}, \mu)$ is highly related to the class of kernels $\mathscr{K}_K(S_\theta \times \Sigma_\delta^{n-1}, \mu)$. This is concretized by the next remark.

Remark 3.59. It is easy to see that

$$\begin{aligned}
[(\lambda, z', x_n, y_n) \mapsto h(\lambda, z', x_n + y_n)] &\in \mathscr{K}_{K+1}(S_\theta \times \Sigma_\delta^n, \mu), \\
[(\lambda, z', x_n, y_n) \mapsto \Phi_{N(\mu)}(\lambda, z')h(\lambda, z', x_n + y_n)] &\in \mathscr{K}_{K+1}(S_\theta \times \Sigma_\delta^n, \mu), \\
[(\lambda, z', x_n, y_n) \mapsto (\partial_n h)(\lambda, z', x_n + y_n)] &\in \mathscr{K}_K(S_\theta \times \Sigma_\delta^n, \mu)
\end{aligned}$$

for all $h \in \mathscr{E}_K(S_\theta \times \Sigma_\delta^n, \mu)$.

Definition 3.60. Let $h \in \mathscr{E}_K(S_\theta \times \Sigma_\delta^n, \mu)$ with $K \geq 1$,

$$\mathcal{W} := L_{p,\varrho}(\mathbb{R}_+, L_q(\mathbb{R}_+, L_q(\mathbb{R}^{n-1}))), \quad \varrho \geq 0.$$

Then we define

$$E^{\mathcal{W}}[h]f := -G^{\mathcal{W}}[\partial_n h]f - G^{\mathcal{W}}[h](\partial_n f)$$

for $f \in L_{p,\varrho}(\mathbb{R}_+, H_q^1(\mathbb{R}_+, L_q(\mathbb{R}^{n-1})))$. Note that this definition is meaningful in view of Remark 3.59.

Remark 3.61. The last definition is related to the so-called "Volevich trick". Here a solution of the half-space problem which is given in the form $u(t, x', x_n) = [h(\mathcal{D}'_+, x_n)g](t, x')$ is written in the form

$$u(t, x', x_n) = -\int_0^\infty \partial_n [h(\mathcal{D}'_+, x_n + y_n)\widetilde{g}(y_n)](t, x') dy_n$$
$$= -\int_0^\infty [(\partial_n h)(\mathcal{D}'_+, x_n + y_n)\widetilde{g}(y_n)](t, x') dy_n$$
$$\quad -\int_0^\infty [h(\mathcal{D}'_+, x_n + y_n)(\partial_n \widetilde{g})(y_n)](t, x') dy_n$$

where $\widetilde{g}$ is an extension of g to the half-space. The precise formulation can be found in the following proposition.

Proposition 3.62. *Let* $h \in \mathcal{E}_K(S_\theta \times \Sigma_\delta^{n-1}, \mu)$ *with* $K \geq 1$ *and*

$$\mathcal{W} := L_{p,\varrho}(\mathbb{R}_+, L_q(\mathbb{R}_+, L_q(\mathbb{R}^{n-1}, X))),$$
$$\mathcal{W}' := L_{p,\varrho}(\mathbb{R}_+, L_q(\mathbb{R}^{n-1}, L_q(\mathbb{R}_+, X))),$$
$$\mathcal{W}'' := L_{p,\varrho}(\mathbb{R}_+, L_q(\mathbb{R}^{n-1}, X)).$$

(i) *For all* $g \in L_{p,\varrho}(\mathbb{R}_+, H_q^1(\mathbb{R}_+, L_q(\mathbb{R}^{n-1}, X)))$ *we have*

$$E^{\mathcal{W}}[h]g = -G^{\mathcal{W}}[\partial_n h]g - G^{\mathcal{W}}[\Phi_{N(\mu)}h](U\Phi_{N(\mu)}^{-1}((\mathcal{D}'_+)^{\mathcal{W}'})U^{-1}\partial_n g).$$

(ii) *We have*

$$[(E^{\mathcal{W}}[h]g)(t)](x_n) = (h((\mathcal{D}'_+)^{\mathcal{W}''}, x_n)(\gamma_{0,n}^+ g))(t) \quad \text{for almost all } t, x_n > 0$$

for all $g \in L_{p,\varrho}(\mathbb{R}_+, H_q^2(\mathbb{R}_+, L_q(\mathbb{R}^{n-1}, X)))$. *Here* $\gamma_{0,n}$ *denotes the classical trace operator on* $H_q^2(\mathbb{R}_+, L_q(\mathbb{R}^{n-1}, X))$ *associated with* $x_n = 0$, *i.e.,* $\gamma_{0,n}u = u|_{x_n=0}$.

(iii) *The operator* $E^{\mathcal{W}}[h]$ *has the mapping property*

$$E^{\mathcal{W}}[h] \in L\Big(X_K(\mu, 0) \cap X_{K-1}(\mu, 1), \bigcap_{j=0}^K X_{K-j}(\mu, j)\Big).$$

Proof. (i) This can be proved in the same way as in the proof of Lemma 3.54.

(ii) Let $g \in L_{p,\varrho}(\mathbb{R}_+, H_q^2(\mathbb{R}_+, L_q(\mathbb{R}^{n-1}, X)))$ and

$$(g_j)_{j \in \mathbb{N}} \subseteq \mathscr{D}(\mathbb{R}_+, H_q^2(\mathbb{R}_+, L_q(\mathbb{R}^{n-1}, X)))$$

3.5. Singular integral operators on L_p-L_q

with $g_j \to g$ in $L_{p,\varrho}(\mathbb{R}_+, H_q^2(\mathbb{R}_+, L_q(\mathbb{R}^{n-1}, X)))$. Due to the representation by Fourier multipliers we obtain

$$\begin{aligned}-(E^{\mathcal{W}}[h]g_j)(t) &= (T_1(\mathcal{D}_t^{\mathcal{W}})g_j + T_2(\mathcal{D}_t^{\mathcal{W}})\partial_n g_j)(t) \\ &= (\mathscr{F}^{-1}(T_1(i\tau)(\mathscr{F}g_j) + T_2(i\tau)(\mathscr{F}\partial_n g_j)))(t) \\ &= (2\pi)^{-1/2}\int_{\mathbb{R}} e^{it\tau}(T_1(i\tau)(\mathscr{F}g_j)(\tau) + T_2(i\tau)(\partial_n \mathscr{F}g_j)(\tau))d\tau\end{aligned}$$

for almost all $t > 0$, where

$$T_1(\lambda) := (G^{\mathcal{N}}[\lambda, \partial_n h])^+, \ T_2(\lambda) := (G^{\mathcal{N}}[\lambda, h])^+, \ \lambda \in S_\theta,$$

with $\mathcal{N} := L_q(\mathbb{R}^{n-1}, X)$. For almost all $x_n > 0$ and $\tau \in \mathbb{R}$ we obtain

$$\begin{aligned}(T_1(i\tau)&(\mathscr{F}g_j)(\tau) + T_2(i\tau)(\partial_n \mathscr{F}g_j)(\tau))(x_n) \\ &= \int_0^\infty (\partial_n h)(i\tau, (\mathcal{D}')^{\mathcal{N}}, x_n + y_n)[(\mathscr{F}g_j)(\tau)](y_n)dy_n \\ &\quad + \int_0^\infty h(i\tau, (\mathcal{D}')^{\mathcal{N}}, x_n + y_n)[\partial_n(\mathscr{F}g_j)(\tau)](y_n)dy_n \\ &= \int_0^\infty \partial_{y_n}\left\{h(i\tau, (\mathcal{D}')^{\mathcal{N}}, x_n + y_n)[(\mathscr{F}g_j)(\tau)](y_n)\right\}dy_n \\ &= -h(i\tau, (\mathcal{D}')^{\mathcal{N}}, x_n)\gamma_{0,n}((\mathscr{F}g_j)(\tau)) \\ &= -h(i\tau, (\mathcal{D}')^{\mathcal{N}}, x_n)[(\mathscr{F}(\gamma_{0,n}^+ g_j))(\tau)]\end{aligned}$$

by the classical fundamental theorem of calculus. Altogether, we get

$$\begin{aligned}[(E[h]g_j)(t)](x_n) &= (2\pi)^{-1/2}\int_{\mathbb{R}} e^{it\tau}h(i\tau, (\mathcal{D}')^{\mathcal{N}}, x_n)[(\mathscr{F}(\gamma_{0,n}^+ g_j))(\tau)]d\tau \\ &= (\mathscr{F}^{-1}(h(i\tau, (\mathcal{D}')^{\mathcal{N}}, x_n)\mathscr{F}(\gamma_{0,n}^+ g_j)))(t) \\ &= (h((\mathcal{D}'_+)^{\mathcal{W}''}, x_n)(\gamma_{0,n}^+ g_j))(t) \quad \text{for almost all } t, x_n > 0.\end{aligned}$$

Due to $G^{\mathcal{W}}[\partial_n h], G^{\mathcal{W}}[h] \in L(L_{p,\varrho}(\mathbb{R}_+, L_q(\mathbb{R}_+, L_q(\mathbb{R}^{n-1}, X))))$ we have

$$((E[h]g_j)(t))(x_n) \to ((E[h]g)(t))(x_n), \quad j \to \infty$$

in $L_q(\mathbb{R}^{n-1}, X)$ for almost all $t, x_n > 0$. With the convergence $\gamma_{0,n}^+ g_j \to \gamma_{0,n}^+ g$ in $L_{p,\varrho}(\mathbb{R}_+, L_q(\mathbb{R}^{n-1}, X))$ and the boundedness of $h((\mathcal{D}'_+)^{\mathcal{W}''}, x_n)$ we obtain

$$h((\mathcal{D}'_+)^{\mathcal{W}''}, x_n)(\gamma_{0,n}^+ g_j) \to h((\mathcal{D}'_+)^{\mathcal{W}''}, x_n)(\gamma_{0,n}^+ g), \quad j \to \infty$$

in $L_{p,\varrho}(\mathbb{R}_+, L_q(\mathbb{R}^{n-1}, X))$.

(iii) Remark 3.59, Proposition 3.56, and the trivial embedding

$$X_K(\mu,0) \cap X_{K-1}(\mu,1) \hookrightarrow X_K(\mu,0)$$

already yield

$$G^{\mathcal{W}}[\partial_n h] \in L\Big(X_K(\mu,0) \cap X_{K-1}(\mu,1), \bigcap_{j=0}^{K} X_{K-j}(\mu,j)\Big).$$

It is easy to see that $U\Phi_{N(\mu)}^{-1}((\mathcal{D}'_+)^{\mathcal{W}'})U^{-1}\partial_n \in L(X_{K-1}(\mu,1), X_K(\mu,0))$ and therefore

$$G^{\mathcal{W}}[\Phi_{N(\mu)}h]U\Phi_{N(\mu)}^{-1}((\mathcal{D}'_+)^{\mathcal{W}'})U^{-1}\partial_n$$

$$\in L\Big(X_K(\mu,0) \cap X_{K-1}(\mu,1), \bigcap_{j=0}^{K} X_{K-j}(\mu,j)\Big)$$

by Remark 3.59, Proposition 3.56, and $X_K(\mu,0) \cap X_{K-1}(\mu,1) \hookrightarrow X_K(\mu,0)$. Using (i) we derive the assertion. $\square$

Corollary 3.63. *Let* $\omega(\lambda,z') := (\rho\lambda+\mu|z'|_-^2)^{1/2}$, $|z'|_- := \left(-\sum_{k=1}^{n-1} z_k^2\right)^{1/2}$, $\rho,\mu > 0$, *and*

$$h(\lambda,z',x_n) := \exp(-\omega(\lambda,z')x_n)$$

for $(\lambda,z') \in S_\theta \times \Sigma_\delta^{n-1}$, $\theta > \pi/2$, $x_n > 0$. *We define*

$$\mathcal{W} := L_{p,\varrho}(\mathbb{R}_+, L_q(\mathbb{R}_+, L_q(\mathbb{R}^{n-1}))),$$
$$\mathcal{W}' := L_{p,\varrho}(\mathbb{R}_+, L_q(\mathbb{R}^{n-1}, L_q(\mathbb{R}_+))),$$
$$\mathcal{W}'' := L_{p,\varrho}(\mathbb{R}_+, L_q(\mathbb{R}^{n-1})).$$

(i) *We have* $h \in \mathscr{E}_2(S_\theta \times \Sigma_\delta^{n-1}, \mu)$ *where* $\mu(\gamma) := \max\{1,\gamma/2\}$ *for* $\gamma > 0$.

(ii) *The symbol h gives rise to a bounded operator*

$$E^{\mathcal{W}}[h] \in L(\mathbb{X}),$$

where

$$\mathbb{X} := {}_0H_{p,\varrho}^1(\mathbb{R}_+, L_q(\mathbb{R}_+^n)) \cap L_{p,\varrho}(\mathbb{R}_+, H_q^2(\mathbb{R}_+^n)).$$

(iii) *Let e_n be the extension operator associated with $\gamma_{0,n}$. Then we have*

$$E^{\mathcal{W}}[h]e_n \in L(\gamma_{0,n}\mathbb{X}, \mathbb{X})$$

3.5. Singular integral operators on L_p-L_q

and $E^{\mathcal{W}}[h]e_n\varphi$ additionally fulfills

$$\gamma_{0,n} E^{\mathcal{W}}[h]e_n\varphi = \varphi,$$
$$\gamma_{0,n} \partial_n E^{\mathcal{W}}[h]e_n\varphi = -\omega((\boldsymbol{D}'_+)^{\mathcal{W}''})\varphi$$

for all $\varphi \in \gamma_{0,n}\mathbb{X} := {}_0F^{1-1/(2q)}_{pq,\varrho}(\mathbb{R}_+, L_q(\mathbb{R}^{n-1})) \cap L_{p,\varrho}(\mathbb{R}_+, B^{2-1/q}_{qq}(\mathbb{R}^{n-1}))$.
Note that the representation of the trace space as well as the existence of an extension operator follow from the discussion in Section 3.2, cf. Remark 3.22 and Proposition 3.23.

(iv) We have

$$\partial_n^2 E^{\mathcal{W}}[h]e_n\varphi = -U\omega^2((\boldsymbol{D}'_+)^{\mathcal{W}'})U^{-1}E^{\mathcal{W}}[h]e_n\varphi$$
$$= -(\partial_t - \Delta')E^{\mathcal{W}}[h]e_n\varphi$$

for all $\varphi \in \gamma_{0,n}\mathbb{X}$.

Proof. (i) This is easy to see.

(ii) According to Remark 3.59 and Proposition 3.62 (iii) we already know

$$E^{\mathcal{W}}[h] \in L\Big(X_2(\mu,0) \cap X_1(\mu,1), \bigcap_{j=0}^{2} X_{2-j}(\mu,j)\Big), \qquad (3.38)$$

where

$$X_0(\mu,2) = L_{p,\varrho}(\mathbb{R}_+, H_q^2(\mathbb{R}_+, L_q(\mathbb{R}^{n-1}))),$$
$$X_1(\mu,1) = {}_0H^{1/2}_{p,\varrho}(\mathbb{R}_+, H_q^1(\mathbb{R}_+, L_q(\mathbb{R}^{n-1}))) \cap L_{p,\varrho}(\mathbb{R}_+, H_q^1(\mathbb{R}_+, H_q^1(\mathbb{R}^{n-1}))),$$
$$X_2(\mu,0) = {}_0H^1_{p,\varrho}(\mathbb{R}_+, L_q(\mathbb{R}_+, L_q(\mathbb{R}^{n-1}))) \cap L_{p,\varrho}(\mathbb{R}_+, L_q(\mathbb{R}_+, H_q^2(\mathbb{R}^{n-1}))).$$

With

$$L_{p,\varrho}(\mathbb{R}_+, H_q^2(\mathbb{R}_+^n)) = L_{p,\varrho}(\mathbb{R}_+, H_q^2(\mathbb{R}_+, L_q(\mathbb{R}^{n-1})))$$
$$\cap L_{p,\varrho}(\mathbb{R}_+, L_q(\mathbb{R}_+, H_q^2(\mathbb{R}^{n-1})))$$
$$\cap L_{p,\varrho}(\mathbb{R}_+, H_q^1(\mathbb{R}_+, H_q^1(\mathbb{R}^{n-1})))$$

and the embedding

$$\mathbb{X} \hookrightarrow {}_0H^{1/2}_{p,\varrho}(\mathbb{R}_+, H_q^1(\mathbb{R}_+^n)) \hookrightarrow {}_0H^{1/2}_{p,\varrho}(\mathbb{R}_+, H_q^1(\mathbb{R}_+, L_q(\mathbb{R}^{n-1}))) \qquad (3.39)$$

(cf. Lemma 2.61 and Remark 2.77) we get

$$\bigcap_{j=0}^{2} X_{2-j}(\mu,j) = \mathbb{X} \cap {}_0H^{1/2}_{p,\varrho}(\mathbb{R}_+, H_q^1(\mathbb{R}_+, L_q(\mathbb{R}^{n-1}))) = \mathbb{X}.$$

From (3.38) we therefore derive $E^{\mathcal{W}}[h] \in L(X_2(\mu,0) \cap X_1(\mu,1), \mathbb{X})$. Again applying (3.39) we obtain $\mathbb{X} \hookrightarrow X_2(\mu,0) \cap X_1(\mu,1)$ and therefore the assertion follows from (3.38).

(iii) Due to (ii) and $e_n \in L(\gamma_{0,n}\mathbb{X}, \mathbb{X})$ we trivially have $E^{\mathcal{W}}[h]e_n \in L(\gamma_{0,n}\mathbb{X}, \mathbb{X})$. Moreover, it holds that $(E^{\mathcal{W}}[h]e_n\varphi)(t) \in C([0,\infty), L_q(\mathbb{R}^{n-1}))$ and we can also show

$$\left[x_n \mapsto (h((\mathcal{D}'_+)^{\mathcal{W}''}, x_n)(\gamma^+_{0,n}g))(t)\right] \in C([0,\infty), L_q(\mathbb{R}^{n-1}))$$

with $\mathcal{W}'' := L_{p,\varrho}(\mathbb{R}_+, L_q(\mathbb{R}^{n-1}, X))$. According to Proposition 3.62 we have, for almost all $t > 0$,

$$\gamma_{0,n}(E^{\mathcal{W}}[h]e_n\varphi)(t) = \lim_{j \to \infty} [(E^{\mathcal{W}}[h]e_n\varphi)(t)](1/j)$$
$$= \lim_{j \to \infty} (h((\mathcal{D}'_+)^{\mathcal{W}''}, 1/j)\varphi)(t)$$
$$= (h((\mathcal{D}'_+)^{\mathcal{W}''}, 0)\varphi)(t) = \varphi(t).$$

For $f \in \mathbb{X}$ we obtain

$$\partial_n E^{\mathcal{W}}[h]f = -G^{\mathcal{W}}[\Phi^{-1}_{N(\mu)}\partial_n^2 h](U\Phi_{N(\mu)}((\mathcal{D}'_+)^{\mathcal{W}'})U^{-1}f)$$
$$- G^{\mathcal{W}}[\Phi^{-1}_{N(\mu)}\partial_n h](U\Phi_{N(\mu)}((\mathcal{D}'_+)^{\mathcal{W}'})U^{-1}\partial_n f)$$
$$= G^{\mathcal{W}}[\omega\Phi^{-1}_{N(\mu)}\partial_n h](U\Phi_{N(\mu)}((\mathcal{D}'_+)^{\mathcal{W}'})U^{-1}f)$$
$$+ G^{\mathcal{W}}[\omega\Phi^{-1}_{N(\mu)}h](U\Phi_{N(\mu)}((\mathcal{D}'_+)^{\mathcal{W}'})U^{-1}\partial_n f)$$

according to Lemma 3.55. Due to the boundedness of $\omega^{-1}\Phi_{N(\mu)}$ one can show that

$$\partial_n E^{\mathcal{W}}[h]f = G^{\mathcal{W}}[\partial_n h](U\omega((\mathcal{D}'_+)^{\mathcal{W}'})U^{-1}f) + G^{\mathcal{W}}[h](U\omega((\mathcal{D}'_+)^{\mathcal{W}'})U^{-1}\partial_n f)$$
$$= -E^{\mathcal{W}}[h](U\omega((\mathcal{D}'_+)^{\mathcal{W}'})U^{-1}f)$$

as in the proof of Lemma 3.55. With

$$\gamma_{0,n}U\omega((\mathcal{D}'_+)^{\mathcal{W}'})U^{-1}f = \omega((\mathcal{D}'_+)^{\mathcal{W}'})\gamma_{0,n}f,$$

Proposition 3.62, and the same arguments as before we get the equality $\gamma^+_{0,n}\partial_n E^{\mathcal{W}}[h]e_n\varphi = -\omega((\mathcal{D}'_+)^{\mathcal{W}'})\varphi$ for all $\varphi \in \gamma_{0,n}\mathbb{X}$. Note that the mapping properties of $\omega((\mathcal{D}'_+)^{\mathcal{W}'})$ are characterized by Theorem 2.62.

(iv) This can be proved by the same techniques used in the proof of part (iii). □

Remark 3.64. By minor modifications the theory of this chapter can also be established for $x_n < 0$. Hence, we can also construct operators $E^{\mathcal{W}-}[h]$ on the ground space $\mathcal{W}_- := L_{p,\varrho}(\mathbb{R}_+, L_q(\mathbb{R}^n_-))$ with the same properties mutatis mutandis as in Proposition 3.62 and Corollary 3.63 with

$$h(\lambda, z', x_n) := \exp(\omega(\lambda, z')x_n), \quad x_n < 0.$$

3.5. Singular integral operators on L_p-L_q

Remark 3.65. For $p = q$ the result of Corollary 3.63 can also be obtained in a vector-valued version where X is of class $\mathcal{HT}$ with property (α).

Using the concepts developed in this section, we can give a representation of the solution of the Dirichlet heat equation in L_p-L_q. One can find L_p-L_q-maximal regularity results for parabolic equations with inhomogeneous boundary conditions in [Wei02] and [DHP07], for instance. In [Wei02] the author considers Dirichlet and conormal boundary conditions under the restriction $q \leq p$. The boundary conditions in [DHP07] are more general but exhibit no time derivatives on the boundary.

The theory developed in this section and in Section 3.4 enables us to give a solution of the linearized two-phase Stefan problem in an L_p-L_q-setting. The treatment of this problem can be found in Section 4.8.

Example 3.66. Let $1 < p, q < \infty$. Consider the L_p-L_q Dirichlet heat equation

$$\begin{cases} (\partial_t - \Delta)u = 0 & \text{in } \mathbb{R}_+ \times \mathbb{R}_+^n, \\ \gamma_{0,n} u = \varphi & \text{on } \mathbb{R}_+ \times \mathbb{R}^{n-1}, \\ u(t=0) = 0 & \text{in } \mathbb{R}_+^n \end{cases}$$

where $\varphi \in {}_0F_{pq,\varrho}^{1-1/(2q)}(\mathbb{R}_+, L_q(\mathbb{R}^{n-1})) \cap L_{p,\varrho}(\mathbb{R}_+, B_{qq}^{2-1/q}(\mathbb{R}^{n-1}))$, $\varrho \geq 0$. Then the solution in ${}_0H_{p,\varrho}^1(\mathbb{R}_+, L_q(\mathbb{R}_+^n)) \cap L_{p,\varrho}(\mathbb{R}_+, H_q^2(\mathbb{R}_+^n))$ is given by

$$u = E^{\mathcal{W}}[h]e_n\varphi, \quad \mathcal{W} := L_{p,\varrho}(\mathbb{R}_+, L_q(\mathbb{R}_+, L_q(\mathbb{R}^{n-1})))$$

where h is defined as in Corollary 3.63.

Chapter 4

Application to parabolic differential equations

In this chapter we present a selection of applications of our main result on mixed-order systems stated in Theorem 2.69, Theorem 3.41, Corollary 2.80, and Corollary 3.44. The applications split into two subgroups of parabolic partial differential equations. In the first part we consider problems on the whole space and in the second part we consider boundary value problems. This seems to be one of the first works which treat parabolic problems on the whole space by a direct approach without semigroup theory and a reduction to a first-order system.

The applications to free boundary problems in the Sections 4.6-4.8 will show the full potential of the developed theory. For example, one can find results on N-parabolic mixed-order systems associated with free boundary problems in [DSS08] and [DV08]. In [DV08] the authors present the L_2-theory of boundary value problems where the boundary operators can also depend on ∂_t. They treat the associated Lopatinskii matrix as a mixed-order system in L_2. In [DSS08] the authors consider symbols on L_p employing a joint H^∞-calculus with some restrictive conditions on the structure of the symbols. With this approach they are able to derive a solution of the problem which occurs as reduction of the Stefan problem. Our work is a generalization and linking of both papers cited above.

Most of the considered problems have been solved in the literature before but our approach allows a unified and systematic treatment of all these problems. Furthermore, our approach yields proofs which are shorter and more direct than in the literature.

In this chapter we always search for solutions of partial differential equations with vanishing time trace (i.e., $u|_{t=0} = 0$) and vanishing right-hand side for equations on the half-space. For a suitable treatment of the associated nonlinear problem the fully inhomogeneous system with non-vanishing time trace is required. In [DSS08, Theorem 4.5] R. Denk, J. Saal, and J. Seiler established the existence of an extension operator for the time trace, where the ground space is given by intersections of Sobolev-Slobodeckij spaces. Using [DSS08, Theorem 4.5] it is possible to give suitable extensions of the time traces appearing in the treatment of fully inhomogeneous systems, see the treatment of the Stefan problem in [DSS08, Section 5] for example. In general the solution of the fully inhomogeneous system can then be derived by the superposition principle.

4.1 The generalized L_p-L_q Stokes problem on $\Omega = \mathbb{R}^n$

Motivation. In this section we want to treat the generalized L_p-L_q Stokes problem on the whole space $\mathbb{R}^n$ ($n \in \mathbb{N}$). This problem occurs in the treatment of the Navier-Stokes system for a class of non-Newtonian fluids. By Fourier and Laplace transform we can write the Stokes equation as a Douglis-Nirenberg system. Theorem 2.69 then yields an existence and uniqueness result for the Stokes problem. This problem was already considered in [BP07] for $p = q$. In [SS08, Theorem 4.1] one can find an L_p-L_q result for the common Stokes equation.

As the natural space for the pressure in the Stokes equation is a homogeneous Sobolev space, we recall the main definitions and some properties of these spaces in part a) of this section.

a) Remarks on homogeneous Sobolev spaces

There are many references that introduce scalar-valued homogenous spaces. For a detailed scalar discussion we refer to [BL76, 6.3], [Tri83, 5.1-5.2], and [RS96].

In this part we consider homogeneous Bessel potential and Besov spaces and state some basic properties. We define

$$\mathscr{Z}(\mathbb{R}^n) := \{\varphi \in \mathscr{S}(\mathbb{R}^n) \colon (\partial^\alpha \mathscr{F}\varphi)(0) = 0 \text{ for all } \alpha \in \mathbb{N}_0^n\}.$$

The space $\mathscr{Z}(\mathbb{R}^n)$ is a complete locally convex space with the subspace topology of $\mathscr{S}(\mathbb{R}^n)$. Therefore $\mathscr{Z}(\mathbb{R}^n)$ is also a Fréchet space. As usual we denote $\mathscr{Z}'(\mathbb{R}^n)$ as the set of all $T \colon \mathscr{Z}(\mathbb{R}^n) \to \mathbb{C}$ which are linear and continuous.

Proposition 4.1 ([RS96, p. 93]). *The linear operator*

$$\Pi \colon \mathscr{S}'(\mathbb{R}^n) \to \mathscr{Z}'(\mathbb{R}^n), \quad T \mapsto T|_{\mathscr{Z}(\mathbb{R}^n)}$$

is onto and $\ker \Pi = P(\mathbb{R}^n)$. *Here* $P(\mathbb{R}^n)$ *is defined as the set of all polynomials* $\sum_{|\alpha| \leq M} a_\alpha x^\alpha$ *with* $M \in \mathbb{N}_0$ *and* $a_\alpha \in \mathbb{C}$. *The operator* Π *gives rise to the isomorphism*

$$\Gamma \colon \mathscr{S}'(\mathbb{R}^n)/P(\mathbb{R}^n) \to \mathscr{Z}'(\mathbb{R}^n), \quad [T]_P \mapsto T|_{\mathscr{Z}(\mathbb{R}^n)}.$$

4.1. The generalized L_p-L_q Stokes problem on $\Omega = \mathbb{R}^n$

Definition 4.2 (Homogeneous Bessel potential and Besov spaces). Let $\{\varphi_j\}_{j \in \mathbb{Z}} \subseteq \mathscr{S}(\mathbb{R}^n)$ be such that

(i) there exist $B, C > 0$ with $\operatorname{supp} \varphi_j \subseteq \{x \in \mathbb{R}^n : B\, 2^{j-1} \leq |x| \leq C\, 2^{j+1}\}$,

(ii) for all $\alpha \in \mathbb{N}_0^n$ there exists $c_\alpha > 0$ with $\sup_{x \in \mathbb{R}^n}\{\sup_{j \in \mathbb{Z}}(2^{j|\alpha|}|D^\alpha \varphi_j(x)|)\} \leq c_\alpha$,

(iii) it holds that $\sum_{j \in \mathbb{Z}} \varphi_j(x) = 1$ for all $x \in \mathbb{R}^n$.

Then we define the *homogeneous Bessel potential spaces* by

$$\dot{H}_p^s(\mathbb{R}^n) := \left\{ T \in \mathscr{Z}'(\mathbb{R}^n) : \|T\|_{\dot{H}_p^s} := \|(2^{sj}(\mathscr{F}^{-1}[\varphi_j \mathscr{F} T]))_{j \in \mathbb{Z}}\|_{L_p(\mathbb{R}^n, \ell_2(\mathbb{Z}))} < \infty \right\}$$

and the *homogeneous Besov spaces* by

$$\dot{B}_{pp}^s(\mathbb{R}^n) := \left\{ T \in \mathscr{Z}'(\mathbb{R}^n) : \|T\|_{\dot{B}_{pp}^s} := \|(2^{sj}(\mathscr{F}^{-1}[\varphi_j \mathscr{F} T]))_{j \in \mathbb{Z}}\|_{\ell_p(\mathbb{Z}, L_p(\mathbb{R}^n))} < \infty \right\}$$

for $s \in \mathbb{R}$ and $p \in (1, \infty)$.

Proposition 4.3 (Homogeneous lift operator, [Tri83, Section 5.2.3]**).** *The operator*

$$\dot{J}_\sigma : \mathscr{Z}(\mathbb{R}^n) \to \mathscr{Z}(\mathbb{R}^n), \quad f \mapsto \mathscr{F}^{-1} |\xi|^\sigma \mathscr{F} f, \quad \sigma \in \mathbb{R}$$

is well-defined, bijective, and continuous. By duality we can extend the operator $\dot{J}_\sigma$ to $\mathscr{Z}'(\mathbb{R}^n)$ and denote this extension by $\dot{J}_\sigma$, too. Hence $\dot{J}_\sigma$ is also bijective and continuous in $\mathscr{Z}'(\mathbb{R}^n)$. We also have

$$\dot{J}_\sigma \in L_{\mathrm{Isom}}(\dot{H}_p^s(\mathbb{R}^n), \dot{H}_p^{s-\sigma}(\mathbb{R}^n)), \quad \dot{J}_\sigma \in L_{\mathrm{Isom}}(\dot{B}_{pp}^s(\mathbb{R}^n), \dot{B}_{pp}^{s-\sigma}(\mathbb{R}^n)).$$

Due to

$$L_p(\mathbb{R}^n) \hookrightarrow \mathscr{S}'(\mathbb{R}^n) \xrightarrow{\Pi} \mathscr{Z}'(\mathbb{R}^n), \quad 1 < p < \infty$$

and the fact that $L_p(\mathbb{R}^n) \cap P(\mathbb{R}^n) = \{0\}$ we can interpret $L_p(\mathbb{R}^n)$ as a subspace of $\mathscr{Z}'(\mathbb{R}^n)$. According to [Tri83, Theorem 5.2.1] *we have $\dot{H}_p^0(\mathbb{R}^n) = L_p(\mathbb{R}^n)$.*

Proposition 4.4 ([Tri92, Section 2.3.3])**.** *Inhomogeneous spaces can be expressed by homogeneous spaces in the form*

$$H_p^s(\mathbb{R}^n) = L_p(\mathbb{R}^n) \cap \dot{H}_p^s(\mathbb{R}^n), \quad B_{pp}^s(\mathbb{R}^n) = L_p(\mathbb{R}^n) \cap \dot{B}_{pp}^s(\mathbb{R}^n), \quad s > 0,\ 1 < p < \infty$$

with equivalent norms. This representation yields

$$\dot{J}_\sigma|_{H_p^s(\mathbb{R}^n)} \in L(H_p^s(\mathbb{R}^n), \dot{H}_p^{-\sigma}(\mathbb{R}^n) \cap \dot{H}_p^{s-\sigma}(\mathbb{R}^n)), \quad s \geq 0, \quad \sigma \in \mathbb{R}.$$

In particular, we obtain that $\dot{J}_1|_{H_p^2(\mathbb{R}^n)} \in L(H_p^2(\mathbb{R}^n), H_p^1(\mathbb{R}^n))$ due to $H_p^2(\mathbb{R}^n) = L_p(\mathbb{R}^n) \cap \dot{H}_p^2(\mathbb{R}^n) \cap \dot{H}_p^1(\mathbb{R}^n)$.

Proposition 4.5 ([Tri83, Theorem 5.2.3]). *We have*
$$\dot{H}_p^1(\mathbb{R}^n) = \{f \in \mathscr{L}'(\mathbb{R}^n) : \nabla f \in [L_p(\mathbb{R}^n)]^n\}, \quad 1 < p < \infty$$
and an equivalent norm on $\dot{H}_p^1(\mathbb{R}^n)$ *is given by* $\|f\| := \left(\sum_{k=1}^n \|\partial_k f\|_{L_p(\mathbb{R}^n)}^p\right)^{1/p}$.

Proposition 4.6 ([Tri83, Theorem 5.15]). *For all* $s \in \mathbb{R}$ *and* $1 < p < \infty$ *we have the dense embeddings*
$$\mathscr{L}(\mathbb{R}^n) \overset{d}{\hookrightarrow} \dot{H}_p^s(\mathbb{R}^n), \quad \mathscr{L}(\mathbb{R}^n) \overset{d}{\hookrightarrow} \dot{B}_{pp}^s(\mathbb{R}^n).$$

b) The generalized Stokes problem

Definition 4.7. We define the formal second order differential operator
$$\mathcal{A}(\partial)u := -\left(\sum_{j,k,l=1}^n a_{ij}^{kl} \partial_k \partial_l u_j\right)_{i=1,\ldots,n}$$

for a function $u\colon \mathbb{R}^n \to \mathbb{C}^n$ with $a_{ij}^{kl} \in \mathbb{C}$ for all $i,j,k,l = 1,\ldots,n$. This operator is called *strongly elliptic* if there exists $C > 0$ such that

$$\operatorname{Re}\langle \mathcal{A}(i\xi)\eta, \eta\rangle \geq C \qquad (4.1)$$

for all $\xi \in \mathbb{R}^n$ and $\eta \in \mathbb{C}^n$ with $|\xi| = |\eta| = 1$.

For the strongly elliptic operator $\mathcal{A}(\partial)$ we first state the generalized Stokes problem in $\mathbb{R}^n$. We consider the partial differential equation
$$\begin{cases} \partial_t u + \mathcal{A}(\partial)u + \nabla \pi = f, & (t,x) \in \mathbb{R}_+ \times \mathbb{R}^n, \\ \operatorname{div} u = g, & (t,x) \in \mathbb{R}_+ \times \mathbb{R}^n, \\ u(t=0) = 0, & x \in \mathbb{R}^n \end{cases} \qquad (4.2)$$

for the unknown functions $u\colon \mathbb{R}^n \to \mathbb{R}^n$ and $\pi\colon \mathbb{R}^n \to \mathbb{R}$. The main problem is to find suitable spaces for the right-hand sides f and g such that (4.2) has a solution with the canonical regularities

$$u \in {}_0H_p^1(\mathbb{R}_+, L_q(\mathbb{R}^n, \mathbb{C}^n)) \cap L_p(\mathbb{R}_+, H_q^2(\mathbb{R}^n, \mathbb{C}^n)), \quad \pi \in L_p(\mathbb{R}_+, \dot{H}_q^1(\mathbb{R}^n, \mathbb{C}^n))$$

where $1 < p,q < \infty$. Note that one obtains the standard Stokes problem by setting $a_{ij}^{kl} = \delta_{ij} \cdot \delta_{kl}$. After formal Fourier transform in the space variables and formal Laplace transform in the time variable we get the formal transformed problem

$$\begin{pmatrix} \lambda + \mathcal{A}(i\xi) & i\xi \\ i\xi^T & 0 \end{pmatrix} \cdot \begin{pmatrix} \hat{u} \\ \hat{\pi} \end{pmatrix} = \begin{pmatrix} \hat{f} \\ \hat{g} \end{pmatrix}, \quad (\lambda, \xi) \in \mathbb{R}_+ \times \mathbb{R}^n. \qquad (4.3)$$

All functions in this matrix can be extended to holomorphic functions in $H_P(S_\theta \times \Sigma_\delta^n)$. So we are able to plug in $\nabla_+ = (\partial_t, \nabla)$ by the joint H^∞-calculus developed in

4.1. The generalized L_p-L_q Stokes problem on $\Omega = \mathbb{R}^n$

Section 1.3. The mapping properties of this mixed-order system can be determined with the aid of Theorem 2.69. This also answers the question of suitable spaces for the right-hand sides f and g.

We can define the complex version of (4.3) by

$$\mathscr{L}(\lambda, z) := \begin{pmatrix} \lambda + \mathcal{A}(z) & z \\ z^T & 0 \end{pmatrix}, \quad (\lambda, z) \in S_\theta \times \Sigma_\delta^n, \tag{4.4}$$

which yields $\mathscr{L} \in [H_P(S_\theta \times \Sigma_\delta^n)]^{(n+1) \times (n+1)}$. Here we choose δ and θ such that the next lemma holds.

Lemma 4.8. *Let $\mathcal{A}(\partial)$ be strongly elliptic as in Definition 4.7. Then there exists $\delta \in (0, \pi/2)$, $\theta > \pi/2$, and $C > 0$ such that the following assertions hold.*

(i) *We have*
$$\mathrm{Re}\,\langle \mathcal{A}(z)\eta, \eta \rangle \geq C$$
for all $z \in \overline{\Sigma}_\delta^n$ and $\eta \in \mathbb{C}^n$ with $|z| = |\eta| = 1$.

(ii) *The matrix $\lambda + \mathcal{A}(z)$ is invertible for all $z \in \overline{\Sigma}_\delta^n$ and $\lambda \in \overline{S}_\theta$.*

Proof. Assume that for all $j \in \mathbb{N}$ there exist $z^{(j)} \in \overline{\Sigma}_{1/j}^n$ and $\eta^{(j)} \in \mathbb{C}^n$ with $|z^{(j)}| = |\eta^{(j)}| = 1$ and
$$\mathrm{Re}\,\left\langle \mathcal{A}(z^{(j)})\eta^{(j)}, \eta^{(j)} \right\rangle = 0.$$
Without loss of generality we assume $(z^{(j)}, \eta^{(j)})_j$ to be convergent with $(z, \eta) := \lim_{j \to \infty}(z^{(j)}, \eta^{(j)})$. We have $|z| = |\eta| = 1$, $z \in (i\mathbb{R})^n$, and $\mathrm{Re}\,\langle \mathcal{A}(z)\eta, \eta \rangle = 0$, which contradicts (4.1). From this the claim in (i) easily follows.

Therefore there exist $\delta, C_\mathcal{A}, C > 0$ such that
$$\mathrm{Re}\,\langle \mathcal{A}(z)\eta, \eta \rangle \geq C_\mathcal{A}, \quad |\langle \mathcal{A}(z)\eta, \eta \rangle| \leq C$$
for all $z \in \overline{\Sigma}_\delta^n$ and $\eta \in \mathbb{C}^n$ with $|z| = |\eta| = 1$. So we can choose $\theta > \pi/2$ such that
$$W(-\mathcal{A}) := \{-\langle \mathcal{A}(z)\eta, \eta \rangle : z \in \overline{\Sigma}_\delta^n, \eta \in \mathbb{C}^n \text{ with } |z| = |\eta| = 1\} \cap \overline{S}_\theta = \emptyset. \tag{4.5}$$
For $z \in \overline{\Sigma}_\delta^n \setminus \{0\}$, $\hat{z} := z/|z|$, and $\lambda \in \overline{S}_\theta$ we have $\sigma_p(-\mathcal{A}(\hat{z})) \subseteq W(-\mathcal{A})$ and therefore
$$\ker(\lambda + \mathcal{A}(z)) = \ker(\lambda/|z|^{-2} + \mathcal{A}(\hat{z})) = \{0\} \tag{4.6}$$
due to the choice of θ. $\square$

Lemma 4.9. *Let $A \in \mathbb{C}^{n \times n}$ be invertible and let $\eta \in \mathbb{C}^n$. Then we have*
$$\det \begin{pmatrix} A & \eta \\ \eta^T & 0 \end{pmatrix} = -\det(A) \cdot \eta^T A^{-1} \eta.$$

Proof. Using Laplace's formula in the last row we obtain

$$\det\begin{pmatrix} A & \eta \\ \eta^T & 0 \end{pmatrix} = \sum_{k=1}^{n+1}(-1)^{n+k+1}\eta_k(-1)^{n-k}\det(A_k(\eta))$$

$$= -\det(A)\sum_{k=1}^{n+1}\eta_k[\det(A_k(\eta))/\det(A)],$$

where $A_k(\eta)$ is defined as the matrix A where column k is substituted by η. By Cramer's rule we get

$$\det(A_k(\eta))/\det(A) = (A^{-1}\eta)_k,$$

which yields the assertion. □

According to Lemma 4.8 and Lemma 4.9 we obtain

$$\det \mathscr{L}(\lambda, z) = -\det(\lambda + \mathcal{A}(z))\cdot z^T(\lambda + \mathcal{A}(z))^{-1}z, \quad (\lambda, z) \in S_\theta \times \Sigma_\delta^n.$$

The corresponding Newton polygon to $\det \mathscr{L}$ is given by $N := N(\{(2n, 0), (2, n-1), (0, n-1)\})$ and therefore is not regular in space. Indeed, we first consider the transformed system

$$\mathscr{L}'(\lambda, z) := \begin{pmatrix} 1 & 0 \\ 0 & |z|_-^{-1} \end{pmatrix}\mathscr{L}(\lambda, z)\begin{pmatrix} 1 & 0 \\ 0 & |z|_-^{-1} \end{pmatrix} = \begin{pmatrix} \lambda + \mathcal{A}(z) & z/|z|_- \\ z^T/|z|_- & 0 \end{pmatrix} \quad (4.7)$$

for $(\lambda, z) \in S_\theta \times \Sigma_\delta^n$ and $|z|_- := \sqrt{-\sum_{k=1}^n z_k^2}$. This transformation is convenient for a shift between homogeneous and non-homogeneous Sobolev spaces. Hence, we get $\mathscr{L}' \in [H_P(S_\theta \times \Sigma_\delta^n)]^{(n+1)\times(n+1)}$ and

$$\det \mathscr{L}'(\lambda, z) = -\det(\lambda + \mathcal{A}(z))\cdot z^T/|z|_-(\lambda + \mathcal{A}(z))^{-1}z/|z|_-.$$

In the following we show that we can apply Theorem 2.69 to the matrix $\mathscr{L}'$. Unfortunately the symbol $\det \mathscr{L}'$ does not fit into the symbol class $S(\overline{S}_\theta \times \overline{\Sigma}_\delta^n)$ because it includes the fractions $z_k/|z|_-$, cf. Example 2.6 (ii). Therefore, we are not able to apply the characterization of Corollary 2.57.

Lemma 4.10. *Let $\mathcal{A}(\partial)$ be a strongly elliptic differential operator as in Definition 4.7. Then there exist $\theta > \pi/2$ and $\delta \in (0, \pi)$ such that the symbol*

$$f\colon S_\theta \times \Sigma_\delta^n \to \mathbb{C}, \quad (\lambda, z) \mapsto -z^T/|z|_-\left[\det(\lambda + \mathcal{A}(z))\cdot(\lambda + \mathcal{A}(z))^{-1}\right]z/|z|_-$$

is well-defined and N-parabolic in the sense of Definition 2.39 with

$$[\mu(f)](\gamma) = \max\{2(n-1), \gamma(n-1)\}, \quad \gamma \geq 0.$$

Proof. First, we define the continuous extension $F\colon \overline{S}_\theta \times \overline{\Sigma}_\delta^n \to \mathbb{C}$ by

$$F(\lambda, z) := \begin{cases} -z^T/|z|_-\left[\det(\lambda + \mathcal{A}(z))\cdot(\lambda + \mathcal{A}(z))^{-1}\right]z/|z|_-, & z \neq 0, \\ -\lambda^{n-1}, & z = 0. \end{cases}$$

4.1. The generalized L_p-L_q Stokes problem on $\Omega = \mathbb{R}^n$

(I) Let $\lambda \in \overline{S}_\theta$, $\zeta \in i\mathbb{R}^n \setminus \{0\}$, $\eta := (\lambda + \mathcal{A}(\zeta))^{-1}\hat{\zeta}$ with $\hat{\zeta} := \zeta/|\zeta|$, and $\hat{\eta} := \eta/|\eta|$. Then we obtain

$$\eta = (\lambda + \mathcal{A}(\zeta))^{-1}\hat{\zeta} = |\zeta|^{-2}(\lambda|\zeta|^{-2} + \mathcal{A}(\hat{\zeta}))^{-1}\hat{\zeta} \neq 0$$

according to (4.6). Hence, taking the scalar product in $\mathbb{C}^n$ and noting that $\overline{\zeta} = -\zeta$, we get

$$\begin{aligned}
F(\lambda, \zeta) &= \det(\lambda + \mathcal{A}(z)) \cdot \left\langle (\lambda + \mathcal{A}(\zeta))^{-1}\hat{\zeta}, \hat{\zeta} \right\rangle \\
&= \det(\lambda + \mathcal{A}(z)) \cdot \langle \eta, (\lambda + \mathcal{A}(\zeta))\eta \rangle \\
&= \det(\lambda + \mathcal{A}(z)) \cdot (\overline{\lambda}|\eta|^2 + \langle \eta, \mathcal{A}(\zeta)\eta \rangle) \\
&= |\eta|^2 \det(\lambda + \mathcal{A}(z)) \cdot \overline{\left(\lambda + |\zeta|^2 \left\langle \mathcal{A}(\hat{\zeta})\hat{\eta}, \hat{\eta} \right\rangle \right)} \\
&= |\eta|^2 |\zeta|^2 \det(\lambda + \mathcal{A}(z)) \cdot \overline{\left(\lambda/|\zeta|^2 + \left\langle \mathcal{A}(\hat{\zeta})\hat{\eta}, \hat{\eta} \right\rangle \right)} \neq 0
\end{aligned}$$

due to $\lambda/|\zeta|^2 \in \overline{S}_\theta$, $-\left\langle \mathcal{A}(\hat{\zeta})\hat{\eta}, \hat{\eta} \right\rangle \in W(-\mathcal{A}(\hat{\zeta}))$, (4.6), and (4.5). Trivially, $F(\lambda, 0) \neq 0$ for all $\lambda \in \overline{S}_\theta \setminus \{0\}$. Altogether we derive

$$|F(\lambda, \zeta)| \neq 0, \quad (\lambda, \zeta) \in \overline{S}_\theta \times (i\mathbb{R})^n, \quad |\lambda| + |\zeta|^2 = 1.$$

(II) Assume that for all $j \in \mathbb{N}$ there exists $(\lambda^{(j)}, z^{(j)}) \in \overline{S}_\theta \times \overline{\Sigma}_{1/j}^n$ with $|\lambda^{(j)}| + |z^{(j)}|^2 = 1$ and $F(\lambda^{(j)}, z^{(j)}) = 0$. Without loss of generality we may assume that $(\lambda^{(j)}, z^{(j)})_j$ is convergent and $(\lambda, \zeta) := \lim_{j \to \infty} (\lambda^{(j)}, z^{(j)})$. Then we have $\zeta \in (i\mathbb{R})^n$, $|\lambda| + |\zeta|^2 = 1$, and $F(\lambda, \zeta) = 0$, which contradicts (I).

This yields that there exists $\delta \leq \delta_0$ such that $F(\lambda, z) \neq 0$ for all $\lambda \in \overline{S}_\theta$ and $z \in \overline{\Sigma}_\delta^n$ with $|\lambda| + |z|^2 = 1$.

It is easy to see that F is 2-homogeneous of degree $2(n-1)$ and therefore there exist $C, C' > 0$ such that

$$C' \cdot (|\lambda|^{n-1} + |z|^{2(n-1)}) \leq |F(\lambda, z)| \leq C(|\lambda|^{n-1} + |z|^{2(n-1)})$$

for $(\lambda, z) \in (\overline{S}_\theta \times \overline{\Sigma}_\delta^n) \setminus \{(0,0)\}$ due to the compactness of $K := \{(\lambda, z) \in \overline{S}_\theta \times \overline{\Sigma}_\delta^n : |\lambda| + |z|^2 = 1\}$. Let $\lambda_0 > 0$ be arbitrary, then we get $|\lambda|^{n-1} + |z|^{2(n-1)} \geq 1/2|\lambda|^{n-1} + \lambda_0^{n-1}/2 + |z|^{2(n-1)} \geq \min\{1/2, \lambda_0^{n-1}/2\}(1 + |\lambda|^{n-1} + |z|^{2(n-1)})$ for $|\lambda| \geq \lambda_0$ and $z \in \overline{\Sigma}_\delta^n$. $\square$

For an application of Theorem 2.69 we define the order functions

$$t_i(\gamma) := \begin{cases} 2, & \gamma \in (0, 2), \\ \gamma, & \gamma \geq 2, \end{cases} \quad i = 1, \ldots n, \quad t_{n+1}(\gamma) := 0,$$

$$s_j(\gamma) := 0, \quad j = 1, \ldots n, \quad s_{n+1}(\gamma) := -\begin{cases} 2, & \gamma \in (0,2), \\ \gamma, & \gamma \geq 2 \end{cases}$$

for $\gamma > 0$. For each $i, j = 1, \ldots, n$ the order function $s_j + t_i$ is a strictly positive upper order function of $\mathscr{L}'_{ji}$. The strictly negative order function $s_{n+1} + r_{n+1}$ is an upper order function of $\mathscr{L}'_{n+1,n+1} = 0$. Trivially, we have

$$[\mu(\det \mathscr{L}')](\gamma) = \max\{2(n-1), \gamma(n-1)\} = \sum_{i=1}^{n+1}(t_i(\gamma) + s_i(\gamma)), \quad \gamma > 0.$$

Therefore the transformed Stokes problem (4.7) is an N-parabolic mixed-order system in the sense of Definition 2.67. If we define $(r'_\ell, s'_\ell) := (0,0)$ for $\ell = 0,1$ and $M := 1$ we obtain (2.50) and that $\mu_{\mathbb{H}_i}$ and $\mu_{\mathbb{F}_j}$ are positive order functions. In the notation of Theorem 2.69 we derive

$$\mathbb{H}_i = {}_0H_p^1(\mathbb{R}_+, L_q(\mathbb{R}^n)) \cap L_p(\mathbb{R}_+, H_q^2(\mathbb{R}^n)),$$
$$\mathbb{H}_{n+1} = L_p(\mathbb{R}_+, L_q(\mathbb{R}^n)),$$
$$\mathbb{F}_j = L_p(\mathbb{R}_+, L_q(\mathbb{R}^n)),$$
$$\mathbb{F}_{n+1} = {}_0H_p^1(\mathbb{R}_+, L_q(\mathbb{R}^n)) \cap L_p(\mathbb{R}_+, H_q^2(\mathbb{R}^n))$$

for $i,j = 1, \ldots, n$ with $(\mathcal{F}_\ell, \mathcal{K}_\ell) := (H_p, H_q)$ and $1 < p, q < \infty$. As mentioned in Proposition 2.72 the embedding conditions (2.51) and (2.52) are fulfilled in this case.

Now we can apply Theorem 2.69 and Remark 2.70 (iv) to $\mathscr{L}'$ and get the following theorem.

Theorem 4.11. *Let $1 < p, q < \infty$. There exists $\varrho_0 > 0$ such that we get*

$$[\mathscr{L}'(\mathcal{D}_+^{(\varrho)})]|_{\mathbb{E}} \in L_{\mathrm{Isom}}(\mathbb{E}, \mathbb{D}), \quad \varrho \geq \varrho_0 \tag{4.8}$$

where

$$\mathbb{E} := \mathbb{E}_0 \times \mathbb{E}_1, \quad \begin{aligned} \mathbb{E}_0 &:= \left[{}_0H_{p,\varrho}^1(\mathbb{R}_+, L_q(\mathbb{R}^n)) \cap L_{p,\varrho}(\mathbb{R}_+, H_q^2(\mathbb{R}^n))\right]^n, \\ \mathbb{E}_1 &:= L_{p,\varrho}(\mathbb{R}_+, L_q(\mathbb{R}^n)), \end{aligned}$$

$$\mathbb{D} := \mathbb{D}_0 \times \mathbb{D}_1, \quad \begin{aligned} \mathbb{D}_0 &:= [L_{p,\varrho}(\mathbb{R}_+, L_q(\mathbb{R}^n))]^n, \\ \mathbb{D}_1 &:= {}_0H_{p,\varrho}^1(\mathbb{R}_+, L_q(\mathbb{R}^n)) \cap L_{p,\varrho}(\mathbb{R}_+, H_q^2(\mathbb{R}^n)). \end{aligned}$$

Returning to the original problem (4.2) and (4.3), we define the operator

$$L_{\mathrm{St}} \colon \mathbb{E}_0 \times \dot{\mathbb{E}}_1 \to \dot{\mathbb{D}}_0 \times \dot{\mathbb{D}}_1,$$

$$\begin{pmatrix} u \\ \pi \end{pmatrix} \mapsto \begin{pmatrix} 1 & 0 \\ 0 & \dot{J}_1 \end{pmatrix} \mathscr{L}'(\nabla_+^{(\varrho)}) \begin{pmatrix} 1 & 0 \\ 0 & \dot{J}_1 \end{pmatrix} \begin{pmatrix} u \\ \pi \end{pmatrix}$$

where the spaces $\dot{\mathbb{E}}_1$ and $\dot{\mathbb{D}}_1$ are given by $\dot{\mathbb{E}}_1 := L_{p,\varrho}(\mathbb{R}_+, \dot{H}_q^1(\mathbb{R}^n))$ and $\dot{\mathbb{D}}_1 := {}_0H_{p,\varrho}^1(\mathbb{R}_+, \dot{H}_q^{-1}(\mathbb{R}^n)) \cap L_{p,\varrho}(\mathbb{R}_+, H_q^1(\mathbb{R}^n))$. Note that $\dot{J}_1$ is the homogeneous shift operator introduced in Proposition 4.3.

4.1. The generalized L_p-L_q Stokes problem on $\Omega = \mathbb{R}^n$

Corollary 4.12. *The operator L_{St} is well-defined, invertible, and bounded due to Theorem 4.11 and Proposition 4.4.*

In the following we want to illuminate that L_{St} indeed represents the original differential equation (4.2). For this we need the next lemma.

Lemma 4.13. (i) *Let $\psi_j(z) := z_j/|z|_-$ and $\varphi_{ij}(\lambda, z) := \delta_{ij}\lambda - \sum_{k,l=1}^{n} a_{ij}^{kl} z_k z_l$, $i, j = 1, \ldots, n$, where $(\lambda, z) \in (S_\theta \times \Sigma_\delta^n)$. Then we have*

$$\psi_j(\boldsymbol{D}^{\mathcal{N}})\dot{J}_1 f = \partial_j f, \tag{4.9}$$

$$\dot{J}_1 \psi_j(\boldsymbol{D}^{\mathcal{N}}) g = \partial_j g, \quad \mathcal{N} := L_q(\mathbb{R}^n), \tag{4.10}$$

$$\varphi_{ij}(\boldsymbol{D}_+^{(\varrho)}) u = \delta_{ij}\partial_t u - \sum_{k,l=1}^{n} a_{ij}^{kl} \partial_k \partial_l u, \quad \varrho \geq \varrho_0 \tag{4.11}$$

for all $f \in \dot{H}_q^1(\mathbb{R}^n)$, $g \in H_q^1(\mathbb{R}^n)$ and all

$$u \in [{}_0H_{p,\varrho}^1(\mathbb{R}_+, L_q(\mathbb{R}^n)) \cap L_{p,\varrho}(\mathbb{R}_+, H_q^2(\mathbb{R}^n))]^n.$$

(ii) *For all $(u, \pi)^T \in \mathbb{E}_0 \times \dot{\mathbb{E}}_1$ we have*

$$\begin{pmatrix} \partial_t + \mathcal{A}(\partial) & \nabla \\ \nabla^T & 0 \end{pmatrix} \begin{pmatrix} u \\ \pi \end{pmatrix} = L_{\text{St}}(u, \pi)^T.$$

Proof. (i) For $f \in \mathscr{Z}(\mathbb{R}^n)$ we get $\dot{J}_1 f = \mathscr{F}^{-1}|\xi|\mathscr{F}f \in \mathscr{Z}(\mathbb{R}^n) \subseteq \mathscr{S}(\mathbb{R}^n)$. Therefore we obtain

$$\psi_j(\boldsymbol{D}^{\mathcal{N}})\dot{J}_1 f = \mathscr{F}^{-1} i\xi_j/|\xi|\mathscr{F}\mathscr{F}^{-1}|\xi|\mathscr{F}f = \mathscr{F}^{-1} i\xi_j \mathscr{F}f = \partial_j f$$

by the representation (1.30). By virtue of Propositions 4.6 and 4.5 we obtain (4.9).

Let $g \in \mathscr{S}(\mathbb{R}^n)$ and $\varphi \in \mathscr{Z}(\mathbb{R}^n)$. Then we get

$$[\dot{J}_1 \psi_j(\boldsymbol{D}^{\mathcal{N}})g](\varphi) = [\psi_j(\boldsymbol{D}^{\mathcal{N}})g](\dot{J}_1\varphi) = [\mathscr{F}^{-1} i\xi_j/|\xi|\mathscr{F}g](\mathscr{F}^{-1}|\xi|\mathscr{F}\varphi)$$
$$= [g](\mathscr{F}i\xi_j(\mathscr{F}\varphi)(-\cdot)) = [g](\mathscr{F}i\xi_j\mathscr{F}^{-1}\varphi)$$
$$= [\partial_j g](\varphi)$$

according to (1.30). Therefore we have $[\dot{J}_1\psi_j(\boldsymbol{D}^{\mathcal{N}})g] = [\partial_j g]$ as an equality in $\mathscr{Z}'(\mathbb{R}^n)$. Using $\dot{J}_1\psi_j(\boldsymbol{D}^{\mathcal{N}})g, \partial_j g \in L_q(\mathbb{R}^n)$ we get $\dot{J}_1\psi_j(\boldsymbol{D}^{\mathcal{N}})g = \partial_j g$ in $L_q(\mathbb{R}^n)$. The density of $\mathscr{S}(\mathbb{R}^n)$ in $H_q^1(\mathbb{R}^n)$ then yields

$$\dot{J}_1\psi_j(\boldsymbol{D}^{\mathcal{N}})g = \partial_j g, \quad g \in H_q^1(\mathbb{R}^n).$$

The assertion in (4.11) follows directly from Theorem 1.26.

(ii) This follows easily from (4.9)–(4.11). $\square$

4.2 The generalized L_p-L_q thermo-elastic plate equations on $\Omega = \mathbb{R}^n$

> *Motivation.* Results on maximal L_p-L_q-regularity of the thermo-elastic plate equation can be found in [DR06], [DRS09], and [Nai09], for instance. The latter work is mainly based on the application of Fourier and Laplace transforms and a consideration of a similar problem in $\mathbb{R}^{n+1}$.
>
> In [MR96] and [DR06] the authors consider the generalized thermo-elastic plate equation also called the α-β-system. In [MR96] the authors study the associated equations in an L_2-setting whereas in [DR06] the authors prove existence and uniqueness results in an L_p-L_q setting on $\Omega = \mathbb{R}^n$. There the used techniques are a reduction to a first-order system, the Newton polygon method, and Michlin's multiplier theorem. The results given there fit into our general theory and can now be seen in a larger context. In particular, here we can avoid a reduction to a first-order system.

For given functions $f, g \colon \mathbb{R}_+ \times \mathbb{R}^n \to \mathbb{R}$ the generalized thermo-elastic plate equations read as follows:

$$\begin{cases} u_{tt} + a \cdot S_\eta u - b \cdot S_\eta^\beta \theta = f, & (t, x) \in \mathbb{R}_+ \times \mathbb{R}^n, \\ d \cdot \theta_t + g \cdot S_\eta^\alpha \theta + b \cdot S_\eta^\beta u_t = g, & (t, x) \in \mathbb{R}_+ \times \mathbb{R}^n, \\ u(t = 0) = 0, & x \in \mathbb{R}^n, \\ u_t(t = 0) = 0, & x \in \mathbb{R}^n, \\ \theta(t = 0) = 0, & x \in \mathbb{R}^n \end{cases} \quad (4.12)$$

for the unknown functions $u \colon \mathbb{R}_+ \times \mathbb{R}^n \to \mathbb{R}$, $\theta \colon \mathbb{R}_+ \times \mathbb{R}^n \to \mathbb{R}$, and constants $\alpha, \beta \in [0, 1]$, $a, b, d, g > 0$. For $\eta > 0$ and $\delta \geq 0$ the operators S_η^δ are given by $S_\eta^\delta := f_{\eta,\delta}(\nabla)$ for $f_{\eta,\delta}(z) := |z|^{2\eta\delta}$, i.e., $S_\eta^\delta = (-\Delta)^{\eta\delta}$. If we set $\eta := 2$ and $\alpha = \beta = 1/2$, then we obtain the usual thermo-elastic plate equations.

As mentioned before, with our method it is not necessary to rewrite (4.12) as a first-order system. A formal Fourier and Laplace transform in the space and time variables yield the formal transformed problem of (4.2):

$$\begin{pmatrix} \lambda^2 + a|\xi|^{2\eta} & -b|\xi|^{2\eta\beta} \\ b\lambda|\xi|^{2\eta\beta} & d\lambda + g|\xi|^{2\eta\alpha} \end{pmatrix} \begin{pmatrix} \hat{u} \\ \hat{\theta} \end{pmatrix} = \begin{pmatrix} \hat{f} \\ \hat{g} \end{pmatrix}, \quad (\lambda, \xi) \in \mathbb{R}_+ \times \mathbb{R}^n. \quad (4.13)$$

Now we define the complex version of (4.13) by

$$\mathscr{L}(\lambda, z) := \begin{pmatrix} \lambda^2 + a|z|_-^{2\eta} & -b|z|_-^{2\eta\beta} \\ b\lambda|z|_-^{2\eta\beta} & d\lambda + g|z|_-^{2\eta\alpha} \end{pmatrix} \in \mathbb{C}^{2\times 2}, \quad (\lambda, z) \in S_\vartheta \times \Sigma_\delta^n$$

and get $\mathscr{L} \in [H_P(S_\vartheta \times \Sigma_\delta^n)]^2$. As the determinant of this system we obtain

$$\det \mathscr{L}(\lambda, z) = d\lambda^3 + g\lambda^2|z|_-^{2\eta\alpha} + ad\lambda|z|_-^{2\eta} + b^2\lambda|z|_-^{4\eta\beta} + ag|z|_-^{2\eta(\alpha+1)}.$$

4.2. The generalized L_p-L_q thermo-elastic plate equations on $\Omega = \mathbb{R}^n$

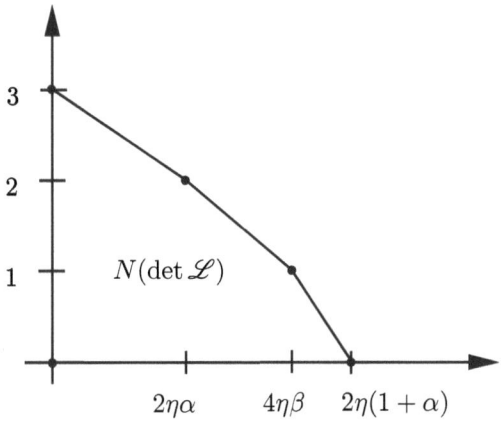

Figure 4.1: Newton polygon of the α-β-system

for $(\lambda, z) \in S_\vartheta \times \Sigma_\delta^n$. In [DR06, Lemma 3.1] it was already proved that $\det \mathscr{L}$ is N-parabolic in $S_\vartheta \times \Sigma_\delta^n$ with

$$[\mu(\det \mathscr{L})](\gamma) = \max\{2\eta(1+\alpha), \gamma + 4\eta\beta, 3\gamma\}, \quad \gamma > 0$$

if α and β fulfill

$$\alpha \geq \beta \quad \text{and} \quad 2\beta - \alpha \geq \frac{1}{2}. \tag{4.14}$$

In the sequel we always assume that (α, β) fulfills (4.14).

For the application of Theorem 2.69 we define $(r'_\ell, s'_\ell) := (0,0)$, $\ell = 0, \ldots, 3$ and $M := 3$, and

$$t_1(\gamma) := \begin{cases} 2\eta + 2\eta(\alpha - \beta), & \gamma \in [0, \gamma_1], \\ \gamma + 2\eta\beta, & \gamma \in (\gamma_1, \gamma_2], \\ 2\gamma + 2\eta(\alpha - \beta), & \gamma \in (\gamma_2, \gamma_3], \\ 2\gamma + 2\eta(\alpha - \beta), & \gamma \in (\gamma_3, \infty), \end{cases} \quad t_2(\gamma) := \begin{cases} 2\eta\alpha, & \gamma \in [0, \gamma_1], \\ 2\eta\alpha, & \gamma \in (\gamma_1, \gamma_2], \\ 2\eta\alpha, & \gamma \in (\gamma_2, \gamma_3], \\ \gamma, & \gamma \in (\gamma_3, \infty), \end{cases}$$

$$s_1(\gamma) := -2\eta(\alpha - \beta), \qquad s_2(\gamma) := 0$$

where $\gamma_1 := 2\eta(1 + \alpha - 2\beta)$, $\gamma_2 := 2\eta(2\beta - \alpha)$, $\gamma_3 := 2\eta\alpha$. Obviously, $s_j + t_i$ is a convex upper order function for $\mathscr{L}_{ji}$ for all $i, j = 1, 2$. Note that $s_1 + t_2$ is a convex increasing order function but not strictly positive. This application is one of the reasons for the formulation of Definition 2.66 with convex / concave order functions instead of strictly positive/negative ones.

It is easy to see that $[\mu(\det \mathscr{L})] = s_1 + s_2 + t_1 + t_2$. Hence, $\mathscr{L}$ is an N-parabolic mixed-order system in the sense of Definition 2.67. With the notation of Theorem

2.69 we have
$$\mu_{\mathbb{H}_i} = t_i, \quad \mu_{\mathbb{F}_j} = -s_j.$$

Therefore, $\mu_{\mathbb{H}_i}$ and $\mu_{\mathbb{F}_j}$ are strictly positive order functions and

$$\begin{aligned}
\mathbb{H}_1 &:= {}_0H_p^2(\mathbb{R}_+, H_q^{2\eta(\alpha-\beta)}(\mathbb{R}^n)) \\
&\quad \cap {}_0H_p^1(\mathbb{R}_+, H_q^{2\eta\beta}(\mathbb{R}^n)) \cap L_p(\mathbb{R}_+, H_q^{2\eta(1+\alpha-\beta)}(\mathbb{R}^n)), \\
\mathbb{H}_2 &:= {}_0H_p^1(\mathbb{R}_+, L_q(\mathbb{R}^n)) \cap L_p(\mathbb{R}_+, H_q^{2\eta\alpha}(\mathbb{R}^n)), \\
\mathbb{F}_1 &:= L_p(\mathbb{R}_+, H_q^{2\eta(\alpha-\beta)}(\mathbb{R}^n)), \\
\mathbb{F}_2 &:= L_p(\mathbb{R}_+, L_q(\mathbb{R}^n)), \quad 1 < p, q < \infty.
\end{aligned}$$

Due to Proposition 2.72 we can apply Theorem 2.69 and Remark 2.70 (iv), respectively, and obtain the following result.

Theorem 4.14. *Let $1 < p, q < \infty$. There exists $\varrho_0 > 0$ such that we get*

$$L := \left[\mathscr{L}(\mathcal{D}_+^{(\varrho)})\right]\Big|_{\mathbb{E}} \in L_{\mathrm{Isom}}(\mathbb{E}, \mathbb{D}), \quad \varrho \geq \varrho_0$$

where

$$\begin{aligned}
\mathbb{E} &:= {}_0H_{p,\varrho}^2(\mathbb{R}_+, H_q^{2\eta(\alpha-\beta)}(\mathbb{R}^n)) \cap {}_0H_{p,\varrho}^1(\mathbb{R}_+, H_q^{2\eta\beta}(\mathbb{R}^n)) \\
&\quad \cap L_{p,\varrho}(\mathbb{R}_+, H_q^{2\eta(1+\alpha-\beta)}(\mathbb{R}^n)) \\
&\quad \times {}_0H_{p,\varrho}^1(\mathbb{R}_+, L_q(\mathbb{R}^n)) \cap L_{p,\varrho}(\mathbb{R}_+, H_q^{2\eta\alpha}(\mathbb{R}^n)), \\
\mathbb{D} &:= L_{p,\varrho}(\mathbb{R}_+, H_q^{2\eta(\alpha-\beta)}(\mathbb{R}^n)) \times L_{p,\varrho}(\mathbb{R}_+, L_q(\mathbb{R}^n)).
\end{aligned}$$

Lemma 4.15. *For all $(u, \theta)^T \in \mathbb{E}$ we have*

$$\begin{pmatrix} \partial_t^2 + aS_\eta & -bS_\eta^\beta \\ bS_\eta^\beta \partial_t & d\partial_t + gS_\eta^\alpha \end{pmatrix} \begin{pmatrix} u \\ \theta \end{pmatrix} = L(\mathcal{D}_+^{(\varrho)}) \begin{pmatrix} u \\ \theta \end{pmatrix}.$$

Proof. This easily follows under frequent use of Theorem 1.26. $\square$

Remark 4.16 (Higher regularity). Using the same techniques it is possible to establish this result on other scales and spaces of higher regularity. Higher regularity results can be obtained by specifying the constants (r'_ℓ, s'_ℓ). Let μ' be an arbitrary strictly positive order function which fits into the order structure of t_1, t_2, s_1, and s_2, i.e.,

$$\mu'(\gamma) = \begin{cases} m_0(\mu')\gamma + b_0(\mu'), & \gamma \in [0, \gamma_1), \\ m_1(\mu')\gamma + b_1(\mu'), & \gamma \in [\gamma_1, \gamma_2), \\ m_2(\mu')\gamma + b_2(\mu'), & \gamma \in [\gamma_2, \gamma_3), \\ m_3(\mu')\gamma + b_3(\mu'), & \gamma \in [\gamma_3, \infty), \end{cases} \quad \gamma \geq 0,$$

with $m_\ell(\mu'), b_\ell(\mu') \geq 0$, $\ell = 0, \ldots, 3$, where $\gamma_1, \ldots, \gamma_3$ are defined as above. Then we can apply Theorem 2.69 with

$$(r'_\ell, s'_\ell) := (b_\ell(\mu'), m_\ell(\mu')), \quad \ell = 0, \ldots, 3,$$

since the sum of strictly positive order functions is also strictly positive. More precisely, we can define $\mu'(\gamma) := \max\{2\kappa\eta\alpha, \kappa\gamma\}$ for some $\kappa \geq 0$ if we want to add regularity in time for the right-hand sides. With this higher regularity order function we then have $(r'_0, s'_0) = (r'_1, s'_1) = (r'_2, s'_2) = (2\kappa\eta\alpha, 0)$, and $(r'_3, s'_3) = (0, \kappa)$. An application of Theorem 2.69, Proposition 2.72, and Remark 2.70 (iv) then yield the existence of $\varrho_0 > 0$ such that

$$\left[\mathscr{L}(\mathcal{D}_+^{(\varrho)})\right]\Big|_{\mathbb{E}} \in L_{\mathrm{Isom}}(\mathbb{E}, \mathbb{D}), \quad \varrho \geq \varrho_0$$

where the spaces of higher regularity are given by $\mathbb{E} := \mathbb{E}_1 \times \mathbb{E}_2$, $\mathbb{D} = \mathbb{D}_1 \times \mathbb{D}_2$, and

$$\mathbb{E}_1 := {}_0H^{2+\kappa}_{p,\varrho}(\mathbb{R}_+, H^{2\eta(\alpha-\beta)}_q(\mathbb{R}^n)) \cap {}_0H^2_{p,\varrho}(\mathbb{R}_+, H^{2\eta((\kappa+1)\alpha-\beta)}_q(\mathbb{R}^n))$$
$$\cap {}_0H^1_{p,\varrho}(\mathbb{R}_+, H^{2\eta(\kappa\alpha+\beta)}_q(\mathbb{R}^n))$$
$$\cap L_{p,\varrho}(\mathbb{R}_+, H^{2\eta(1+(\kappa+1)\alpha-\beta)}_q(\mathbb{R}^n)),$$
$$\mathbb{E}_2 := {}_0H^{1+\kappa}_{p,\varrho}(\mathbb{R}_+, L_q(\mathbb{R}^n)) \cap L_{p,\varrho}(\mathbb{R}_+, H^{2(\kappa+1)\eta\alpha}_q(\mathbb{R}^n)),$$
$$\mathbb{D}_1 := {}_0H^\kappa_{p,\varrho}(\mathbb{R}_+, H^{2\eta(\alpha-\beta)}_q(\mathbb{R}^n)) \cap L_{p,\varrho}(\mathbb{R}_+, H^{2\eta((\kappa+1)\alpha-\beta)}_q(\mathbb{R}^n)),$$
$$\mathbb{D}_2 = {}_0H^\kappa_{p,\varrho}(\mathbb{R}_+, L_q(\mathbb{R}^n)) \cap L_{p,\varrho}(\mathbb{R}_+, H^{2\kappa\eta\alpha}_q(\mathbb{R}^n)), \quad \kappa \geq 0.$$

For $\kappa = 0$ these spaces coincide with the spaces in Theorem 4.14. An analog result can also be derived by using the Besov scale instead of the Bessel potential scale.

4.3 A linear L_p-L_q Cahn-Hilliard-Gurtin problem in $\Omega = \mathbb{R}^n$

Motivation. In this section we want to treat the linear Cahn-Hilliard-Gurtin problem in an L_p-L_q-setting. For $p = q$ this can be found in [Wil07], for example. This problem fits into our developed theory and we can give a proof of the well-posedness in a short and elegant way.

For the unknown functions $u: \mathbb{R}_+ \times \mathbb{R} \to \mathbb{R}$ and $\mu: \mathbb{R}_+ \times \mathbb{R} \to \mathbb{R}$ the Cahn-Hilliard-Gurtin problem reads as follows:

$$\begin{cases} \partial_t u - \mathrm{div}\,(a\partial_t u) - \mathrm{div}\,(B\nabla\mu) = f, & (t, x) \in \mathbb{R}_+ \times \mathbb{R}^n, \\ \Delta u - \beta\partial_t u + \mu - c \cdot \nabla\mu = g, & (t, x) \in \mathbb{R}_+ \times \mathbb{R}^n, \\ u(t=0) = 0, & x \in \mathbb{R}^n \end{cases} \quad (4.15)$$

with $\beta > 0$, $a, c \in \mathbb{R}^n$, and a symmetric and positive definite matrix $B \in \mathbb{R}^{n \times n}$. As in [Wil07, Theorem 4.2.1] we assume that $A := \beta B - \frac{1}{2}(a \otimes c + c \otimes a)$ ($x \otimes y := x \cdot y^T$) is also positive definite. After formal Laplace and partial Fourier transform we obtain the mixed-order system

$$\mathscr{L}(\lambda, z) := \begin{pmatrix} \lambda - \lambda \langle z, a \rangle & -z^T B z \\ -\beta \lambda - |z|_-^2 & 1 - \langle z, c \rangle \end{pmatrix} \in \mathbb{C}^{2 \times 2}, \quad (\lambda, z) \in S_\theta \times \Sigma_\delta^n$$

and get $\mathscr{L} \in [H_P(S_\theta \times \Sigma_\delta^n)]^{2 \times 2}$. For the determinant we easily calculate that

$$\det \mathscr{L}(\lambda, z) = \lambda(1 - \langle z, a \rangle - \langle z, c \rangle) - \lambda z^T A z - |z|_-^2 z^T B z$$

due to $\frac{1}{2} z^T (a \otimes c + c \otimes a) z = \langle z, a \rangle \langle z, c \rangle$. The symbol $\det \mathscr{L}$ fits into the class $S(S_\theta \times \Sigma_\delta^n)$ defined in Definition 2.11. We have $d_\gamma(\det \mathscr{L}) = \max\{4, 2 + \gamma\}$ for $\gamma > 0$ and therefore the Newton polygon of $\det \mathscr{L}$ is not regular in space, cf. Figure 4.2. Now we can benefit from the generalized characterization presented in Corollary 2.57 for symbols with a Newton polygon which is not regular in space. Thus, we also have to consider $\pi_\infty \det \mathscr{L}$ and get

$$\pi_\gamma \det \mathscr{L}(\lambda, z) = \begin{cases} -|z|_-^2 z^T B z, & \gamma < 2, \\ -|z|_-^2 z^T B z - \lambda z^T A z, & \gamma = 2, \\ -\lambda z^T A z, & \gamma \in (2, \infty), \\ \lambda[1 - \langle z, a \rangle - \langle z, c \rangle - z^T A z], & \gamma = \infty. \end{cases}$$

Lemma 4.17. *Let the matrix $M \in \mathbb{R}^{n \times n}$ be symmetric and positive definite.*

(i) *There exists $\delta > 0$ and $C > 0$ such that*

$$|z^T M z| \geq C |z|^2, \quad z \in \overline{\Sigma}_\delta^n \setminus \{0\}.$$

(ii) *For all $\varepsilon > 0$ there exists $\delta > 0$ such that*

$$|\arg(-z^T M z)| \leq \varepsilon, \quad z \in \overline{\Sigma}_\delta^n \setminus \{0\}.$$

Proof. Both assertions can be proved by contradiction. □

Using the last lemma it is obvious that there exist $\theta > \pi/2$ and $\delta > 0$ such that $\pi_\gamma \det \mathscr{L}(\lambda, z) \neq 0$ for all $(\lambda, z) \in (\overline{S}_\theta \setminus \{0\}) \times (\overline{\Sigma}_\delta^n \setminus \{0\})$ and $\gamma \in (0, \infty]$. Due to $\pi_\infty \det \mathscr{L}(\lambda, 0) = \lambda$ we can now apply Corollary 2.57 and obtain the N-parabolicity for $\det \mathscr{L}$. Hence, we have

$$\det \mathscr{L} \in S_N(S_\theta \times \Sigma_\delta^n)$$

for some θ and δ.

4.3. A linear L_p-L_q Cahn-Hilliard-Gurtin problem in $\Omega = \mathbb{R}^n$

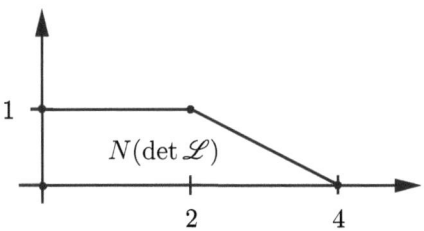

Figure 4.2: Newton polygon of the Cahn-Hilliard-Gurtin problem

Next, we define $(r'_\ell, s'_\ell) := (0,0)$, $\ell = 0, 1 =: M$, and

$$t_1(\gamma) := \max\{3, 1+\gamma\}, \qquad t_2(\gamma) := 2,$$
$$s_1(\gamma) := 0, \qquad s_2(\gamma) := -1$$

and derive $\mu(\det \mathscr{L}) = \sum_{i=1}^{2}(t_i + s_i)$. Altogether we obtain that $\mathscr{L}$ is an N-parabolic mixed-order system in the sense of Definition 2.67. An application of Theorem 2.69, Proposition 2.72, and Remark 2.70 (iv) then yields the following result.

Theorem 4.18. Let $1 < p, q < \infty$. There exists $\varrho_0 > 0$ such that we get

$$\left[\mathscr{L}(\mathcal{D}_+^{(\varrho)})\right]\bigg|_{\mathbb{H}} \in L_{\text{Isom}}(\mathbb{H}, \mathbb{F}), \quad \varrho \geq \varrho_0$$

where

$$\mathbb{H}_1 := {}_0H^1_{p,\varrho}(\mathbb{R}_+, H^1_q(\mathbb{R}^n)) \cap L_{p,\varrho}(\mathbb{R}_+, H^3_q(\mathbb{R}^n)),$$
$$\mathbb{H}_2 := L_{p,\varrho}(\mathbb{R}_+, H^2_q(\mathbb{R}^n)),$$
$$\mathbb{F}_1 := L_{p,\varrho}(\mathbb{R}_+, L_q(\mathbb{R}^n)),$$
$$\mathbb{F}_2 := L_{p,\varrho}(\mathbb{R}_+, H^1_q(\mathbb{R}^n)),$$

and $\mathbb{H} := \mathbb{H}_1 \times \mathbb{H}_2$, $\mathbb{F} := \mathbb{F}_1 \times \mathbb{F}_2$.

For $p = q$ the same spaces as in Theorem 4.18 also occur in [Wil07, Theorem 4.2.1].

4.4 A compressible fluid model of Korteweg type on $\Omega = \mathbb{R}^n$

> *Motivation.* Here we want to consider the linearized model problem on the whole space which occurs in the treatment of a fluid model of Korteweg type. This model describes the dynamics of a non-thermal, compressible fluid with viscosity and capillarity.

The linearized compressible fluid model of Korteweg type on $\mathbb{R}^n$ is given by (cf. [Kot08, Theorem 2.2] where the case $p = q$ is treated)

$$\begin{cases} \partial_t u - \mu_0 \Delta u - (\lambda_0 + \mu_0)\nabla \operatorname{div} u - \kappa_0 \Delta \nabla \rho = f, & (t,x) \in \mathbb{R}_+ \times \mathbb{R}^n, \\ \partial_t \rho + \beta_0 \operatorname{div} u = g, & (t,x) \in \mathbb{R}_+ \times \mathbb{R}^n, \\ u(t=0) = 0, & x \in \mathbb{R}^n, \\ \rho(t=0) = 0, & x \in \mathbb{R}^n \end{cases}$$

for the unknown functions $u \colon \mathbb{R}_+ \times \mathbb{R}^n \to \mathbb{R}^n$, $\rho \colon \mathbb{R}_+ \times \mathbb{R}^n \to \mathbb{R}$. As in [Kot08] we assume that the constants μ_0, $2\mu_0 + \lambda_0$, κ_0, and β_0 are positive. The constants λ_0 and μ_0 are viscosity coefficients and κ_0 is a capillarity coefficient. The associated $(n+1) \times (n+1)$ mixed-order system is given by

$$\mathscr{L}(\lambda, z) := \begin{pmatrix} (\lambda + \mu_0 |z|_-^2)\operatorname{id}_n - (\lambda_0 + \mu_0)(z \otimes z) & \kappa_0 |z|_-^2 z \\ \beta z^T & \lambda \end{pmatrix} \quad (4.16)$$

for $(\lambda, z) \in \overline{S}_\theta \times \overline{\Sigma}_\delta^n$, $\theta > \pi/2$, $\delta > 0$. To compute the determinant of this system we will use the next lemma.

Lemma 4.19. *Let $(x_k)_{k=1,\ldots,n} \subseteq \mathbb{C}$ and define $A := (A_{ik})_{i,k=1,\ldots,n}$ where*

$$A_{ik} := \begin{cases} x_k + 1, & i = k, \\ 1, & i \neq k. \end{cases} \quad (4.17)$$

Then we have

$$\det A = \prod_{k=1}^n x_k + \sum_{k=1}^n \prod_{\substack{j=1, \\ j \neq k}}^n x_j.$$

Proof. The assertion is obvious for $n = 1$. By induction we obtain, for $n > 1$,

$$\det A = (x_1 + 1)\Big[\prod_{k=2}^n x_k + \sum_{k=2}^n \prod_{\substack{j=2, \\ j \neq k}}^n x_j\Big] - \sum_{k=2}^n \Big[0 + \prod_{\substack{j=2, \\ j \neq k}}^n x_j\Big]$$

with Laplace's formula. Note that all $(n-1) \times (n-1)$-matrices appearing in Laplace's formula have the same structure as A. This proves the assertion. $\square$

4.4. A compressible fluid model of Korteweg type on $\Omega = \mathbb{R}^n$

Proposition 4.20. (i) *We have*
$$\det \mathscr{L}(\lambda, z) = (\lambda + \mu_0 |z|_-^2)^{n-1} (\lambda^2 + (\lambda_0 + 2\mu_0)\lambda |z|_-^2 + \beta \kappa_0 |z|_-^4)$$
for all $(\lambda, z) \in S_\theta \times \Sigma_\delta^n$.

(ii) *There exist* $\theta > \pi/2$ *and* $\delta > 0$ *such that* $\det \mathscr{L}$ *is N-parabolic with*
$$d_\gamma(\det \mathscr{L}) = [\mu(\det \mathscr{L})](\gamma) = \max\{2(n+1), (n+1)\gamma\}, \quad \gamma \geq 0.$$

Proof. (i) To apply Lemma 4.19 we assume $z_k \neq 0$ for all $k = 1, \ldots, n$ and get
$$\det \mathscr{L}(\lambda, z) = \beta\kappa_0 (\lambda_0 + \mu_0)^{n-1} |z|_-^2 \cdot \det \begin{pmatrix} \frac{\lambda + \mu_0 |z|^2}{\lambda_0 + \mu_0} \mathrm{id}_n - z \otimes z & z \\ z^T & \frac{\lambda_0 + \mu_0}{\beta\kappa_0 |z|_-^2}\lambda \end{pmatrix}$$
$$= (-1)^{n+1} \beta\kappa_0 (\lambda_0 + \mu_0)^{n-1} |z|_-^2 \prod_{\ell=1}^n z_\ell^2$$
$$\cdot \det \begin{pmatrix} \mathrm{diag}((x_k)_{k=1,\ldots,n}) + \mathbb{1}\otimes\mathbb{1} & \mathbb{1} \\ \mathbb{1}^T & x_{n+1} + 1 \end{pmatrix}$$
where $\mathbb{1} := (1, \ldots, 1)^T$, $x_k := -\frac{\lambda + \mu_0|z|^2}{\lambda_0 + \mu_0} \frac{1}{z_k^2}$ $(k = 1, \ldots, n)$, and $x_{n+1} := -\frac{\lambda_0 + \mu_0}{\beta\kappa_0 |z|_-^2}\lambda - 1$. With this we derive
$$\det \mathscr{L}(\lambda, z) = (-1)^{n+1} \beta\kappa_0 (\lambda_0 + \mu_0)^{n-1} |z|_-^2 \prod_{\ell=1}^n z_\ell^2 \left[\prod_{k=1}^{n+1} x_k + \sum_{k=1}^{n+1} \prod_{\substack{j=1, \\ j \neq k}}^{n+1} x_j \right]$$
$$= (-1)^{n+1} \beta\kappa_0 (\lambda_0 + \mu_0)^{n-1} |z|_-^2 \Big[(-1)^n(\lambda_0 + \mu_0)^{-n}(\lambda + \mu_0|z|_-^2)^n x_{n+1}$$
$$+ (-1)^n (\lambda_0 + \mu_0)^{-n} (\lambda + \mu_0|z|_-^2)^n$$
$$+ (-1)^n (\lambda_0 + \mu_0)^{-n+1} (\lambda + \mu_0|z|_-^2)^{n-1} |z|_-^2 x_{n+1} \Big]$$
$$= -\beta\kappa_0(\lambda_0+\mu_0)^{-1}|z|_-^2 (\lambda+\mu_0|z|_-^2)^{n-1}\Big[(\lambda+\mu_0|z|_-^2)x_{n+1}$$
$$+ (\lambda+\mu_0|z|_-^2) + (\lambda_0+\mu_0)|z|_-^2 x_{n+1}\Big]$$
$$= \beta\kappa_0(\lambda_0+\mu_0)^{-1}|z|_-^2(\lambda+\mu_0|z|_-^2)^{n-1}\Big[(\lambda+\mu_0|z|_-^2)\frac{\lambda_0+\mu_0}{\beta\kappa_0|z|_-^2}\lambda$$
$$+ (\lambda_0+\mu_0)|z|_-^2\left(\frac{\lambda_0+\mu_0}{\beta\kappa_0|z|_-^2}\lambda + 1\right)\Big]$$
$$= (\lambda+\mu_0|z|_-^2)^{n-1}\Big[(\lambda+\mu_0|z|_-^2)\lambda + |z|_-^2\big((\lambda_0+\mu_0)\lambda + \beta\kappa_0|z|_-^2\big)\Big]$$
$$= (\lambda+\mu_0|z|_-^2)^{n-1} P(\lambda, z)$$
with $P(\lambda, z) := \lambda^2 + (\lambda_0 + 2\mu_0)\lambda|z|_-^2 + \beta\kappa_0|z|_-^4$. By continuity this formula also holds for $z \in \mathbb{C}^n$.

(ii) First we consider P. For all $(\lambda, z) \in \mathbb{C} \times \mathbb{C}^n$ we get
$$P(\lambda, z) = 0 \Leftrightarrow \lambda = \frac{|z|_-^2}{2}\left(-(\lambda_0 + \mu_0) \pm \sqrt{(\lambda_0 + \mu_0)^2 - 4\beta\kappa_0}\right).$$
We always have
$$\operatorname{Re}\left(-(\lambda_0 + \mu_0) \pm \sqrt{(\lambda_0 + \mu_0)^2 - 4\beta\kappa_0}\right) < 0,$$
$$\left|\operatorname{Im}\left(-(\lambda_0 + \mu_0) \pm \sqrt{(\lambda_0 + \mu_0)^2 - 4\beta\kappa_0}\right)\right| \leq \sqrt{|(\lambda_0 + \mu_0)^2 - 4\beta\kappa_0|}.$$
Hence we can choose $\theta > \pi/2$ and $\delta > 0$ such that $P(\lambda, z) \neq 0$ for all $(\lambda, z) \in (\overline{S}_\theta \setminus \{0\}) \times (\overline{\Sigma}_\delta^n \setminus \{0\})$. By homogeneity we obtain $P \in S_N(\overline{S}_\theta \times \overline{\Sigma}_\delta^n)$ and
$$d_\gamma(P) = \max\{4, 2\gamma\}.$$
Altogether we derive $\det \mathscr{L} \in S_N(\overline{S}_\theta \times \overline{\Sigma}_\delta^n)$ due to the N-parabolicity of the symbol $(\lambda + \mu_0 |z|_-^2)$. The associated order function is then given by
$$[\mu(\det \mathscr{L})](\gamma) = (n-1)\max\{2, \gamma\} + d_\gamma(P) = \max\{2(n+1), (n+1)\gamma\}, \quad \gamma \geq 0.$$
$\square$

Next, we define the order functions
$$t_1(\gamma), \ldots, t_n(\gamma) := \max\{2, \gamma\}, \qquad t_{n+1}(\gamma) := \max\{3, 3/2 \cdot \gamma\},$$
$$s_1(\gamma), \ldots, s_n(\gamma) := 0, \qquad s_{n+1}(\gamma) := -\max\{1, 1/2 \cdot \gamma\}$$
and easily obtain $[\mu(\det \mathscr{L})] = \sum_{k=1}^{n+1}(s_k + t_k)$. For all $j, k = 1, \ldots, n$ the order function $s_j + t_k$ is an upper order function of the component $\mathscr{L}_{jk}$. So we have proved the following proposition:

Proposition 4.21. *The complex matrix $\mathscr{L}$ defined in (4.16) is an N-parabolic mixed-order system in the sense of Definition 2.67.*

With $(r'_\ell, s'_\ell) := (0, 0)$, $\ell = 0, 1$ and $M := 1$, we can apply Theorem 2.69, Proposition 2.72, and Remark 2.70 (iv). Using the notation of Theorem 2.69 we derive the following result.

Theorem 4.22. *Let $1 < p, q < \infty$. There exists $\varrho_0 > 0$ such that we have*
$$\left[\mathscr{L}(\mathcal{D}_+^{(\varrho)})\right]\Big|_{\mathbb{H}} \in L_{\operatorname{Isom}}(\mathbb{H}, \mathbb{F}), \quad \varrho \geq \varrho_0$$
where
$$\mathbb{H}_1, \ldots, \mathbb{H}_n := {}_0H^1_{p,\varrho}(\mathbb{R}_+, L_q(\mathbb{R}^n)) \cap L_{p,\varrho}(\mathbb{R}_+, H^2_q(\mathbb{R}^n)),$$
$$\mathbb{H}_{n+1} := {}_0H^{3/2}_{p,\varrho}(\mathbb{R}_+, L_q(\mathbb{R}^n)) \cap L_{p,\varrho}(\mathbb{R}_+, H^3_q(\mathbb{R}^n)),$$
$$\mathbb{F}_1, \ldots, \mathbb{F}_n := L_{p,\varrho}(\mathbb{R}_+, L_q(\mathbb{R}^n)),$$
$$\mathbb{F}_{n+1} := {}_0H^{1/2}_{p,\varrho}(\mathbb{R}_+, L_q(\mathbb{R}^n)) \cap L_{p,\varrho}(\mathbb{R}_+, H^1_q(\mathbb{R}^n)),$$
and $\mathbb{H} := \prod_{k=1}^{n+1} \mathbb{H}_k$, $\mathbb{F} := \prod_{k=1}^{n+1} \mathbb{F}_k$.

For $p = q$ this result coincides with [Kot08, Theorem 2.2].

4.5 A linear three-phase problem on $\Omega = \mathbb{R}^n$

Motivation. In [Kot10] the author considers a linear three-phase problem which appears in the treatment of a nonlinear chemical reaction system with electromigration. We can treat this system easily with the Newton polygon approach. Note that this is an example for a non-regular Newton polygon.

In [Kot10, Theorem 3.2] the author considers a special case of the following full space problem in an L_p-setting:

$$\begin{cases} \partial_t w - D[\Delta - 1]w - Mu[\Delta - 1]\psi = f, & (t,x) \in \mathbb{R}_+ \times \mathbb{R}^n, \\ \langle \zeta, w \rangle = g, & (t,x) \in \mathbb{R}_+ \times \mathbb{R}^n, \\ w(t=0) = 0, & x \in \mathbb{R}^n, \\ \psi(t=0) = 0, & x \in \mathbb{R}^n \end{cases} \quad (4.18)$$

for the unknown functions $w\colon \mathbb{R}_+ \times \mathbb{R}^n \to \mathbb{R}^N$ ($N \geq 2$) and $\psi\colon \mathbb{R}_+ \times \mathbb{R}^n \to \mathbb{R}$ with the charges of the species $\zeta \in \mathbb{R}^N \setminus \{0\}$, $u \in \mathbb{R}^N$ ($u_k > 0$), $D := \operatorname{diag}((d_k)_k) \in \mathbb{R}^{N \times N}$ ($d_k > 0$), $\lambda_0 > 0$, $Z := \operatorname{diag}((\zeta_k)_k) \in \mathbb{R}^{N \times N}$, and the electrochemical mobility $M := \lambda_0 DZ \in \mathbb{R}^{N \times N}$. Note that (4.18) consists of one partial differential equation and one purely algebraic equation. To treat this problem we consider the associated complex mixed-order system

$$\mathscr{L}(\lambda, z) := \begin{pmatrix} \lambda \operatorname{id}_N + (1+|z|_-^2)D & (1+|z|_-^2)Mu \\ \zeta^T & 0 \end{pmatrix} \in \mathbb{C}^{(N+1) \times (N+1)} \quad (4.19)$$

where $(\lambda, z) \in \overline{S}_\theta \times \overline{\Sigma}_\delta^n$.

Proposition 4.23. (i) *We have*

$$\det \mathscr{L}(\lambda, z) = (-1)^{N+1} \lambda_0 (1+|z|_-^2) \left[\prod_{j=1}^N (\lambda + d_j(1+|z|_-^2)) \right]$$

$$\cdot \left[\sum_{k=1}^N \frac{\zeta_k^2 d_k u_k}{\lambda + d_k(1+|z|_-^2)} \right]$$

for all $(\lambda, z) \in \overline{S}_\theta \times \overline{\Sigma}_\delta^n$.

(ii) *There exist* $\theta > \pi/2$ *and* $\delta > 0$ *such that* $\det \mathscr{L}$ *is* N-*parabolic with*

$$[\mu(\det \mathscr{L})](\gamma) = \max\{2N, (N-1)\gamma + 2\},$$

cf. Figure 4.3.

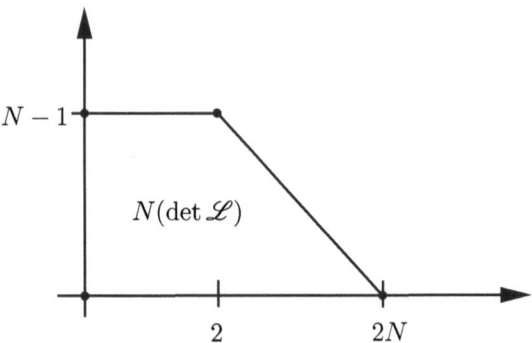

Figure 4.3: Newton polygon of det $\mathscr{L}$ for the three-phase problem

Proof. The first assertion is easy to verify. It is clear that there exist $\theta > \pi/2$ and $\delta > 0$ such that the two-sided estimates

$$\left|\prod_{j=1}^{N}(\lambda + d_j(1+|z|_-^2))\right| \approx (1+|\lambda|^N + |z|^{2N}), \quad |1+|z|_-^2| \approx 1+|z|^2$$

hold for all $(\lambda, z) \in S_\theta \times \Sigma_\delta^n$ with $|\lambda|$ sufficiently large. So we only have to show the two-sided estimate

$$\left|\sum_{k=1}^{N} \frac{\zeta_k^2 d_k u_k}{\lambda + d_k(1+|z|_-^2)}\right| \approx (1+|\lambda|+|z|^2)^{-1}.$$

This can be shown easily by homogeneity and the fact that $\zeta_k^2 d_k u_k \geq 0$ for all $k = 1, \ldots, N$. □

To show that $\mathscr{L}$ is a Douglis-Nirenberg system in the sense of Definition 2.66 we define the order functions

$$t_1(\gamma), \ldots, t_N(\gamma) := \max\{2, \gamma\}, \qquad t_{N+1}(\gamma) := 2,$$
$$s_1(\gamma), \ldots, s_N(\gamma) := 0, \qquad s_{N+1}(\gamma) := -\max\{2, \gamma\}$$

and easily obtain that $s_j + t_k$ is a convex or concave, respectively, upper order function of $\mathscr{L}_{jk}$. Additionally we have $[\mu(\det \mathscr{L})] = \sum_{k=1}^{N+1}(s_k + t_k)$. According to Proposition 4.23 there exist $\theta > \pi/2$ and $\delta > 0$ such that the mixed-order system $\mathscr{L}$, given in (4.19), is N-parabolic. Applying Theorem 2.69, Remark 2.70 (iv), and Proposition 2.72 with $(r_\ell', s_\ell') := (0,0)$, $\ell = 0, 1 =: M$, we obtain the following theorem.

Theorem 4.24. *Let $1 < p, q < \infty$. There exists $\varrho_0 > 0$ such that we get*

$$\left[\mathscr{L}(\mathcal{D}_+^{(\varrho)})\right]\Big|_{\mathbb{H}} \in L_{\text{Isom}}(\mathbb{H}, \mathbb{F}), \qquad \varrho \geq \varrho_0$$

where

$$\mathbb{H}_1, \ldots, \mathbb{H}_N := {}_0H^1_{p,\varrho}(\mathbb{R}_+, L_q(\mathbb{R}^n)) \cap L_{p,\varrho}(\mathbb{R}_+, H^2_q(\mathbb{R}^n)),$$
$$\mathbb{H}_{N+1} := L_{p,\varrho}(\mathbb{R}_+, H^2_q(\mathbb{R}^n)),$$
$$\mathbb{F}_1, \ldots, \mathbb{F}_N := L_{p,\varrho}(\mathbb{R}_+, L_q(\mathbb{R}^n)),$$
$$\mathbb{F}_{N+1} := {}_0H^1_{p,\varrho}(\mathbb{R}_+, L_q(\mathbb{R}^n)) \cap L_{p,\varrho}(\mathbb{R}_+, H^2_q(\mathbb{R}^n)),$$

and $\mathbb{H} := \prod_{i=1}^{N+1} \mathbb{H}_i$, $\mathbb{F} := \prod_{j=1}^{N+1} \mathbb{F}_j$.

Remark 4.25. In [Kot10, Theorem 3.2] the regularity of ψ is better than in Theorem 4.24 because only the case of $g = 0$ and $\zeta^T f \in {}_0H^{1/2}_{p,\varrho}(\mathbb{R}_+, H^{-1}_q(\mathbb{R}^n))$ is considered there. This can be seen by multiplying ζ^T from the left to the first equation of (4.18) followed by an application of $(\Delta - 1)^{-1}$.

4.6 The spin-coating process

> *Motivation.* In [DGH+11], R. Denk, M. Geissert, M. Hieber, J. Saal, and O. Sawada present existence and uniqueness results on the equations modeling the spin-coating process in an L_p-setting. These equations may serve as an example of a free boundary value problem where here one of the main difficulties lies in the presence of surface tension in the model and the fact that the resulting Newton polygon has no triangular structure. The key for well-posedness results in [DGH+11] is the analysis of a mixed-order system which fits into the Newton polygon approach.

Without going into detail, the equations for the spin coating model read as follows:

$$\begin{cases} \rho(\partial_t u + (u \cdot \nabla)u) = \mu \Delta u - \nabla \pi - \rho[2\widetilde{\omega} \times u \\ \qquad\qquad\qquad\qquad + \widetilde{\omega} \times (\widetilde{\omega} \times (\chi_R(x)(x,y)^T))] & \text{in } \Omega(t), \\ \operatorname{div} u = 0 & \text{in } \Omega(t), \\ -T\nu = \sigma \kappa \nu & \text{on } \Gamma^+(t), \\ V = u \cdot v & \text{on } \Gamma^+(t), \\ v = c(h+\delta)^\alpha \partial_y c & \text{on } \Gamma^-(t), \\ w = 0 & \text{on } \Gamma^-(t), \\ u(t=0) = u_0 & \text{in } \Omega(0), \\ \Gamma^+(t=0) = \Gamma_0^+ & \text{in } \Gamma_0^+. \end{cases} \quad (4.20)$$

(i) The region $\Omega(t)$ is filled with a Newtonian fluid. It is assumed that there exists an unknown height function $h \colon \mathbb{R}_+ \times \mathbb{R}^2 \to [0, \infty)$ and a constant $\delta > 0$ such that

$$\Omega(t) = \{(x,y) \in \mathbb{R}^3 \colon x \in \mathbb{R}^2, y \in (0, h(t,x) + \delta)\}.$$

(ii) The surface $\Gamma(t)$ is the boundary of $\Omega(t)$ at time t. It splits into the free surface $\Gamma^+(t)$ which is given by $\{(x,y) \in \mathbb{R}^3 : x \in \mathbb{R}^2, y = h(t,x) + \delta\}$ and the bottom part $\Gamma^-(t) := \mathbb{R}^2 \times \{0\}$. The outer normal on $\Gamma(t)$ is denoted by $\nu(t,\cdot)$. Furthermore $V(t,\cdot)$ and $\kappa(t,\cdot)$ denote the normal velocity and the mean curvature of $\Gamma(t)$.

(iii) The constants $\rho, \mu > 0$ and $\sigma > 0$ denote density, viscosity and surface tension of the fluid. For simplicity we set $\rho = \mu = 1$ in the following. The unknown function $u(t,\cdot) : \Omega(t) \to \mathbb{R}^3$ models the velocity of the fluid at time t and splits into $u = (v,w)^T$ with $v := (u_1, u_2)^T$ and $w := u_3$. The pressure in $\Omega(t)$ is denoted by the unknown function $\pi(t,\cdot) : \Omega(t) \to [0,\infty)$.

(vi) The constant vector $\widetilde{\omega} = |\widetilde{\omega}|e_z \in \mathbb{R}^3$ describes the speed of rotation and $\chi_R : \mathbb{R}^2 \to [0,\infty)$ is a smooth cut-off function ($R > 0$ large) such that one can neglect the centrifugal force outside of $B(0,R)$. The stress tensor T is given by $T := \mu(\nabla u + (\nabla u)^T) - \pi I$.

(v) The constants $c, \alpha > 0$ are slip parameters of the Navier slip condition.

Roughly speaking, the authors of [DGH$^+$11] apply the Hanzawa transform to obtain a problem on a time-independent strip. After linearization, this problem can be treated as so-called reduced Stokes problems in the half-spaces $D^- := \mathbb{R}^3_+$ and $D^+ := \mathbb{R}^2 \times (-\infty, \delta)$. Altogether, it is crucial to solve the model problem (cf. [DGH$^+$11, Proposition 5.3])

$$\begin{cases} \partial_t u - \Delta u + \nabla \pi = 0 & \text{in } \mathbb{R}_+ \times \mathbb{R}^3_+, \\ \operatorname{div} u = 0 & \text{in } \mathbb{R}_+ \times \mathbb{R}^3_+, \\ \partial_t h + \gamma_{y=0} w = f_h & \text{on } \mathbb{R}_+ \times \mathbb{R}^2, \\ \gamma_{y=0} \partial_y v + \gamma_{y=0} \nabla_x w = g_1 & \text{on } \mathbb{R}_+ \times \mathbb{R}^2, \\ 2\gamma_{y=0} \partial_y w - \gamma_{y=0} \pi - \sigma \Delta_x h = g_2 & \text{on } \mathbb{R}_+ \times \mathbb{R}^2, \\ u(t=0) = 0 & \text{in } \mathbb{R}^3_+, \\ h(t=0) = 0 & \text{in } \mathbb{R}^2. \end{cases} \quad (4.21)$$

In the following we use the standard approach to reduce this problem to the boundary as in [DGH$^+$11]. After formal application of the Laplace and Fourier transform to (4.21) we deduce the following system of ordinary differential equations:

$$\tau(\lambda,\xi)^2 \hat{u}(\lambda,\xi,y) - \partial_y^2 \hat{u}(\lambda,\xi,y) + (i\xi, \partial_y)^T \hat{\pi}(\lambda,\xi,y) = 0, \quad y > 0, \quad (4.22)$$

$$i\xi \cdot \hat{v}(\lambda,\xi,y) + \partial_y \hat{w}(\lambda,\xi,y) = 0, \quad y > 0, \quad (4.23)$$

$$\lambda \hat{h}(\lambda,\xi) + \hat{w}(\lambda,\xi,0) = \hat{f}_h(\lambda,\xi), \quad (4.24)$$

$$(\partial_y \hat{v})(\lambda,\xi,0) + i\xi \hat{w}(\lambda,\xi,0) = \hat{g}_1(\lambda,\xi), \quad (4.25)$$

$$2(\partial_y \hat{w})(\lambda,\xi,0) - \hat{\pi}(\lambda,\xi,0) + \sigma|\xi|^2 \hat{h}(\lambda,\xi) = \hat{g}_2(\lambda,\xi) \quad (4.26)$$

for fixed (λ,ξ), $\tau = \tau(\lambda,\xi) := (\lambda + |\xi|^2)^{1/2}$, and the transformed functions $\hat{u} = (\hat{v}, \hat{w})^T$, $\hat{\pi}$, $\hat{h}$, and $\hat{f}_h, \hat{g}_1, \hat{g}_2$. By a multiplication of (4.22) with $(i\xi, \partial_y)$ and with use

4.6. The spin-coating process

of (4.23) we can uncouple the equations for $\hat{\pi}$ and $\hat{u}$. We obtain $(\partial_y - |\xi|^2)\hat{\pi} = 0$, which yields the stable solution

$$\hat{\pi}(\lambda, \xi, y) = \hat{p}(\lambda, \xi) \cdot \exp(-|\xi|y), \quad y > 0 \tag{4.27}$$

for an unknown function $\hat{p}$. With Green's functions $k_\pm$ we can now give a formula for $\hat{v}$ and $\hat{w}$ by

$$\hat{v}(\lambda, \xi, y) = -\int_0^\infty k_-(\lambda, \xi, y, s) \cdot i\xi \hat{\pi}(\lambda, \xi, s) ds + \hat{\Phi}_v(\lambda, \xi) \exp(-\tau y), \tag{4.28}$$

$$\hat{w}(\lambda, \xi, y) = -\int_0^\infty k_+(\lambda, \xi, y, s) \cdot \partial_y \hat{\pi}(\lambda, \xi, s) ds + \hat{\Phi}_w(\lambda, \xi) \exp(-\tau y) \tag{4.29}$$

where $\hat{\Phi}_v, \hat{\Phi}_w$ are unknown and

$$k_-(\lambda, \xi, y, s) := \begin{cases} \frac{\exp(-\tau s)}{\tau} \sinh(\tau y), & y \leq s, \\ \frac{\exp(-\tau y)}{\tau} \sinh(\tau s), & y \geq s, \end{cases}$$

$$k_+(\lambda, \xi, y, s) := \begin{cases} \frac{\exp(-\tau s)}{\tau} \cosh(\tau y), & y \leq s, \\ \frac{\exp(-\tau y)}{\tau} \cosh(\tau s), & y \geq s. \end{cases}$$

Inserting (4.27)-(4.29) into (4.23)-(4.26) we obtain the same system of linear equations

$$\begin{pmatrix} i\xi^T & -\tau & 0 & 0 \\ 0 & 1 & \lambda & \frac{|\xi|}{\tau(\tau+|\xi|)} \\ -\tau & i\xi & 0 & -i\xi \frac{\tau-|\xi|}{\tau(\tau+|\xi|)} \\ 0 & -2\tau & \sigma|\xi|^2 & -1 \end{pmatrix} \begin{pmatrix} \hat{\Phi}_v \\ \hat{\Phi}_w \\ \hat{h} \\ \hat{p} \end{pmatrix} = \begin{pmatrix} 0 \\ \hat{f}_h \\ \hat{g}_1 \\ \hat{g}_2 \end{pmatrix} \tag{4.30}$$

for the unknown functions $\hat{\Phi}_v, \hat{\Phi}_w, \hat{h}$, and $\hat{p}$ as in [DGH+11, Proposition 5.3]. In contrast to the argumentation in [DGH+11] we solve this mixed-order system at one stroke with Theorem 2.69. In order to do this we define the complex version of (4.30) by

$$\mathscr{L}(\lambda, z) := \begin{pmatrix} z^T & -\omega & 0 & 0 \\ 0 & 1 & \lambda & \frac{|z|_-}{\omega(\omega+|z|_-)} \\ -\omega & z & 0 & -z \cdot \frac{\omega-|z|_-}{\omega(\omega+|z|_-)} \\ 0 & -2\omega & \sigma|z|_-^2 & -1 \end{pmatrix} \in \mathbb{C}^{5 \times 5} \tag{4.31}$$

for $(\lambda, z) \in \overline{S}_\theta \times \overline{\Sigma}_\delta^2$ where $\omega = \omega(\lambda, z) := (\lambda + |z|_-^2)^{1/2}$. A straightforward calculation shows that

$$\det \mathscr{L}(\lambda, z) = \frac{P(\lambda, z)}{\omega + |z|_-},$$

$$P(\lambda, z) := \lambda^3 + \lambda^2 |z|_- \omega + 5\lambda^2 |z|_-^2 + 4\lambda |z|_-^4 + \sigma \lambda |z|_-^3 + \sigma |z|_-^4 \omega + \sigma |z|_-^5.$$

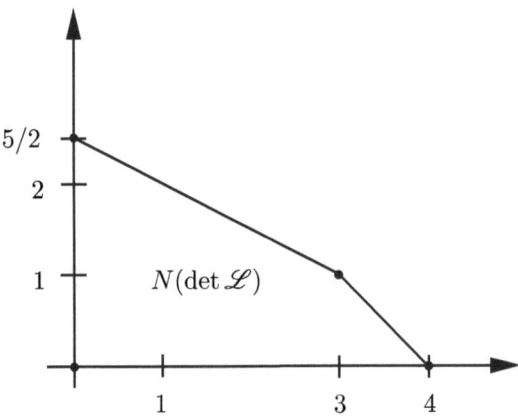

Figure 4.4: Newton polygon for the spin-coating process

Lemma 4.26. *For sufficiently small $\delta > 0$ and $\theta > \pi/2$ we obtain the following results:*

(i) *We get*
$$P \in S_N(\overline{S}_\theta \times \overline{\Sigma}_\delta^2)$$
and $[\mu(P)](\gamma) = d_\gamma(P) = \max\{5, \gamma + 4, 3\gamma\}$, $\gamma > 0$.

(ii) *The symbol* $\det \mathscr{L}$ *is N-parabolic with*
$$[\mu(\det \mathscr{L})](\gamma) = \max\{4, \gamma + 3, 5/2\gamma\}, \quad \gamma > 0,$$
cf. Figure 4.4.

Proof. (i) It is easy to see that $P \in S(\overline{S}_\theta \times \overline{\Sigma}_\delta^2)$. We obtain the γ-principal part

$$\pi_\gamma P(\lambda, z) = \begin{cases} 2\sigma |z|_-^5, & \gamma \in (0,1), \\ 2|z|_-^4 (\sigma |z|_- + 2\lambda), & \gamma = 1, \\ 4\lambda |z|_-^4, & \gamma \in (1,2), \\ 4\lambda |z|_-^4 + 5\lambda^2 |z|_-^2 + \lambda^2 |z|_{-\omega} + \lambda^3, & \gamma = 2, \\ \lambda^3, & \gamma > 2, \end{cases}$$

which is non-vanishing for $\gamma \neq 2$ and $(\lambda, z) \in [\overline{S}_\theta \setminus \{0\}] \times [\overline{\Sigma}_\delta^2 \setminus \{0\}]$. For $\gamma = 2$ we obtain

4.6. The spin-coating process

$$\pi_2 P(\lambda, z) = \lambda[\lambda^2 + \lambda\omega|z|_- + 5\lambda|z|_-^2 + 4|z|_-^4]$$
$$= \lambda[\lambda^2 + \lambda|z|_-^2 + \lambda\omega|z|_- + 4\omega^2|z|_-^2]$$
$$= \lambda[\lambda(\lambda + |z|_-^2) + \lambda\omega|z|_- + 4\omega^2|z|_-^2]$$
$$= \lambda\omega^2\left[\lambda\left(1 + \frac{|z|_-}{\omega}\right) + 4|z|_-^2\right].$$

We have $|z|_- \in \overline{S}_\delta$, $\omega^{-1} \in \overline{S}_{\theta/2}$ and therefore $1 + |z|_-/\omega \in \overline{S}_{\delta+\theta/2}$. This yields

$$\lambda(1 + |z|_-/\omega) \in \overline{S}_{\delta+3/2\cdot\theta}.$$

Let θ and δ satisfy $\pi - (\delta + 3/2 \cdot \theta) > 2\delta$. Then we have

$$\lambda\left(1 + \frac{|z|_-}{\omega}\right) + 4|z|_-^2 \neq 0$$

for all $(\lambda, z) \in [\overline{S}_\theta \setminus \{0\}] \times [\overline{\Sigma}_\delta^2 \setminus \{0\}]$. Corollary 2.57 then yields the assertion in (i).

(ii) First, we define $Q(\lambda, z) := \omega + |z|_-$ and the strictly positive order functions $\mu_1(\gamma) := d_\gamma(Q)$ and $\mu_2(\gamma) := d_\gamma(P) - d_\gamma(Q) = \max\{4, \gamma+3, 3/2\cdot\gamma\}$. Lemma 2.33 then directly yields

$$|\det \mathscr{L}(\lambda, z)| \geq \frac{W_{\mu_1}(\lambda, z) \cdot W_{\mu_2}(\lambda, z)}{W_{\mu_1}(\lambda, z)} = W_{\mu_2}(\lambda, z)$$

for $(\lambda, z) \in S_\theta \times \Sigma_\delta^2$ with $|\lambda| \geq \lambda_0$. □

In order to apply Corollary 2.80 we define $M := 2$, $\gamma_1 := 1$, $\gamma_2 := 2$, and

$$t_1(\gamma) := \begin{cases} 1, & \gamma \in (0, \gamma_1], \\ 1, & \gamma \in (\gamma_1, \gamma_2], \\ \frac{1}{2}\gamma, & \gamma \in (\gamma_2, \infty), \end{cases}$$

$$t_4(\gamma) := \begin{cases} 2, & \gamma \in (0, \gamma_1], \\ \gamma+1, & \gamma \in (\gamma_1, \gamma_2], \\ \frac{3}{2}\gamma, & \gamma \in (\gamma_2, \infty), \end{cases}$$

$$t_2 := t_3 := t_1, \quad t_5 := 0,$$
$$s_1 := s_3 := s_4 := s_5 := 0, \quad s_2 := -t_1.$$

We have $(s_2 + t_4) = \max\{1, \gamma\}$, $|\mathscr{L}_{2,5}(\lambda, z)| \leq C|\omega|^{-1}$, $|\mathscr{L}_{k,5}(\lambda, z)| \leq C$ for $|\lambda| \geq \lambda_0$ and $k \in \{3, 4\}$. Thus, $s_j + t_i$ is a strictly positive or strictly negative upper order function of $\mathscr{L}_{j,i}$ for all $i, j = 1, \ldots, 5$. It is also easy to verify that

$$\sum_{i=1}^{5}[s_i(\gamma) + t_i(\gamma)] = 2 \cdot t_1(\gamma) + t_4(\gamma) = \max\{4, \gamma+3, 5/2\gamma\} = [\mu(\det \mathscr{L})](\gamma)$$

for $\gamma > 0$. So we have proved:

Proposition 4.27. *The complex matrix $\mathscr{L}$ defined in (4.31) is an N-parabolic mixed-order system in the sense of Definition 2.67.*

For $p \in (1, \infty)$ we define

$$(r_0', s_0') := (1 - 1/p, 0), \qquad (\mathcal{F}_0, \mathcal{K}_0) := (H_p, B_{pp}),$$
$$(r_1', s_1') := (1 - 1/p, 0), \qquad (\mathcal{F}_1, \mathcal{K}_1) := (H_p, B_{pp}),$$
$$(r_2', s_2') := (0, 1/2 - 1/(2p)), \qquad (\mathcal{F}_2, \mathcal{K}_2) := (B_{pp}, H_p).$$

With the notation of Theorem 2.69 and Corollary 2.80, respectively, we get for $i \in \{1, 2, 3\}$

$$\mu_{\mathbb{H}_i}(\gamma) = \begin{cases} 2 - 1/p, & \gamma \in (0, \gamma_1], \\ 2 - 1/p, & \gamma \in (\gamma_1, \gamma_2], \\ (1 - 1/(2p)) \cdot \gamma, & \gamma > \gamma_2, \end{cases}$$

$$\mu_{\mathbb{H}_4}(\gamma) = \begin{cases} 3 - 1/p, & \gamma \in (0, \gamma_1], \\ \gamma + 2 - 1/p, & \gamma \in (\gamma_1, \gamma_2], \\ (2 - 1/(2p)) \cdot \gamma, & \gamma > \gamma_2, \end{cases}$$

$$\mu_{\mathbb{H}_5}(\gamma) = \begin{cases} 1 - 1/p, & \gamma \in (0, \gamma_1], \\ 1 - 1/p, & \gamma \in (\gamma_1, \gamma_2], \\ (1/2 - 1/(2p)) \cdot \gamma, & \gamma > \gamma_2 \end{cases}$$

and for $j \in \{1, 3, 4, 5\}$,

$$\mu_{\mathbb{F}_j}(\gamma) = \mu_{\mathbb{H}_5}, \qquad \mu_{\mathbb{F}_2}(\gamma) = \mu_{\mathbb{H}_1}.$$

So it is easy to see that for all $i, j \in \{1, \ldots, 5\}$ the scale $(\mathcal{F}_\ell, \mathcal{K}_\ell)_\ell$ is $(\mu_{\mathbb{H}_i}, \mu_{\mathbb{F}_j})$-admissible in the sense of Definition 2.78. So we can apply Corollary 2.80 and Remark 2.70 (iv).

Theorem 4.28. *Let $1 < p < \infty$. There exists $\varrho_0 > 0$ such that we get*

$$\left[\mathscr{L}(\boldsymbol{\mathcal{D}}_+^{(\varrho)})\right]\Big|_{\mathbb{H}} \in L_{\text{Isom}}(\mathbb{H}, \mathbb{F}), \quad \varrho \geq \varrho_0$$

where

$$\mathbb{H}_1 = \mathbb{H}_2 = \mathbb{H}_3 = {}_0B_{p,\varrho}^{1 - 1/(2p)}(\mathbb{R}_+, L_p(\mathbb{R}^2)) \cap L_{p,\varrho}(\mathbb{R}_+, B_{pp}^{2 - 1/p}(\mathbb{R}^2)),$$

$$\mathbb{H}_4 = {}_0B_{p,\varrho}^{2 - 1/(2p)}(\mathbb{R}_+, L_p(\mathbb{R}^2)) \cap {}_0H_{p,\varrho}^1(\mathbb{R}_+, B_{pp}^{2 - 1/p}(\mathbb{R}^2))$$
$$\cap L_{p,\varrho}(\mathbb{R}_+, B_{pp}^{3 - 1/p}(\mathbb{R}^2)),$$

$$\mathbb{H}_5 = {}_0B_{p,\varrho}^{1/2 - 1/(2p)}(\mathbb{R}_+, L_p(\mathbb{R}^2)) \cap L_{p,\varrho}(\mathbb{R}_+, B_{pp}^{1 - 1/p}(\mathbb{R}^2)),$$

$$\mathbb{F}_1 = \mathbb{F}_3 = \mathbb{F}_4 = \mathbb{F}_5 = {}_0B_{p,\varrho}^{1/2 - 1/(2p)}(\mathbb{R}_+, L_p(\mathbb{R}^2)) \cap L_{p,\varrho}(\mathbb{R}_+, B_{pp}^{1 - 1/p}(\mathbb{R}^2)),$$

$$\mathbb{F}_2 = {}_0B_{p,\varrho}^{1 - 1/(2p)}(\mathbb{R}_+, L_p(\mathbb{R}^2)) \cap L_{p,\varrho}(\mathbb{R}_+, B_{pp}^{2 - 1/p}(\mathbb{R}^2))$$

and $\mathbb{H} := \prod_{i=1}^5 \mathbb{H}_i$, $\mathbb{F} := \prod_{j=1}^5 \mathbb{F}_j$.

4.6. The spin-coating process

So we can give a solution to the reduced problem (4.30) with right-hand side $(0, f_h, g_1, g_2)^T \in \mathbb{F}$ by

$$(\Phi_v, \Phi_w, h, p)^T := [\mathscr{L}(\mathcal{D}_+^{(\varrho)})]^{-1}(0, f_h, g_1, g_2)^T \in \mathbb{H}. \tag{4.32}$$

Remark 4.29. Using (4.32) the solution of (4.21) can be determined by standard arguments. For clarity we want to give some remarks and references. The function h is already determined by (4.32). Furthermore, we have $p \in \mathbb{H}_5 \hookrightarrow L_{p,\varrho}(\mathbb{R}_+, \dot{B}_p^{1-1/p}(\mathbb{R}^2))$ and the formal identity $\hat{\pi} = \exp(-|\xi|y)\hat{p}$. Here we consider the symbol $\exp(-|\xi|y)$ of the Cauchy-Poisson semigroup $\{P(y)\}_{y \geq 0}$, cf. [Tri78, Section 2.5.3], [Tri97, Section 12.2]. Using this we can define the pressure term by

$$\pi(t) := P(\cdot)p(t), \quad t > 0.$$

According to [Tri82, Corollary 3 (i)] and [Tri83, Theorem 5.2.3.2, Remark 5.2.3.4], there exists an equivalent norm on the homogeneous Besov space $\dot{B}_p^{1-1/p}(\mathbb{R}^n)$ in terms of the Cauchy-Poisson semigroup. This can be used to show that $\pi \in L_{p,\varrho}(\mathbb{R}_+, \dot{H}_p^1(\mathbb{R}_+^3))$, cf. [PS10]. Following (4.28) and (4.29) we can now define $u := (v, w)$ in an obvious way with the operators $k_\pm(\mathcal{D}_+^{(\varrho)}, y, s)$ and Corollary 3.63.

Remark 4.30 (Solution on a finite time interval). With the approach described above we can also derive a solution of (4.21) on a finite time interval $(0, T)$ without an exponential weight. For the given data

$$f_h \in {_0}B_p^{1-1/(2p)}((0,T), L_p(\mathbb{R}^2)) \cap L_p((0,T), B_{pp}^{2-1/p}(\mathbb{R}^2)),$$
$$g_1 \in {_0}B_p^{1/2-1/(2p)}((0,T), L_p(\mathbb{R}^2)) \cap L_p((0,T), B_{pp}^{1-1/p}(\mathbb{R}^2)),$$
$$g_2 \in {_0}B_p^{1/2-1/(2p)}((0,T), L_p(\mathbb{R}^2)) \cap L_p((0,T), B_{pp}^{1-1/p}(\mathbb{R}^2))$$

we can define extensions $(\widetilde{f_h}, \widetilde{g_1}, \widetilde{g_2})$ of (f_h, g_1, g_2) with

$$(0, \widetilde{f_h}, \widetilde{g_1}, \widetilde{g_2}) \in \prod_{j=1}^{5} \mathbb{F}_j$$

where the exponentially weighted spaces $\mathbb{F}_j$ are given above (cf. [DSS08, Lemma 2.2]). Thus the previous results yield a corresponding solution

$$\widetilde{u} \in \left[{_0}H_{p,\varrho}^1(\mathbb{R}_+, L_p(\mathbb{R}_+^3)) \cap L_{p,\varrho}(\mathbb{R}_+, H_p^2(\mathbb{R}_+^3))\right]^3,$$
$$\widetilde{h} \in {_0}B_{p,\varrho}^{2-1/(2p)}(\mathbb{R}_+, L_p(\mathbb{R}^2)) \cap {_0}H_{p,\varrho}^1(\mathbb{R}_+, B_{pp}^{2-1/p}(\mathbb{R}^2)) \cap L_{p,\varrho}(\mathbb{R}_+, B_{pp}^{3-1/p}(\mathbb{R}^2)),$$
$$\widetilde{\pi} \in L_{p,\varrho}(\mathbb{R}_+, \dot{H}^1(\mathbb{R}_+^3)).$$

Now we define the restriction $(u, h, \pi) := (\widetilde{u}, \widetilde{h}, \widetilde{\pi})|_{(0,T)}$ to the time interval $(0, T)$ and get

$$u \in \left[{}_0H_p^1((0,T), L_p(\mathbb{R}_+^3)) \cap L_p((0,T), H_p^2(\mathbb{R}_+^3))\right]^3,$$
$$h \in {}_0B_p^{2-1/(2p)}((0,T), L_p(\mathbb{R}^2)) \cap {}_0H_p^1((0,T), B_{pp}^{2-1/p}(\mathbb{R}^2))$$
$$\cap L_p((0,T), B_{pp}^{3-1/p}(\mathbb{R}^2)),$$
$$\pi \in L_p((0,T), \dot{H}^1(\mathbb{R}_+^3))$$

(cf. [DSS08, Lemma 2.2]). All operators in (4.21) are differential operators and therefore (u, h, π) is also a solution on the ground spaces with finite time interval.

4.7 Two-phase Navier-Stokes equations with Boussinesq-Scriven surface and gravity

Motivation. In this section we want to concentrate on the two-phase Navier-Stokes problem with surface viscosity (Boussinesq-Scriven surface), gravity and free boundary. It describes the motion of two viscous incompressible fluids separated by a closed free interface. The two-phase Navier-Stokes equations were treated by many other authors (cf. [PS09], [PS10], [PS11], [SS11a], and the given references therein) for the case $\lambda_s, \mu_s = 0$ (i.e., in the absence of surface viscosity). A rigorous derivation of the mathematical model of the two-phase Navier-Stokes equations with Boussinesq-Scriven surface and the associated linearization can be found in a work of D. Bothe and J. Prüss, cf. [BP10]. In [PS11] J. Prüss and G. Simonett proved the well-posedness of the linearized two-phase Navier-Stokes problem with surface tension and gravity but without surface viscosity (i.e., $\lambda_s, \mu_s = 0$). Up to now the well-posedness for $\lambda_s, \mu_s > 0$ has not been considered before in an L_p-setting.

Here we present a unified treatment, which can handle both cases $\lambda_s, \mu_s = 0$ and $\lambda_s, \mu_s > 0$ at one stroke. In addition our approach simplifies the proofs given in [PS09], [PS10], and [PS11].

We consider the linearized problem for the pressure $\pi \colon \mathbb{R}_+ \times \dot{\mathbb{R}}^{n+1} \to \mathbb{R}$, the velocity field $u = (v, w) \colon \mathbb{R}_+ \times \dot{\mathbb{R}}^{n+1} \to \mathbb{R}$, and the function $h \colon \mathbb{R}_+ \times \mathbb{R}^n \to \mathbb{R}$, which models the boundary dynamics. This problem reads as follows:

$$\begin{cases} \rho\partial_t u - \mu \Delta u + \nabla \pi = f & \text{in } \mathbb{R}_+ \times \dot{\mathbb{R}}^{n+1}, \\ \operatorname{div} u = f_d & \text{in } \mathbb{R}_+ \times \dot{\mathbb{R}}^{n+1}, \\ -\mu_s \Delta_x v - \lambda_s \nabla_x \operatorname{div}_x v - [\![\mu \partial_y v]\!] - [\![\mu \nabla_x w]\!] = g_v & \text{on } \mathbb{R}_+ \times \mathbb{R}^n, \\ -2[\![\mu \partial_y w]\!] + [\![\pi]\!] - \sigma \Delta h - Gh = g_w & \text{on } \mathbb{R}_+ \times \mathbb{R}^n, \\ [\![u]\!] = 0 & \text{on } \mathbb{R}_+ \times \mathbb{R}^n, \\ \partial_t h - w + \langle b_0, \nabla \rangle h = g_h & \text{on } \mathbb{R}_+ \times \mathbb{R}^n, \\ u(t=0) = u_0 & \text{in } \dot{\mathbb{R}}^{n+1}, \\ h(t=0) = h_0 & \text{in } \mathbb{R}^n. \end{cases} \quad (4.33)$$

4.7. Two-phase Navier-Stokes equations with Boussinesq-Scriven surface

where we define $\dot{\mathbb{R}}^{n+1} := \mathbb{R}^n \times (\mathbb{R} \setminus \{0\})$, $\rho := \rho_1 \chi_{\mathbb{R}_-^{n+1}} + \rho_2 \chi_{\mathbb{R}_+^{n+1}}$, $\mu := \mu_1 \chi_{\mathbb{R}_-^{n+1}} + \mu_2 \chi_{\mathbb{R}_+^{n+1}}$ with $\rho_i, \mu_i, \sigma > 0$, $\lambda_s, \mu_s \geq 0$, and $G := [\![\rho]\!]\gamma_a$ with $\gamma_a \geq 0$. Due to the fact that gravitation has a potential we have added the gravitational force on the right-hand side to the pressure π. Here $[\![\varphi]\!]$ always denotes the jump of the function φ across the boundary interface $\mathbb{R}^n$, i.e., $[\![\varphi]\!] := \gamma_{0,n+1}\varphi_{\mathbb{R}_-^{n+1}} - \gamma_{0,n+1}\varphi_{\mathbb{R}_+^{n+1}}$.

Let $b_0 \in \mathbb{R}^n$ with $|b_0| \leq \beta$ for fixed $\beta > 0$. Due to the localization argument in [PS11, Theorem 3.1 (v)] it is crucial to solve (4.33) uniformly in b_0, i.e., the norm of the solution operator should be estimated from above uniformly for all $b_0 \in \mathbb{R}^n$ with $|b_0| \leq \beta$.

Remark 4.31. We consider the two cases $\lambda_s, \mu_s = 0$ and $\lambda_s \geq 0, \mu_s > 0$ due to the fact that the highest order on the boundary is not elliptic if $\lambda_s > 0, \mu_s = 0$.

As in the proof of [PS10, Theorem 3.1] the system (4.33) can be reduced to

$$\begin{cases} \rho \partial_t u - \mu \Delta u + \nabla \pi = 0 & \text{in } \mathbb{R}_+ \times \dot{\mathbb{R}}^{n+1}, \\ \operatorname{div} u = 0 & \text{in } \mathbb{R}_+ \times \dot{\mathbb{R}}^{n+1}, \\ -\mu_s \Delta_x v - \lambda_s \nabla_x \operatorname{div}_x v - [\![\mu \partial_y v]\!] - [\![\mu \nabla_x w]\!] = g_v & \text{on } \mathbb{R}_+ \times \mathbb{R}^n, \\ -2[\![\mu \partial_y w]\!] + [\![\pi]\!] - \sigma \Delta h - Gh = g_w & \text{on } \mathbb{R}_+ \times \mathbb{R}^n, \\ [\![u]\!] = 0 & \text{on } \mathbb{R}_+ \times \mathbb{R}^n, \\ \partial_t h - w + \langle b_0, \nabla \rangle h = g_h & \text{on } \mathbb{R}_+ \times \mathbb{R}^n, \\ u(t=0) = 0 & \text{in } \dot{\mathbb{R}}^{n+1}, \\ h(t=0) = 0 & \text{in } \mathbb{R}^n. \end{cases} \quad (4.34)$$

After formal application of the Laplace and Fourier transform to (4.34) we deduce the system of ordinary differential equations for $(\hat{u}, \hat{\pi}) = (\hat{u}_1, \hat{\pi}_1)\chi_{\mathbb{R}_-^{n+1}} + (\hat{u}_2, \hat{\pi}_2)\chi_{\mathbb{R}_+^{n+1}}$ and $\hat{h}$

$$\mu_j \omega_j^2 \hat{u}_j(\lambda, \xi, y) - \mu_j \partial_y^2 \hat{u}_j(\lambda, \xi, y) + (i\xi, \partial_y)^T \hat{\pi}_j(\lambda, \xi, y) = 0, \quad (4.35)$$
$$(-1)^j y > 0, \ j = 1, 2,$$

$$i\xi \cdot \hat{v}_j(\lambda, \xi, y) + \partial_y \hat{w}_j(\lambda, \xi, y) = 0, \quad (4.36)$$
$$(-1)^j y > 0, \ j = 1, 2,$$

$$(\mu_s|\xi|^2 + \lambda_s(\xi \otimes \xi))\hat{v}(\lambda, \xi, 0) - [\![\mu \partial_y \hat{v}]\!](\lambda, \xi) - i\xi [\![\mu \hat{w}]\!](\lambda, \xi) = \hat{g}_v(\lambda, \xi), \quad (4.37)$$

$$-2[\![\mu \partial_y \hat{w}]\!](\lambda, \xi) + [\![\hat{\pi}]\!](\lambda, \xi) + \sigma|\xi|^2 \hat{h}(\lambda, \xi) - G\hat{h}(\lambda, \xi) = \hat{g}_w(\lambda, \xi), \quad (4.38)$$

$$[\![\hat{u}]\!](\lambda, \xi) = 0, \quad (4.39)$$

$$\lambda \hat{h}(\lambda, \xi) - \hat{w}(\lambda, \xi, 0) + \langle b_0, i\xi \rangle \hat{h} = \hat{g}_h(\lambda, \xi) \quad (4.40)$$

for fixed (λ, ξ), $\omega_j = \omega_j(\lambda, \xi) := \mu_j^{-1/2}(\rho_j \lambda + \mu_j |\xi|^2)^{1/2}$. In contrast to [PS09], [PS10], and [PS11] we use another representation formula for solution of the ordinary differential equations. Similar to Section 4.6 we use the ansatz

$$\hat{\pi}_j(\lambda, \xi, y) = \hat{p}_j(\lambda, \xi) \cdot \exp(-(-1)^j |\xi| y), \quad (-1)^j y > 0, \ j = 1, 2, \quad (4.41)$$

$$\hat{v}_j(\lambda,\xi,y) = -\int_0^\infty k_-^{(j)}(\lambda,\xi,y,s) \cdot i\xi \hat{\pi}_j(\lambda,\xi,s) ds$$
$$+ \hat{\Phi}_v^{(j)}(\lambda,\xi) \exp(-(-1)^j \omega_j y), \quad j=1,2, \tag{4.42}$$

$$\hat{w}_j(\lambda,\xi,y) = -\int_0^\infty k_+^{(j)}(\lambda,\xi,y,s) \cdot \partial_y \hat{\pi}_j(\lambda,\xi,s) ds$$
$$+ \hat{\Phi}_w^{(j)}(\lambda,\xi) \exp(-(-1)^j \omega_j y), \quad j=1,2 \tag{4.43}$$

where $\hat{p}_j, \hat{\Phi}_v^{(j)}, \hat{\Phi}_w^{(j)}$ are unknown functions and $k_\pm^{(j)}$ are the Green's functions

$$k_-^{(1)}(\lambda,\xi,y,s) := -\frac{1}{\mu_1 \omega_1} \begin{cases} e^{\omega_1 y} \sinh(\omega_1 s), & y \leq s, \\ e^{\omega_1 s} \sinh(\omega_1 y), & y \geq s, \end{cases}$$

$$k_+^{(1)}(\lambda,\xi,y,s) := \frac{1}{\mu_1 \omega_1} \begin{cases} e^{\omega_1 y} \cosh(\omega_1 s), & y \leq s, \\ e^{\omega_1 s} \cosh(\omega_1 y), & y \geq s, \end{cases}$$

$$k_-^{(2)}(\lambda,\xi,y,s) := \frac{1}{\mu_2 \omega_2} \begin{cases} \sinh(\omega_2 y) e^{-\omega_2 s}, & y \leq s, \\ \sinh(\omega_2 s) e^{-\omega_2 y}, & y \geq s, \end{cases}$$

$$k_+^{(2)}(\lambda,\xi,y,s) := \frac{1}{\mu_2 \omega_2} \begin{cases} \cosh(\omega_2 y) e^{-\omega_2 s}, & y \leq s, \\ \cosh(\omega_2 s) e^{-\omega_2 y}, & y \geq s. \end{cases}$$

Easy calculations show

$$\hat{\pi}_j(\lambda,\xi,0) = \hat{p}_j(\lambda,\xi), \tag{4.44}$$
$$\hat{v}_j(\lambda,\xi,0) = \hat{\Phi}_v^{(j)}(\lambda,\xi), \tag{4.45}$$
$$\hat{w}_j(\lambda,\xi,0) = \frac{(-1)^j}{\mu_j} \cdot \frac{|\xi|}{\omega_j \gamma_j^+} \hat{p}_j(\lambda,\xi) + \hat{\Phi}_w^{(j)}(\lambda,\xi), \tag{4.46}$$
$$\partial_y \hat{v}_j(\lambda,\xi,0) = -\frac{(-1)^j}{\mu_j} \frac{i\xi}{\gamma_j^+} \hat{p}_j(\lambda,\xi) - (-1)^j \omega_j \hat{\Phi}_v^{(j)}(\lambda,\xi), \tag{4.47}$$
$$\partial_y \hat{w}_j(\lambda,\xi,0) = -(-1)^j \omega_j \hat{\Phi}_w^{(j)}(\lambda,\xi), \tag{4.48}$$

where $\gamma_j^\pm := \omega_j \pm |\xi|$. Note that $\hat{p}_1 = \hat{p}_2 - (\hat{p}_2 - \hat{p}_1) = \hat{p}_2 - [\![\hat{\pi}]\!]$ and $\hat{\Phi}_v^{(1)} = \hat{\Phi}_v^{(2)}$. Inserting (4.44)-(4.48) into the boundary conditions (4.36)-(4.40), we obtain the linear system of equations

$$\mathcal{A}(\lambda,\xi) \begin{pmatrix} \hat{\Phi}_v^{(2)} \\ \hat{\Phi}_w^{(2)} \\ \hat{\Phi}_w^{(1)} \\ \hat{h} \\ [\![\hat{\pi}]\!] \\ |\xi|\hat{p}_2 \end{pmatrix} = \begin{pmatrix} 0 \\ 0 \\ \hat{g}_h \\ 0 \\ \hat{g}_w \\ \hat{g}_v \end{pmatrix} \tag{4.49}$$

4.7. Two-phase Navier-Stokes equations with Boussinesq-Scriven surface

for the unknowns $(\hat{\Phi}_v^{(2)}, \hat{\Phi}_w^{(2)}, \hat{\Phi}_w^{(1)}, \hat{h}, [\![\hat{\pi}]\!], |\xi|\hat{p}_2)^T$. Here the matrix $\mathcal{A}(\lambda, \xi)$ is given by

$$\begin{pmatrix} i\xi^T & -\omega_2 & 0 & 0 & 0 & 0 \\ i\xi^T & 0 & \omega_1 & 0 & 0 & 0 \\ 0 & -1 & 0 & \lambda + \langle b_0, i\xi \rangle & 0 & -\dfrac{1}{\mu_2 \omega_2 \gamma_2^+} \\ 0 & 1 & -1 & 0 & -\dfrac{|\xi|}{\mu_1 \omega_1 \gamma_1^+} & \dfrac{1}{\mu_2 \omega_2 \gamma_2^+} + \dfrac{1}{\mu_1 \omega_1 \gamma_1^+} \\ 0 & 2\mu_2\omega_2 & 2\mu_1\omega_1 & \sigma|\xi|^2 - G & 1 & 0 \\ \widetilde{\Omega}\,\mathrm{id}_n & -\mu_2 i\xi & \mu_1 i\xi & 0 & -i\xi\dfrac{\gamma_1^-}{\omega_1 \gamma_1^+} & \dfrac{i\xi}{|\xi|}\left[\dfrac{\gamma_2^-}{\omega_2\gamma_2^+} + \dfrac{\gamma_1^-}{\omega_1\gamma_1^+}\right] \end{pmatrix}$$

where $\widetilde{\Omega} := \Omega' \mathrm{id}_n + \lambda_s (\xi \otimes \xi)$, $\Omega' := \Omega + \mu_s |\xi|^2$, and $\Omega := \mu_1 \omega_1 + \mu_2 \omega_2$. Solving the fourth equation $[\![\hat{w}]\!] = 0$ we get

$$|\xi|\hat{p}_2 = -\delta \hat{\Phi}_w^{(2)} + \delta \hat{\Phi}_w^{(1)} + \dfrac{1}{\mu_1} \cdot \dfrac{\delta|\xi|}{\omega_1 \gamma_1^+} [\![\hat{\pi}]\!], \quad \delta := \dfrac{\mu_1 \mu_2 \omega_1 \omega_2 \gamma_1^+ \gamma_2^+}{\Omega_+}$$

where $\Omega_+ := \mu_1 \omega_1 \gamma_1^+ + \mu_2 \omega_2 \gamma_2^+$. Plugging in this representation for $|\xi|\hat{p}_2$ we obtain the reduced system

$$\tilde{A}(\lambda, \xi) \begin{pmatrix} \hat{\Phi}_v^{(2)} \\ \hat{\Phi}_w^{(2)} \\ \hat{\Phi}_w^{(1)} \\ \hat{h} \\ [\![\hat{\pi}]\!] \end{pmatrix} = \begin{pmatrix} 0 \\ 0 \\ \hat{g}_h \\ \hat{g}_w \\ \hat{g}_v \end{pmatrix}.$$

Here the matrix $\tilde{A}(\lambda, \xi)$ is defined as

$$\begin{pmatrix} i\xi^T & -\omega_2 & 0 & 0 & 0 \\ i\xi^T & 0 & \omega_1 & 0 & 0 \\ 0 & -\dfrac{\mu_2 \omega_2 \gamma_2^+}{\Omega_+} & -\dfrac{\mu_1 \omega_1 \gamma_1^+}{\Omega_+} & \lambda + |\xi|\left\langle b_0, \dfrac{i\xi}{|\xi|}\right\rangle & -\dfrac{|\xi|}{\Omega_+} \\ 0 & 2\mu_2\omega_2 & 2\mu_1\omega_1 & \sigma|\xi|^2 - G & 1 \\ \widetilde{\Omega}\,\mathrm{id}_n & -i\xi(\mu_2 + \kappa) & i\xi(\mu_1 + \kappa) & 0 & i\xi\dfrac{\mu_2 \gamma_2^- - \mu_1 \gamma_1^-}{\Omega_+} \end{pmatrix}$$

where $\kappa := \dfrac{\mu_1 \mu_2}{|\xi|} \cdot (\omega_1 \gamma_1^+ \gamma_2^- + \omega_2 \gamma_2^+ \gamma_1^-) \Omega_+^{-1}$. In the following we consider this mixed-order system and its mapping properties. In contrast to [PS09], [PS10], and [PS11] we can avoid many auxiliary problems by our approach. Therefore our proof is shorter and more direct.

Here we have to apply the parameter-dependent generalization of the theory developed in the last chapters, see Remark 2.71. To treat this mixed-order system we have to define a complex parameter-dependent matrix. We use the same abbreviations as before in a complex version (i.e., substitute $i\xi \rightsquigarrow z$, $|\xi| \rightsquigarrow |z|_-$, $b_0 \rightsquigarrow \vartheta_0$, and $i\xi/|\xi| \rightsquigarrow \vartheta_1$) to define the parameter-dependent $(n+4) \times (n+4)$-matrix

$$\mathscr{L}[\vartheta](\lambda, z) \tag{4.50}$$

$$:= \begin{pmatrix} z^T & -\omega_2 & 0 & 0 & 0 \\ z^T & 0 & \omega_1 & 0 & 0 \\ 0 & -\frac{\mu_2\omega_2\gamma_2^+}{\Omega_+} & -\frac{\mu_1\omega_1\gamma_1^+}{\Omega_+} & \lambda + |z|_- \langle \vartheta_0, \vartheta_1 \rangle & -\frac{|z|_-}{\Omega_+} \\ 0 & 2\mu_2\omega_2 & 2\mu_1\omega_1 & \sigma|z|_-^2 - G & 1 \\ \widetilde{\Omega}\mathrm{id}_n & -z(\mu_2 + \kappa) & z(\mu_1 + \kappa) & 0 & z\frac{\mu_2\gamma_2^- - \mu_1\gamma_1^-}{\Omega_+} \end{pmatrix}$$

for $(\lambda, z) \in \overline{S}_\theta \times \overline{\Sigma}_\delta^n$ and $\vartheta = (\vartheta_0, \vartheta_1) \in K(\beta, \varepsilon) := K_1(\beta) \times K_2(\varepsilon)$ where

$$K_1(\beta) := \{b_0 \in \mathbb{R}^n : |b_0| \leq \beta\}, \quad K_2(\varepsilon) := \{\zeta \in \mathbb{C}^n : |\operatorname{Re}\zeta| \leq \varepsilon \text{ and } |\operatorname{Im}\zeta| \leq 3/2\}.$$

The constants θ, δ, and ε will be determined by Lemma 4.33. In the sequel we show that for arbitrary $\beta > 0$ there exists θ, δ, and ε such that the matrix $\mathscr{L}$ is an N-parabolic mixed-order system with compact parameter in the sense of Definition 2.67 and Remark 2.71. The symbol $i\xi/|\xi|$, which has been replaced by the compact parameter ϑ_1, is related to the Riesz transform, cf. Definition 4.36. At the end of this section we use the Dunford calculus to plug in the Riesz transform for ϑ_1 to return to the original system. First, we have to determine $\det \mathscr{L}[\vartheta]$ in the next lemma, which shows that we can describe the determinant of a special $(n+4) \times (n+4)$-matrix by the determinant of a 4×4-matrix.

Lemma 4.32. *Let $n \in \mathbb{N}$, $A := (a_{ij})_{i,j=1,\ldots,4} \in \mathbb{C}^{4\times 4}$, $(\alpha_j)_{j=1,\ldots,4} \in \mathbb{C}$, $z \in \mathbb{C}^n$, $C > 0$, and $\Omega' \in \mathbb{C} \setminus \{0\}$. Then we get*

$$\det \begin{pmatrix} z^T & a_{11} & a_{12} & a_{13} & a_{14} \\ 0 & a_{21} & a_{22} & a_{23} & a_{24} \\ 0 & a_{31} & a_{32} & a_{33} & a_{34} \\ 0 & a_{41} & a_{42} & a_{43} & a_{44} \\ \Omega'\mathrm{id}_n - C \cdot z \otimes z & \alpha_1 z & \alpha_2 z & \alpha_3 z & \alpha_4 z \end{pmatrix} = \pm \Omega'^n \det B$$

where $B = (b_{ij})_{i,j=1,\ldots,4}$ is given by

$$b_{1j} = a_{1j} + \frac{\alpha_j + Ca_{1j}}{\Omega'}|z|_-^2, \quad j = 1,\ldots,4$$

and $b_{ij} = a_{ij}$ for $i = 2, 3, 4$ and $j = 1,\ldots,4$.

Proof. This can be verified easily. □

By an application of this lemma and straightforward calculations, we obtain

$$|\det \mathscr{L}[\vartheta](\lambda, z)| = \frac{|\omega_1\omega_2|}{|\Omega_+|}|\Omega'|^{n-1}|P[\vartheta](\lambda, z)|$$

4.7. Two-phase Navier-Stokes equations with Boussinesq-Scriven surface

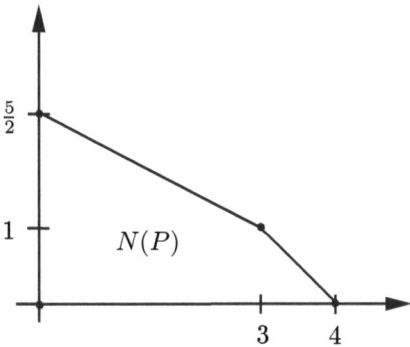

Figure 4.5: Newton polygon of P, $\mathcal{B} = 0$

where we defined, for $(\lambda, z) \in \overline{S}_\theta \times \overline{\Sigma}_\delta^n$ and $\vartheta \in K(\beta, \varepsilon)$,

$$\begin{aligned}
P[\vartheta](\lambda, z) := & (\mu_1 \omega_1^2 + \mu_2 \omega_2^2)(\mu_1 \omega_1 + \mu_2 \omega_2)\lambda \\
& + \left[(\mu_1 \omega_1 + \mu_2 \omega_2)^2 + \mu_1 \mu_2 (\omega_1 + \omega_2)^2\right] \lambda |z|_- \\
& + (\mu_1 \omega_1 + \mu_2 \omega_2)(\mu_1 \omega_1^2 + \mu_2 \omega_2^2) \langle \vartheta \rangle |z|_- \\
& - (\mu_1 \omega_1 + \mu_2 \omega_2) G |z|_- \\
& + \left[3(\mu_2^2 \omega_2 + \mu_1^2 \omega_1) - \mu_1 \mu_2 (\omega_1 + \omega_2)\right] \lambda |z|_-^2 + \mathcal{B}(\mu_1 \omega_1 + \mu_2 \omega_2) \lambda |z|_-^2 \\
& + \left[(\mu_1 \omega_1 + \mu_2 \omega_2)^2 + \mu_1 \mu_2 (\omega_1 + \omega_2)^2\right] \langle \vartheta \rangle |z|_-^2 - \bar{\mu} G |z|_-^2 \\
& - [\![\mu]\!]^2 \lambda |z|_-^3 + \mathcal{B}(\mu_1 \omega_1 + \mu_2 \omega_2) \lambda |z|_-^3 \\
& + \sigma(\mu_1 \omega_1 + \mu_2 \omega_2)|z|_-^3 + \left[3(\mu_1^2 \omega_1 + \mu_2^2 \omega_2) - \mu_1 \mu_2 (\omega_1 + \omega_2)\right] \langle \vartheta \rangle |z|_-^3 \\
& + \mathcal{B} \left[(\mu_1 \omega_1^2 + \mu_2 \omega_2^2) \langle \vartheta \rangle - G\right] |z|_-^3 + \left[\bar{\mu} \sigma - [\![\mu]\!]^2 \langle \vartheta \rangle\right] |z|_-^4 \\
& + \mathcal{B}(\mu_1 \omega_1 + \mu_2 \omega_2) \langle \vartheta \rangle |z|_-^4 + \sigma \mathcal{B} |z|_-^5.
\end{aligned}$$

Here we used the abbreviations $\langle \vartheta \rangle := \langle \vartheta_0, \vartheta_1 \rangle$, $\vartheta \in K(\beta, \varepsilon)$, $\bar{\mu} := \mu_1 + \mu_2$, $\bar{\rho} := \rho_1 + \rho_1$, and $\mathcal{B} := \lambda_s + \mu_s$. In the sequel we consider the cases $\mathcal{B} = 0$ and $\mathcal{B} \neq 0$ separately. If $\mathcal{B} \neq 0$, we have boundary conditions of order 2. Therefore it is obvious that the order structure of P strongly depends on the existence of surface viscosity, i.e., on the constant $\mathcal{B}$.

We can show that

$$d_\gamma(P) = \begin{cases} \max\{4, \gamma + 3, 5/2 \cdot \gamma\}, & \mathcal{B} = 0, \\ \max\{5, \gamma + 4, 2\gamma + 2, 5/2 \cdot \gamma\}, & \mathcal{B} \neq 0, \end{cases} \quad \gamma > 0.$$

Lemma 4.33. *Let* $\lambda_s, \mu_s = 0$ *or* $\lambda_s \geq 0$, $\mu_s > 0$.

(i) *For all* $\beta > 0$ *there exist* $\theta > \pi/2$, $\delta > 0$, *and* $\varepsilon > 0$ *such that the symbol* P

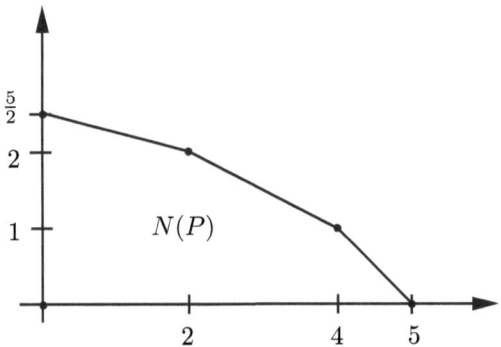

Figure 4.6: Newton polygon of P, $\mathcal{B} \neq 0$

defined above is N-parabolic with compact parameter (see Remark 2.71), i.e.,

$$P \in S_N[K(\beta,\varepsilon)](\overline{S}_\theta \times \overline{\Sigma}_\delta^n).$$

(ii) If θ, δ, and ε are determined by (i), then $\det \mathscr{L}$ is N-parabolic with compact parameter. For $\gamma > 0$, we have

$$[\mu(\det \mathscr{L}[\vartheta])](\gamma) = \begin{cases} \max\{n+3, \gamma+n+2, [n+4]/2\gamma\}, \\ \qquad\qquad\qquad\qquad\qquad \lambda_s, \mu_s = 0, \\ \max\{3+2n, \gamma+2+2n, 2\gamma+2n, (n+4)/2\cdot\gamma\}, \\ \qquad\qquad\qquad\qquad\qquad \lambda_s \geq 0, \mu_s > 0, \end{cases}$$

for all $\vartheta \in K(\beta,\varepsilon)$.

Proof. (i) Here we can apply the characterization of Corollary 2.57. This is why we will consider the principal part of P in the next lines.

(1) Let $\lambda_s, \mu_s = 0$. Then we get

$$\pi_\gamma P[\vartheta](\lambda, z) = \begin{cases} 2\bar{\mu}\left[\sigma + 2\bar{\mu}\langle\vartheta\rangle\right]|z|_-^4, & \gamma \in (0,1), \\ 2\bar{\mu}(2\bar{\mu}\lambda + [\sigma + 2\bar{\mu}\langle\vartheta\rangle]|z|_-)|z|_-^3, & \gamma = 1, \\ 4\bar{\mu}^2\lambda|z|_-^3, & \gamma \in (1,2), \\ \bar{\rho}(\sqrt{\mu_1\rho_1} + \sqrt{\mu_2\rho_2})\lambda^{5/2}, & \gamma > 2 \end{cases}$$

and

$$\pi_{\gamma=2}P[\vartheta](\lambda, z) = \bar{\rho}(\sqrt{\mu_1}\omega_1 + \sqrt{\mu_2}\omega_2)\lambda + 4\sqrt{\mu_1\mu_2}\omega_1\omega_2|z|_- + \bar{\rho}\bar{\mu}\lambda|z|_- \\ + 4(\mu_1^{3/2}\omega_1 + \mu_2^{3/2}\omega_2)|z|_-^2 + 4\mu_1\mu_2|z|_-^3.$$

4.7. Two-phase Navier-Stokes equations with Boussinesq-Scriven surface

It is easily seen that for all $\gamma \in (1, \infty) \setminus \{2\}$ we have $\pi_\gamma P(\lambda, z) \neq 0$ for all non-vanishing tuples $(\lambda, z) \in (\overline{S}_\theta \setminus \{0\}) \times (\overline{\Sigma}_\delta^n \setminus \{0\})$ and $\vartheta \in K(\beta, \varepsilon)$, $\theta > \pi/2$, and $\delta > 0$. But we have to take a closer look at $\pi_\gamma P$ for $\gamma \in (0, 1] \cup \{2\}$. In the following we choose $\varepsilon > 0$ such that $\varepsilon\beta < \sigma/[2\bar\mu]$. Hence we get

$$\operatorname{Re}(\sigma + 2\bar\mu \langle \vartheta \rangle) = \sigma + 2\bar\mu \langle \vartheta_0, \operatorname{Re} \vartheta_1 \rangle > 0, \quad \vartheta \in K(\beta, \varepsilon) \qquad (4.51)$$

due to $|\langle \vartheta_0, \operatorname{Re} \vartheta_1 \rangle| \leq \varepsilon\beta < \sigma/[2\bar\mu]$. Therefore we can choose $\delta > 0$ small and $\theta \in (\pi/2, 2/3\pi)$ such that

$$\sigma + 2\bar\mu \langle \vartheta \rangle \in S_{\pi - \theta - \delta} \text{ for all } \vartheta \in K(\beta, \varepsilon). \qquad (4.52)$$

(I) Let $\gamma < 1$. Then we directly get $\pi_\gamma P[\vartheta](\lambda, z) \neq 0$ for all $(\lambda, z) \in (\overline{S}_\theta \setminus \{0\}) \times (\overline{\Sigma}_\delta^n \setminus \{0\})$ and $\vartheta \in K(\beta, \varepsilon)$ by (4.51).

(II) Let $\gamma = 1$. Then we have $[\sigma + 2\bar\mu \langle \vartheta \rangle]|z|_- \in S_{\pi-\theta}$ according to (4.52). In particular, this yields

$$2\bar\mu \lambda + [\sigma + 2\bar\mu \langle \vartheta \rangle]|z|_- \in \overline{S}_\theta \setminus \{0\} \qquad (4.53)$$

for all $(\lambda, z, \vartheta) \in (\overline{S}_\theta \setminus \{0\}) \times (\overline{\Sigma}_\delta^n \setminus \{0\}) \times K(\beta, \varepsilon)$. We obtain $\pi_1 P[\vartheta](\lambda, z) \neq 0$ for all $(\lambda, z, \vartheta) \in (\overline{S}_\theta \setminus \{0\}) \times (\overline{\Sigma}_\delta^n \setminus \{0\}) \times K(\beta, \varepsilon)$.

(III) Let $\gamma = 2$ and $\theta \in (\pi/2, 2/3\pi)$ be fixed. For simplicity we define the continuous symbol $S \colon X \to \mathbb{C}$ by

$$S(\lambda, r, \varphi) := \bar\rho(\sqrt{\mu_1}\omega_1' + \sqrt{\mu_2}\omega_2')\lambda + 4\sqrt{\mu_1\mu_2}\omega_1'\omega_2' re^{i\varphi} + \bar\rho\bar\mu\lambda re^{i\varphi}$$
$$+ 4(\mu_1^{3/2}\omega_1' + \mu_2^{3/2}\omega_2')(re^{i\varphi})^2 + 4\mu_1\mu_2(re^{i\varphi})^3$$

where $X := \overline{S}_\theta \times [0, \infty) \times [-1, 1]$ and $\omega_j' = \omega_j'(\lambda, r, \varphi) = \mu_j^{-1/2}(\rho_j\lambda + \mu_j re^{i\varphi})^{-1/2}$. Let $\lambda \in \overline{S}_\theta$ with $\operatorname{Im} \lambda > 0$ and $\varphi = 0$. Then we also have $\operatorname{Im} \omega_j > 0$ and $\operatorname{Im}(\omega_j\lambda) > 0$ and therefore $\operatorname{Im} S(\lambda, r, 0) > 0$. Analogously for $\lambda \in \overline{S}_\theta$ with $\operatorname{Im} \lambda < 0$ we also get $\operatorname{Im} S(\lambda, r, 0) < 0$. For $(\lambda, r) \in [0, \infty)^2 \setminus \{0\}$ we trivially have $S(\lambda, r, 0) > 0$. So we have proved $S(\lambda, r, 0) \neq 0$ for all $(\lambda, r) \in [\overline{S}_\theta \times [0, \infty)] \setminus \{(0, 0)\}$. Thus, there exists $C_0 > 0$ such that

$$|S(\lambda, r, 0)| \geq C_0 \qquad (4.54)$$

for all $(\lambda, r) \in V := \{(\tau, s) \in \overline{S}_\theta \times [0, \infty) \colon |\tau| + s^2 = 1\}$. Due to the compactness of $V \times [-1, 1]$ there exists $\delta = \delta(C_0) > 0$ such that $|S(\lambda, r, 0) - S(\lambda, r, \varphi)| < C_0/2$ for all $(\lambda, r, \varphi) \in V \times [-1, 1]$ with $|\varphi| \leq \delta$. Due to (4.54) we obtain

$$|S(\lambda, r, \varphi)| \geq |S(\lambda, r, 0)| - |S(\lambda, r, \varphi) - S(\lambda, r, 0)| > C_0/2$$

for all $(\lambda, r, \varphi) \in V$ with $|\varphi| \leq \delta$. From this we derive $|S(\lambda, r, \varphi)| \geq C_0/2$ for all tuples $(\lambda, r, \varphi) \in \overline{S}_\theta \times [0, \infty) \times [-\delta, \delta]$ because of $S(\cdot, \cdot, \varphi) \in S^{(2,1)}(\overline{S}_\theta \times [0, \infty))$. For all $(\lambda, z) \in \overline{S}_\theta \times \overline{\Sigma}_\delta^n$ we have $|z|_- \in \overline{S}_\delta$ and therefore

$$|\pi_2 P[\vartheta](\lambda, z)| = |S(\lambda, ||z|_-|, \arg|z|_-)| > C_0/2,$$

which especially yields $\pi_2 P(\lambda, z) \neq 0$.

(2) Let $\lambda_s \geq 0$ and $\mu_s > 0$. Then $\pi_\gamma P[\vartheta](\lambda, z)$ is given by

$$\begin{cases} \mathcal{B}\left[\sigma + 2\bar{\mu}\langle\vartheta\rangle\right]|z|_-^5, & \gamma \in (0, 1), \\ \mathcal{B}\left[(\sigma + 2\bar{\mu}\langle\vartheta\rangle)|z|_- + 2\bar{\mu}\lambda\right]|z|_-^4, & \gamma = 1, \\ 2\mathcal{B}\bar{\mu}\lambda|z|_-^4, & \gamma \in (1, 2), \\ \mathcal{B}\left[(\mu_1\omega_1 + \mu_2\omega_2)|z|_- + (\mu_1\omega_1^2 + \mu_2\omega_2^2)\right]\lambda|z|_-^2, & \gamma = 2, \\ \mathcal{B}\bar{\rho}\lambda^2|z|_-^2, & \gamma \in (2, 4), \\ \bar{\rho}\left[\mathcal{B}|z|_-^2 + (\sqrt{\mu_1\rho_1} + \sqrt{\mu_2\rho_2})\lambda^{1/2}\right]\lambda^2, & \gamma = 4, \\ \bar{\rho}(\sqrt{\mu_1\rho_1} + \sqrt{\mu_2\rho_2})\lambda^{5/2}, & \gamma > 4. \end{cases}$$

Using (4.52) and (4.53) we obtain $\pi_\gamma P(\lambda, z) \neq 0$ for all $(\lambda, z) \in (\overline{S}_\theta \setminus \{0\}) \times (\overline{\Sigma}_\delta^n \setminus \{0\})$, $\vartheta \in K(\beta, \varepsilon)$, and $\gamma \in (0, \infty) \setminus \{2\}$ with $\beta, \varepsilon, \theta$, and δ as in part (1).

For $\gamma = 2$ we obtain

$$\pi_{\gamma=2}[\vartheta]P(\lambda, z) = \mathcal{B}(\mu_1\omega_1 + \mu_2\omega_2)\left[|z|_- + \underbrace{\frac{\mu_1\omega_1^2 + \mu_2\omega_2^2}{\mu_1\omega_1 + \mu_2\omega_2}}_{\in S_{3\theta/2}}\right]\lambda|z|_-^2 \neq 0$$

for all $(\lambda, z) \in (\overline{S}_\theta \setminus \{0\}) \times (\overline{\Sigma}_\delta^n \setminus \{0\})$.

So we have proved that the symbol P is N-parabolic with compact parameter in both cases.

(ii) According to (i) we have $P \in S_N[K(\beta, \varepsilon)](\overline{S}_\theta \times \overline{\Sigma}_\delta^n)$. It is easy to see that $\Omega' \in S_N(\overline{S}_\theta \times \overline{\Sigma}_\delta^n)$ with

$$d_\gamma(\Omega') = \begin{cases} \max\{1, 1/2\gamma\}, & \mu_s = 0, \\ \max\{2, 1/2\gamma\}, & \mu_s \neq 0 \end{cases}$$

and $\omega_1\omega_2\Omega_+^{-1} \in S^{(2,0)}(\overline{S}_\theta \times \overline{\Sigma}_\delta^n)$. According to Lemma 2.33 we get, for $\gamma > 0$,

$$[\mu(\det \mathscr{L}[\vartheta])](\gamma) = (n-1) \cdot d_\gamma(\Omega') + d_\gamma(P)$$
$$= \begin{cases} \max\{n+3, \gamma+n+2, [n+4]/2 \cdot \gamma\}, & \lambda_s, \mu_s = 0, \\ \max\{3+2n, \gamma+2+2n, 2\gamma+2n, (n+4)/2 \cdot \gamma\}, & \lambda_s \geq 0, \mu_s > 0. \end{cases} \qquad \square$$

4.7. Two-phase Navier-Stokes equations with Boussinesq-Scriven surface

In order to apply Corollary 2.80 we define the order functions

$$t_1(\gamma) := \ldots := t_n(\gamma) := \begin{cases} \max\{1, 1/2 \cdot \gamma\}, & \lambda_s, \mu_s = 0, \\ \max\{2, 1/2 \cdot \gamma\}, & \lambda_s \geq 0, \mu_s > 0, \end{cases}$$

$$t_{n+1}(\gamma) := t_{n+2}(\gamma) := \max\{1, 1/2 \cdot \gamma\},$$
$$t_{n+3}(\gamma) := \max\{2, \gamma+1, 3/2 \cdot \gamma\},$$
$$t_{n+4} := 0,$$

$$s_1 := s_2 := 0, \quad s_3(\gamma) := -\max\{1, 1/2 \cdot \gamma\}, \quad s_4 := \ldots := s_{n+4} := 0.$$

We easily obtain $\sum_{k=1}^{n+4}(s_k(\gamma) + t_k(\gamma)) = [\mu(\det \mathscr{L}[\vartheta])](\gamma)$ for $\gamma > 0$ in both cases. It is easy to verify that $s_j + t_i$ is an upper convex or concave, respectively, order function of $\mathscr{L}_{ji}$ for all $i, j = 1, \ldots, n+4$.

So we have proved the following proposition.

Proposition 4.34. *The complex matrix $\mathscr{L}$ defined in (4.50) is an N-parabolic mixed-order system with compact parameter in the sense of Definition 2.67 and Remark 2.71.*

For the application of Corollary 2.80 we define, for $p \in (1, \infty)$,

$$(r'_0, s'_0) := (r'_1, s'_1) := (1 - 1/p, 0), \quad (r'_2, s'_2) := (0, 1/2 - 1/(2p)),$$
$$(\mathcal{F}_0, \mathcal{K}_0) := (\mathcal{F}_1, \mathcal{K}_1) := (H_p, B_{pp}), \quad (\mathcal{F}_2, \mathcal{K}_2) := (B_{pp}, H_p)$$

if $\lambda_s, \mu_s = 0$, and

$$(r'_0, s'_0) := (r'_1, s'_1) := (1 - 1/p, 0), \quad (r'_2, s'_2) := (r'_3, s'_3) := (0, 1/2 - 1/(2p)),$$
$$(\mathcal{F}_0, \mathcal{K}_0) := (\mathcal{F}_1, \mathcal{K}_1) := (H_p, B_{pp}), \quad (\mathcal{F}_2, \mathcal{K}_2) := (\mathcal{F}_3, \mathcal{K}_3) := (B_{pp}, H_p)$$

if $\lambda_s \geq 0, \mu_s > 0$. It is easy to see that this scale fulfills all admissibility conditions. Let $\lambda_s, \mu_s = 0$ or $\lambda_s \geq 0, \mu_s > 0$. Then Corollary 2.80 and Remark 2.70 (iv) yield the following result.

Theorem 4.35. *Let $1 < p < \infty$ and $\beta > 0$. There exist $\varrho_0 > 0$ and $\varepsilon > 0$ such that*

$$L := \left[\mathscr{L}[\vartheta](\mathcal{D}_+^{(\varrho)})\right]\Big|_{\mathbb{H}} \in L_{\text{Isom}}(\mathbb{H}, \mathbb{F}), \quad \vartheta \in K(\beta, \varepsilon), \ \varrho \geq \varrho_0,$$

and $\|(L[\vartheta])^{-1}\|_{L(\mathbb{F}, \mathbb{H})} \leq C$ for all $\vartheta \in K(\beta, \varepsilon)$. The associated spaces are given by $\mathbb{H} := \prod_{i=1}^{n+4} \mathbb{H}_i$ and $\mathbb{F} := \prod_{j=1}^{n+4} \mathbb{F}_j$ where

$$\mathbb{H}_i = \begin{cases} {}_0 B_{p,\varrho}^{1-1/(2p)}(\mathbb{R}_+, L_p(\mathbb{R}^n)) \cap L_{p,\varrho}(\mathbb{R}_+, B_{pp}^{2-1/p}(\mathbb{R}^n)), & \lambda_s, \mu_s = 0, \\ {}_0 B_{p,\varrho}^{1-1/(2p)}(\mathbb{R}_+, L_p(\mathbb{R}^n)) \cap {}_0 B_{p,\varrho}^{1/2-1/(2p)}(\mathbb{R}_+, H_p^2(\mathbb{R}^n)) \\ \quad \cap L_{p,\varrho}(\mathbb{R}_+, B_{pp}^{3-1/p}(\mathbb{R}^n)), & \lambda_s \geq 0, \mu_s > 0 \end{cases}$$

for $i = 1, \ldots, n$ and

$$\mathbb{H}_{n+1} = \mathbb{H}_{n+2} = {}_0B_{p,\varrho}^{1-1/(2p)}(\mathbb{R}_+, L_p(\mathbb{R}^n)) \cap L_{p,\varrho}(\mathbb{R}_+, B_{pp}^{2-1/p}(\mathbb{R}^n)),$$
$$\mathbb{H}_{n+3} = {}_0B_{p,\varrho}^{2-1/(2p)}(\mathbb{R}_+, L_p(\mathbb{R}^n)) \cap {}_0H_{p,\varrho}^{1}(\mathbb{R}_+, B_{pp}^{2-1/p}(\mathbb{R}^n))$$
$$\cap L_{p,\varrho}(\mathbb{R}_+, B_{pp}^{3-1/p}(\mathbb{R}^n)),$$
$$\mathbb{H}_{n+4} = {}_0B_{p,\varrho}^{1/2-1/(2p)}(\mathbb{R}_+, L_p(\mathbb{R}^n)) \cap L_{p,\varrho}(\mathbb{R}_+, B_{pp}^{1-1/p}(\mathbb{R}^n)),$$
$$\mathbb{F}_1 = \mathbb{F}_2 = {}_0B_{p,\varrho}^{1/2-1/(2p)}(\mathbb{R}_+, L_p(\mathbb{R}^n)) \cap L_{p,\varrho}(\mathbb{R}_+, B_{pp}^{1-1/p}(\mathbb{R}^n)),$$
$$\mathbb{F}_3 = {}_0B_{p,\varrho}^{1-1/(2p)}(\mathbb{R}_+, L_p(\mathbb{R}^n)) \cap L_{p,\varrho}(\mathbb{R}_+, B_{pp}^{2-1/p}(\mathbb{R}^n)),$$
$$\mathbb{F}_4 = \ldots = \mathbb{F}_{n+4} = {}_0B_{p,\varrho}^{1/2-1/(2p)}(\mathbb{R}_+, L_p(\mathbb{R}^n)) \cap L_{p,\varrho}(\mathbb{R}_+, B_{pp}^{1-1/p}(\mathbb{R}^n)).$$

In the derivation of the matrix $\mathscr{L}$ in (4.50) we replaced the symbol $i\xi/|\xi|$ by the compact parameter ϑ_1. The symbol $i\xi/|\xi|$ is related to the Riesz transform, which is introduced below. To return to the original problem we use the Dunford calculus to plug in the Riesz transform for ϑ_1.

Definition 4.36 (Riesz transform). We define the *Riesz transform* on $H_p^{-\infty}(\mathbb{R}^n)$ by

$$R \colon H_p^{-\infty}(\mathbb{R}^n) \to [H_p^{-\infty}(\mathbb{R}^n)]^n,$$
$$f \mapsto (R_j f)_{j=1,\ldots,n}$$

where $R_j := \mathrm{op}[m_j]$, $m_j(z) := -i \cdot z_j/|z|_-$ for $j \in \{1, \ldots, n\}$ and $z \in \Sigma_\delta^n$ (cf. Remark 1.77). For further details we refer to [Ste70, Ch. III].

Remark 4.37. (i) Due to Remark 1.77 we have

$$R|_{\mathcal{K}^r(\mathbb{R}^n)} \in L(\mathcal{K}^r(\mathbb{R}^n), [\mathcal{K}^r(\mathbb{R}^n)]^n)$$

for $\mathcal{K} \in \{H_p, B_{pp}\}$.

(ii) It is easy to verify that $\sigma(R_j) = [-1, 1]$ for all $j \in \{1, \ldots, n\}$.

We have $\prod_{j=1}^n \sigma(iR_j) \subseteq K_2(\varepsilon)$ and therefore we find paths of integration in $K_2(\varepsilon)$ which envelop the spectrum. So we can use the Dunford calculus to define the operators

$$L(b_0) := L[b_0, iR^+] \in L(\mathbb{H}, \mathbb{F}),$$
$$S(b_0) := (L[b_0, \cdot])^{-1}(iR^+) \in L(\mathbb{F}, \mathbb{H})$$

for all $b_0 \in K_1(\beta)$, where R^+ denotes the natural extension of the Riesz operator R.

Corollary 4.38. *We have*

$$L(b_0) \in L_{\mathrm{Isom}}(\mathbb{H}, \mathbb{F}), \quad S(b_0) \in L_{\mathrm{Isom}}(\mathbb{F}, \mathbb{H}).$$

and $L(b_0)^{-1} = S(b_0)$. For all $\beta > 0$ there exists $C = C(\beta) > 0$ such that
$$\|L(b_0)^{-1}\|_{L(\mathbb{F},\mathbb{H})} \leq C(\beta)$$
for all $b_0 \in \mathbb{R}^n$ with $|b_0| \leq \beta$.

At this point we want to mention Remarks 4.29 and 4.30 again. The arguments discussed there can also be applied to the two-phase Navier-Stokes equations in (4.34).

4.8 The L_p-L_q two-phase Stefan problem with Gibbs-Thomson correction

Motivation. Here we want to consider the L_p-L_q two-phase Stefan problem with Gibbs-Thomson correction, which was treated by J. Escher and J. Prüss in [EPS03] in the case $p = q$. With our approach we can simplify the proof of the well-posedness of the corresponding linearized problem in the case of $p = q$ (cf. [EPS03, Theorem 6.1]) and we can also give a proof in the case of $p \neq q$. For the one-phase Stefan problem with Gibbs-Thomson correction and $p = q$ we also want to refer to a result of R. Denk, J. Saal. and J. Seiler in [DSS08]. Here we essentially use the same techniques as in the latter reference.

It seems that for $p \neq q$ the L_p-L_q two-phase Stefan problem with Gibbs-Thomson correction has not been considered in the literature before. Furthermore, we are able to give an explicit characterization of the spaces on the boundary.

We consider the following system of equations:
$$\begin{cases} \partial_t u - \Delta u = 0 & \text{in } \mathbb{R}_+ \times \dot{\mathbb{R}}^n, \\ [\![u]\!] = 0 & \text{on } \mathbb{R}_+ \times \mathbb{R}^{n-1}, \\ u + \Delta' h = g_1 & \text{on } \mathbb{R}_+ \times \mathbb{R}^{n-1}, \\ \partial_t h - [\![\partial_n u]\!] = g_2 & \text{on } \mathbb{R}_+ \times \mathbb{R}^{n-1}, \\ u(t=0) = 0 & \text{in } \dot{\mathbb{R}}^n, \\ h(t=0) = 0 & \text{in } \mathbb{R}^{n-1} \end{cases} \quad (4.55)$$

for the unknown functions $u \colon \dot{\mathbb{R}}^n \to \mathbb{R}$ and $h \colon \mathbb{R}^{n-1} \to \mathbb{R}$, where $\dot{\mathbb{R}}^n := \mathbb{R}^{n-1} \times (\mathbb{R} \setminus \{0\})$. After formal Laplace and Fourier transform we obtain the following system of ordinary differential equations for $\hat{u} = \hat{u}_1 \chi_{\mathbb{R}^n_-} + \hat{u}_2 \chi_{\mathbb{R}^n_+}$ and $\hat{h}$,

$$\begin{cases} \omega(\lambda, \xi)^2 \hat{u}_j(\lambda, \xi, x_n) - \partial_n^2 \hat{u}_j(\lambda, \xi, x_n) = 0, & (-1)^j x_n > 0, \, j = 1, 2, \\ [\![\hat{u}]\!](\lambda, \xi) = 0, \\ \hat{u}(\lambda, \xi, 0) - |\xi|^2 \hat{h}(\lambda, \xi) = \hat{g}_1(\lambda, \xi), \\ \lambda \hat{h}(\lambda, \xi) - [\![\partial_n \hat{u}]\!](\lambda, \xi) = \hat{g}_2(\lambda, \xi) \end{cases} \quad (4.56)$$

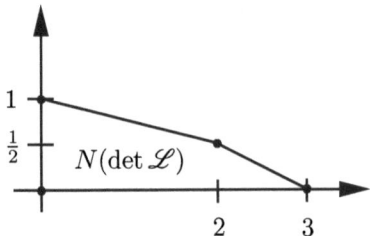

Figure 4.7: Newton polygon for the two-phase Stefan problem

for fixed (λ, ξ) and $\omega(\lambda, \xi) := \sqrt{\lambda + |\xi|^2}$. Solving the first two lines of (4.56) we get
$$\hat{u}_j(\lambda, \xi, x_n) = \hat{\Phi}(\lambda, \xi) \exp(-(-1)^j \omega \cdot x_n), \quad (-1)^j x_n > 0, \ j = 1, 2$$
for an unknown function $\hat{\Phi}$. Due to $\partial_n \hat{u}_j(\lambda, \xi, 0) = -(-1)^j \omega \hat{\Phi}(\lambda, \xi)$ $(j = 1, 2)$ we deduce the following system of linear equations from the boundary conditions in (4.56),
$$\begin{pmatrix} 1 & -|\xi|^2 \\ 2\omega & \lambda \end{pmatrix} \begin{pmatrix} \hat{\Phi} \\ \hat{h} \end{pmatrix} = \begin{pmatrix} \hat{g}_1 \\ \hat{g}_2 \end{pmatrix}.$$

Therefore we define the complex matrix
$$\mathscr{L}(\lambda, z) := \begin{pmatrix} 1 & -|z|_-^2 \\ 2\omega & \lambda \end{pmatrix} \tag{4.57}$$
with $(\lambda, z) \in \overline{S}_\theta \times \overline{\Sigma}_\delta^{n-1}$ and $\omega(\lambda, z) = \sqrt{\lambda + |z|_-^2}$. So we easily obtain
$$(\det \mathscr{L})(\lambda, z) = \lambda + 2\omega |z|_-^2,$$
$$[\mu(\det \mathscr{L})](\gamma) = \max\{3, 1/2 \cdot \gamma + 2, \gamma\}, \quad \gamma > 0,$$

and
$$\pi_\gamma(\det \mathscr{L})(\lambda, z) = \begin{cases} 2|z|_-^3, & \gamma \in (0, 2), \\ 2|z|_-^2 \omega, & \gamma = 2, \\ 2|z|_-^2 \lambda^{1/2}, & \gamma \in (2, 4), \\ \lambda^{1/2}(\lambda^{1/2} + 2|z|_-^2), & \gamma = 4, \\ \lambda, & \gamma > 4, \end{cases} \quad (\lambda, z) \in \overline{S}_\theta \times \overline{\Sigma}_\delta^{n-1}.$$

According to Corollary 2.57 it is obvious that $\det \mathscr{L}$ is N-parabolic (cf. Figure 4.7). In order to apply Corollary 3.44 we define for $1 < p < \infty$ and $2p/(1+p) < q < 2p$ (see Remark 3.28)

4.8. The L_p-L_q two-phase Stefan problem with Gibbs-Thomson correction

$$(r_0', s_0') := (1 - 1/q, 0), \qquad (\mathcal{F}_0, \mathcal{K}_0) := (H_p, B_{qq}),$$
$$(r_1', s_1') := (0, 1/2 - 1/(2q)), \qquad (\mathcal{F}_1, \mathcal{K}_1) := (F_{pq}, H_q),$$
$$(r_2', s_2') := (0, 1/2 - 1/(2q)), \qquad (\mathcal{F}_2, \mathcal{K}_2) := (F_{pq}, H_q),$$

$$t_1(\gamma) := \begin{cases} 1, & \gamma \in (0, 2], \\ 1/2 \cdot \gamma, & \gamma \in (2, 4], \\ 1/2 \cdot \gamma, & \gamma > 4, \end{cases} \quad t_2(\gamma) := \begin{cases} 3, & \gamma \in (0, 2], \\ 1/2 \cdot \gamma + 2, & \gamma \in (2, 4], \\ \gamma, & \gamma > 4, \end{cases}$$

$$s_1(\gamma) := -t_1(\gamma), \qquad s_2(\gamma) := 0.$$

We also obtain $\sum_{k=1}^{2}(s_k(\gamma) + t_k(\gamma)) = [\mu(\det \mathscr{L})](\gamma)$ for $\gamma > 0$. Obviously, $s_j + t_i$ is an upper convex or concave, respectively, order function of $\mathscr{L}_{ji}$ for all $i, j = 1, 2$. It is easy to see that the scale defined above fulfills all admissibility conditions of Corollary 3.44. This yields the next theorem.

Theorem 4.39. *The complex matrix $\mathscr{L}$ defined in (4.57) is an N-parabolic mixed-order system in the sense of Definition 2.67. Let $1 < p < \infty$ and $2p/(1+p) < q < 2p$. Then there exists $\varrho_0 > 0$ such that the matrix $\mathscr{L}$ gives rise to the isomorphism*

$$\left[\mathscr{L}(\mathcal{D}_+^{(\varrho)})\right]\Big|_{\mathbb{H}} \in L_{\text{Isom}}(\mathbb{H}, \mathbb{F}), \quad \varrho \geq \varrho_0$$

where $\mathbb{H} := \mathbb{H}_1 \times \mathbb{H}_2$ and $\mathbb{F} := \mathbb{F}_1 \times \mathbb{F}_2$ with the spaces

$$\mathbb{H}_1 := {}_0F_{pq,\varrho}^{1-1/(2q)}(\mathbb{R}_+, L_q(\mathbb{R}^{n-1})) \cap L_{p,\varrho}(\mathbb{R}_+, B_{qq}^{2-1/q}(\mathbb{R}^{n-1})),$$
$$\mathbb{H}_2 := {}_0F_{pq,\varrho}^{3/2-1/(2q)}(\mathbb{R}_+, L_q(\mathbb{R}^{n-1})) \cap {}_0F_{pq,\varrho}^{1-1/(2q)}(\mathbb{R}_+, H_q^2(\mathbb{R}^{n-1}))$$
$$\cap L_{p,\varrho}(\mathbb{R}_+, B_{qq}^{4-1/q}(\mathbb{R}^{n-1})),$$
$$\mathbb{F}_1 := {}_0F_{pq,\varrho}^{1-1/(2q)}(\mathbb{R}_+, L_q(\mathbb{R}^{n-1})) \cap L_{p,\varrho}(\mathbb{R}_+, B_{qq}^{2-1/q}(\mathbb{R}^{n-1})),$$
$$\mathbb{F}_2 := {}_0F_{pq,\varrho}^{1/2-1/(2q)}(\mathbb{R}_+, L_q(\mathbb{R}^{n-1})) \cap L_{p,\varrho}(\mathbb{R}_+, B_{qq}^{1-1/q}(\mathbb{R}^{n-1})).$$

Proof. This follows from Corollary 3.44, Remark 2.70 (iv), and the lines above. $\square$

Theorem 4.40. *For $1 < p < \infty$, $2p/(1+p) < q < 2p$, and $(g_1, g_2) \in \mathbb{F}_1 \times \mathbb{F}_2$ we derive a solution of the two-phase Stefan problem with Gibbs-Thomson correction (4.55) by*

$$u := u_1 \chi_{\mathbb{R}_-^n} + u_2 \chi_{\mathbb{R}_+^n} \in {}_0H_{p,\varrho}^1(\mathbb{R}_+, L_q(\dot{\mathbb{R}}^n)) \cap L_{p,\varrho}(\mathbb{R}_+, H_q^2(\dot{\mathbb{R}}^n)),$$
$$u_1 := E^{\mathcal{W}_-}[h_1]e_n \Phi \in {}_0H_{p,\varrho}^1(\mathbb{R}_+, L_q(\mathbb{R}_-^n)) \cap L_{p,\varrho}(\mathbb{R}_+, H_q^2(\mathbb{R}_-^n)),$$
$$u_2 := E^{\mathcal{W}_+}[h_2]e_n \Phi \in {}_0H_{p,\varrho}^1(\mathbb{R}_+, L_q(\mathbb{R}_+^n)) \cap L_{p,\varrho}(\mathbb{R}_+, H_q^2(\mathbb{R}_+^n)),$$
$$h \in \mathbb{H}_2$$

where $h_j(\lambda, z, x_n) := \exp(-(-1)^j \omega(\lambda, z) x_n)$, $\mathcal{W}_\pm := L_{p,\varrho}(\mathbb{R}_+, L_q(\mathbb{R}_\pm^n))$, and

$$\begin{pmatrix} \Phi \\ h \end{pmatrix} := \mathscr{L}(\mathcal{D}_+^{(\varrho)})^{-1} \begin{pmatrix} g_1 \\ g_2 \end{pmatrix} \in \mathbb{H}_1 \times \mathbb{H}_2.$$

Furthermore there exists $C = C(\mathscr{L}^{-1}, p, q, h) > 0$ such that

$$\|(u, h)\|_{\mathbb{X} \times \mathbb{H}_2} \leq C \|(g_1, g_2)\|_{\mathbb{F}}$$

where $\mathbb{X} := {}_0H^1_{p,\varrho}(\mathbb{R}_+, L_q(\dot{\mathbb{R}}^n)) \cap L_{p,\varrho}(\mathbb{R}_+, H^2_q(\dot{\mathbb{R}}^n))$.

Proof. According to Example 3.66 and Remark 3.64 we deduce that $u := u_1 \chi_{\mathbb{R}^n_-} + u_2 \chi_{\mathbb{R}^n_+}$ solves the heat equation in $\mathbb{R}_+ \times \dot{\mathbb{R}}^n$ with

$$\gamma_{0,n} u_j = \Phi,$$
$$\gamma_{0,n} \partial_n u_j = -(-1)^j \omega(\nabla^{\mathcal{W}''}_+) \Phi, \quad j = 1, 2$$

where $\mathcal{W}'' := L_{p,\varrho}(\mathbb{R}_+, L_q(\mathbb{R}^{n-1}))$. So we obtain

$$\begin{pmatrix} \gamma_{0,n} u + \Delta' h \\ \partial_t h - [\![\partial_n u]\!] \end{pmatrix} = \begin{pmatrix} 1 & \Delta' \\ 2\omega(\boldsymbol{D}^{\mathcal{W}''}_+) & \partial_t \end{pmatrix} \begin{pmatrix} \Phi \\ h \end{pmatrix} = \mathscr{L}(\boldsymbol{D}^{(\varrho)}_+) \begin{pmatrix} \Phi \\ h \end{pmatrix}$$
$$= \begin{pmatrix} g_1 \\ g_2 \end{pmatrix},$$

which finishes the proof. The claimed boundedness of the solution operator easily follows from the boundedness of $\mathscr{L}(\boldsymbol{D}^{(\varrho)}_+)^{-1}$ and the boundedness of $E[h_j]$, cf. Corollary 3.63 and Remark 3.64. $\square$

Remark 4.41. For the construction of solutions on a finite time interval we can adapt the argumentation in Remark 4.30.

List of Figures

1	Newton polygon for the Stefan problem with Gibbs-Thomson correction	7
1.1	Spectrum, S_θ, and admissible curve Γ_φ	14
1.2	Spectrum, Σ_δ, and admissable curve $\Gamma_\varphi \cup (-\Gamma_\varphi)$	14
1.3	Path of integration $\Gamma(\varepsilon, R)$	49
2.1	Regular Newton polygon N	78
2.2	Newton polygon that is not regular in time	80
2.3	Newton polygon that is not regular in space	80
2.4	Illustration of the partition	94
3.1	Illustration of the set of tuples (p, q) satisfying (3.4)	147
4.1	Newton polygon of the α-β-system	197
4.2	Newton polygon of the Cahn-Hilliard-Gurtin problem	201
4.3	Newton polygon of $\det \mathscr{L}$ for the three-phase problem	206
4.4	Newton polygon for the spin-coating process	210
4.5	Newton polygon of P, $\mathcal{B} = 0$	219
4.6	Newton polygon of P, $\mathcal{B} \neq 0$	220
4.7	Newton polygon for the two-phase Stefan problem	226

Bibliography

[ADN59] S. Agmon, A. Douglis, and L. Nirenberg. Estimates near the boundary for solutions of elliptic partial differential equations satisfying general boundary conditions. I. *Comm. Pure Appl. Math.*, 12:623–727, 1959.

[ADN64] S. Agmon, A. Douglis, and L. Nirenberg. Estimates near the boundary for solutions of elliptic partial differential equations satisfying general boundary conditions. II. *Comm. Pure Appl. Math.*, 17:35–92, 1964.

[AE09] H. Amann and J. Escher. *Analysis III*. Birkhäuser Verlag, Basel, 2009.

[AF03] R.A. Adams and J.J.F. Fournier. *Sobolev Spaces*. Pure and Applied Mathematics. Academic Press, 2003. Second edition.

[Agm62] S. Agmon. On the eigenfunctions and on the eigenvalues of general elliptic boundary value problems. *Comm. Pure Appl. Math.*, 15:119–147, 1962.

[Ama95] H. Amann. *Linear and Quasilinear Parabolic Problems. Vol. I.* Birkhäuser Verlag, 1995.

[Ama00] H. Amann. Compact embeddings of vector-valued Sobolev and Besov spaces. *Glas. Mat. Ser. III*, 35(55)(1):161–177, 2000. Dedicated to the memory of Branko Najman.

[Ama03] H. Amann. Vector-valued distributions and Fourier multipliers. Unpublished manuscript, 2003.

[Ama09] H. Amann. Anisotropic function spaces and maximal regularity for parabolic problems. Part 1: Function spaces. *Jindrich Necas Center for Mathematical Modeling Lecture Notes*, 6, 2009.

[AV64] M.S. Agranovich and M.I. Vishik. Elliptic problems with a parameter and parabolic problems of general form. *Russ. Math. Surv.*, 19(3):53–157, 1964. Engl. transl. of Usp. Mat. Nauk. 19(3):53-161, 1964.

[Ber85] M.Z. Berkolaĭko. Traces theorems on coordinate subspaces for some spaces of differentiable functions with anisotropic mixed norm. *Dokl. Akad. Nauk SSSR*, 282(5):1042–1046, 1985. (Russian), English transl. in Soviet Math. Dokl. 31 (1985).

[Ber87a]　M.Z. Berkolaĭko. Traces of functions in generalized Sobolev spaces with a mixed norm on an arbitrary coordinate subspace. I (Russian). *Trudy Inst. Mat. (Novosibirsk)*, 7(Issled. Geom. Mat. Anal.):30–44, 199, 1987.

[Ber87b]　M.Z. Berkolaĭko. Traces of functions in generalized Sobolev spaces with a mixed norm on an arbitrary coordinate subspace. II (Russian). *Trudy Inst. Mat. (Novosibirsk)*, 9(Issled. Geom. "v tselom" i Mat. Anal.):34–41, 206, 1987.

[BIN78]　O.V. Besov, V.P. Il'in, and Sergey M. Nikol'skiĭ. *Integral Representations of Functions and Imbedding Theorems. Vol. I.* V. H. Winston & Sons, Washington, D.C., 1978. Translated from the Russian, Scripta Series in Mathematics, Edited by Mitchell H. Taibleson.

[BIN79]　O.V. Besov, V.P. Il'in, and Sergey M. Nikol'skiĭ. *Integral Representations of Functions and Imbedding Theorems. Vol. II.* V. H. Winston & Sons, Washington, D.C., 1979. Scripta Series in Mathematics, Edited by Mitchell H. Taibleson.

[BK05]　S. Bu and J. Kim. Operator-valued Fourier multiplier theorems on Triebel spaces. *Acta Math. Sci. Ser. B Engl. Ed.*, 25(4):599–609, 2005.

[BK09]　S. Bu and J. Kim. Some remarks about operator-valued Fourier multiplier theorems on Triebel spaces. *Acta Anal. Funct. Appl.*, 11(1):1–8, 2009.

[BL76]　J. Bergh and J. Löfström. *Interpolation Spaces. An Introduction.* Springer-Verlag, Berlin, 1976.

[Bou83]　J. Bourgain. Some remarks on Banach spaces in which martingale difference sequences are unconditional. *Ark. Mat.*, 21(2):163–168, 1983.

[BP61]　A. Benedek and R. Panzone. The space L^p, with mixed norm. *Duke Math. J.*, 28:301–324, 1961.

[BP07]　D. Bothe and J. Prüss. L_P-theory for a class of non-Newtonian fluids. *SIAM J. Math. Anal.*, 39(2):379–421 (electronic), 2007.

[BP10]　D. Bothe and J. Prüss. On the two-phase Navier-Stokes equations with Boussinesq-Scriven surface fluid. *J. Math. Fluid Mech.*, 12(1):133–150, 2010.

[Bug71]　Ja.S. Bugrov. Functional spaces with mixed norm. *Izv. Akad. Nauk SSSR Ser. Mat.*, 35:1137–1158, 1971.

[Bur81]　D.L. Burkholder. A geometrical characterization of Banach spaces in which martingale difference sequences are unconditional. *Ann. Probab.*, 9(6):997–1011, 1981.

[CDMY96] M. Cowling, I. Doust, A. McIntosh, and A. Yagi. Banach space operators with a bounded H^∞ functional calculus. *J. Austral. Math. Soc. Ser. A*, 60(1):51–89, 1996.

[DD11] R. Denk and M. Dreher. Resolvent estimates for elliptic systems in function spaces of higher regularity. *Electron. J. Differ. Equ.*, 2011(109):1–12, 2011.

[DF10] R. Denk and M. Faierman. Estimates for solutions of a parameter-elliptic multi-order system of differential equations. *Integral Equations Operator Theory*, 66:327–365, 2010.

[DGH+11] R. Denk, M. Geissert, M. Hieber, J. Saal, and O. Sawada. The spin-coating process: Analysis of the free boundary value problem. *Commun. Partial Differ. Equations*, 36(7):1145–1192, 2011.

[DHP03] R. Denk, M. Hieber, and J. Prüss. $\mathcal{R}$-boundedness, Fourier multipliers and problems of elliptic and parabolic type. *Mem. Amer. Math. Soc.*, 166(788):viii+114, 2003.

[DHP07] R. Denk, M. Hieber, and J. Prüss. Optimal L^p-L^q-estimates for parabolic boundary value problems with inhomogeneous data. *Math. Z.*, 257(1):193–224, 2007.

[DMV98] R. Denk, R. Mennicken, and L.R. Volevich. The Newton polygon and elliptic problems with parameter. *Math. Nachr.*, 192:125–157, 1998.

[Don74] T. Donaldson. *A Laplace transform calculus for partial differential operators*. American Mathematical Society, Providence, R.I., 1974. Memoirs of the American Mathematical Society, No. 143.

[DPZ08] R. Denk, J. Prüss, and R. Zacher. Maximal L_p-regularity of parabolic problems with boundary dynamics of relaxation type. *J. Funct. Anal.*, 255(11):3149–3187, 2008.

[DR06] R. Denk and R. Racke. L^p-resolvent estimates and time decay for generalized thermoelastic plate equations. *Electron. J. Differ. Equ.*, 2006(48):1–16, 2006.

[DRS09] R. Denk, R. Racke, and Y. Shibata. L_p theory for the linear thermoelastic plate equations in bounded and exterior domains. *Adv. Differ. Equ.*, 14(7-8):685–715, 2009.

[DS11] R. Denk and J. Seiler. On the maximal L_p-regularity of parabolic mixed-order systems. *J. Evol. Equ.*, 11(2):371–404, 2011.

[DSS08] R. Denk, J. Saal, and J. Seiler. Inhomogeneous symbols, the Newton polygon, and maximal L^p-regularity. *Russ. J. Math. Phys.*, 15(2):171–192, 2008.

[DSS09]   R. Denk, J. Saal, and J. Seiler. Bounded H_∞-calculus for pseudo-differential Douglis-Nirenberg systems of mild regularity. *Math. Nachr.*, 282(3):386–407, 2009.

[DV02a]   R. Denk and L.R. Volevich. Elliptic boundary value problems with large parameter for mixed order systems. In *Partial Differential Equations*, volume 206 of *Amer. Math. Soc. Transl. Ser. 2*, pages 29–64. Amer. Math. Soc., Providence, RI, 2002.

[DV02b]   G. Dore and A. Venni. H^∞-calculus for an elliptic operator on a halfspace with general boundary conditions. *Ann. Scuola Norm. Sup. Pisa Cl. Sci. (5)*, 1(3):487–543, 2002.

[DV05]   G. Dore and A. Venni. H^∞ functional calculus for sectorial and bisectorial operators. *Studia Math.*, 166(3):221–241, 2005.

[DV08]   R. Denk and L.R. Volevich. Parabolic boundary value problems connected with the Newtons's polygon and some problems of crystallization. *J. Evol. Equ.*, 8:523–556, 2008.

[EPS03]   J. Escher, J. Prüss, and G. Simonett. Analytic solutions for a Stefan problem with Gibbs-Thomson correction. *J. Reine Angew. Math.*, 563:1–52, 2003.

[Fer87]   D.L. Fernandez. Vector-valued singular integral operators on L^p-spaces with mixed norms and applications. *Pacific J. Math.*, 129(2):257–275, 1987.

[GGHR06]   M. Geissert, B. Grec, M. Hieber, and E. Radkevich. The model-problem associated to the Stefan problem with surface tension: an approach via Fourier-Laplace multipliers. In *Differential Equations: Inverse and Direct Problems*, volume 251 of *Lect. Notes Pure Appl. Math.*, pages 171–182. Chapman & Hall/CRC, Boca Raton, FL, 2006.

[GR07]   B. Grec and E.V. Radkevich. Newton's polygon method and the local solvability of free boundary problems. *Journal of mathematical sciences*, 143(4):3253–3292, 2007.

[Gri72]   P. Grisvard. Interpolation non commutative. *Atti Accad. Naz. Lincei Rend. Cl. Sci. Fis. Mat. Natur.*, 52:11–15, 1972.

[GV92]   S. Gindikin and L.R. Volevich. *The Method of Newton's Polyhedron in the Theory of Partial Differential Equations*, volume 86 of *Mathematics and its Applications (Soviet Series)*. Kluwer Academic Publishers Group, Dordrecht, 1992. Translated from the Russian manuscript by V. M. Volosov.

[Haa06]   M. Haase. *The Functional Calculus for Sectorial Operators*, volume 169 of *Operator Theory: Advances and Applications*. Birkhäuser Verlag, Basel, 2006.

Bibliography

[Han81] E.I. Hanzawa. Classical solutions of the Stefan problem. *Tôhoku Math. J. (2)*, 33(3):297–335, 1981.

[Hyt05] T.P. Hytönen. On operator-multipliers for mixed-norm $L^{\overline{p}}$ spaces. *Arch. Math. (Basel)*, 85(2):151–155, 2005.

[Jef04] B. Jefferies. *Spectral Properties of Noncommuting Operators*, volume 1843 of *Lecture Notes in Mathematics*. Springer-Verlag, Berlin, 2004.

[JS08] J. Johnsen and W. Sickel. On the trace problem for Lizorkin-Triebel spaces with mixed norms. *Math. Nachr.*, 281(5):669–696, 2008.

[Kai12] M. Kaip. *General parabolic mixed order systems in L_p and applications*. Ph. D. Thesis. University of Konstanz, 2012.

[Kat76] T. Kato. *Perturbation Theory for Linear Operators*. Springer-Verlag, Berlin, second edition, 1976. Grundlehren der Mathematischen Wissenschaften, Band 132.

[KKW06] N.J. Kalton, P.C. Kunstmann, and L. Weis. Perturbation and interpolation theorems for the H^∞-calculus with applications to differential operators. *Math. Ann.*, 336(4):747–801, 2006.

[KMM07] N.J. Kalton, S. Mayboroda, and M. Mitrea. Interpolation of Hardy-Sobolev-Besov-Triebel-Lizorkin spaces and applications to problems in partial differential equations. In *Interpolation Theory and Applications*, volume 445 of *Contemp. Math.*, pages 121–177. Amer. Math. Soc., Providence, RI, 2007.

[Kot08] M. Kotschote. Strong solutions for a compressible fluid model of Korteweg type. *Ann. Inst. H. Poincaré Anal. Non Linéaire*, 25(4):679–696, 2008.

[Kot10] M. Kotschote. Maximal L_p-regularity for a linear three-phase problem of parabolic-elliptic type. *J. Evol. Equ.*, 10(2):293–318, 2010.

[Koz96] A. Kozhevnikov. Asymptotics of the spectrum of Douglis-Nirenberg elliptic operators on a compact manifold. *Math. Nachr.*, 182:261–293, 1996.

[Kra04] T. Krainer. On the inverse of parabolic boundary value problems for large times. *Japanese J. Math.*, 30:91–163, 2004.

[KS12] M. Kaip and J. Saal. The permanence of $\mathcal{R}$-boundedness and property (α) under interpolation and applications to parabolic systems. *J. Math. Sci. Univ. Tokyo*, 19:359–407, 2012.

[KW01] N.J. Kalton and L. Weis. The H^∞-calculus and sums of closed operators. *Math. Ann.*, 321:319–345, 2001.

[KW04]   P.C. Kunstmann and L. Weis. *Maximal L_p-regularity for Parabolic Equations, Fourier Multiplier Theorems, and H^∞-functional Calculus*, volume 1855 of *Functional Analytic Methods for Evolution Equations, Lecture notes in Math.* Springer, 2004.

[LM68]   J.L. Lions and E. Magenes. *Problèmes aux Limites non Homogènes et Applications. Vol. 1.* Travaux et Recherches Mathématiques, No. 17. Dunod, Paris, 1968.

[McI86] A. McIntosh. Operators which have an H_∞ functional calculus. In *Miniconference on Operator Theory and Partial Differential Equations (North Ryde, 1986)*, volume 14 of *Proc. Centre Math. Anal. Austral. Nat. Univ.*, pages 210–231. Austral. Nat. Univ., Canberra, 1986.

[MR96]   J.E. Muñoz Rivera and R. Racke. Large solutions and smoothing properties for nonlinear thermoelastic systems. *J. Differential Equations*, 127(2):454–483, 1996.

[MS12]   M. Meyries and R. Schnaubelt. Interpolation, embeddings and traces of anisotropic fractional sobolev spaces with temporal weights. *J. Funct. Anal.*, 262:1200–1229, 2012.

[MV13]   M. Meyries and M. Veraar. Traces and embeddings of anisotropic function spaces. Submitted, 2013.

[Nai09] Y. Naito. On the L_p-L_q maximal regularity for the linear thermoelastic plate equation in a bounded domain. *Math. Methods Appl. Sci.*, 32(13):1609–1637, 2009.

[New81] I. Newton. De methodis serierum et fluxionum. In *H.W. Turnbull (ed.): The mathematical papers of Isaac Newton*, volume 3, pages 43–71. Cambridge University Press, 1967-1981.

[Pee74] J. Peetre. Über den Durchschnitt von Interpolationsräumen. *Arch. Math. (Basel)*, 25:511–513, 1974.

[Pee71] J. Peetre. Zur Interpolation von Operatorenräumen. *Arch. Math. (Basel)*, 21:601–608, 1970/71.

[PS09]   J. Prüss and G. Simonett. Analysis of the boundary symbol for the two-phase Navier-Stokes equations with surface tension. In *Nonlocal and Abstract Parabolic Equations and their Applications*, volume 86 of *Banach Center Publ.*, pages 265–285. Polish Acad. Sci. Inst. Math., Warsaw, 2009.

[PS10]   J. Prüss and G. Simonett. On the two-phase Navier-Stokes equations with surface tension. *Interfaces Free Bound.*, 12(3):311–345, 2010.

[PS11]   J. Prüss and G. Simonett. Analytic solutions for the two-phase Navier-Stokes equations with surface tension and gravity. In *Parabolic Prob-*

lems. *The Herbert Amann Festschrift*, volume 80 of *Progress in Nonlinear Differential Equations and Their Applications*, pages 507–540. Birkhäuser Verlag, Basel, 2011.

[PSS07] J. Prüss, J. Saal, and G. Simonett. Existence of analytic solutions for the classical Stefan problem. *Math. Ann.*, 338:703–755, 2007.

[RdF86] J.L. Rubio de Francia. Martingale and integral transforms of Banach space valued functions. In *Probability and Banach spaces (Zaragoza, 1985)*, volume 1221 of *Lecture Notes in Math.*, pages 195–222. Springer, Berlin, 1986.

[RdFT87] J.L. Rubio de Francia and J.L. Torrea. Some Banach techniques in vector-valued Fourier analysis. *Colloq. Math.*, 54(2):273–284, 1987.

[RS96]   T. Runst and W. Sickel. *Sobolev Spaces of Fractional Order, Nemytskij Operators, and Nonlinear Partial Differential Equations*, volume 3 of *de Gruyter Series in Nonlinear Analysis and Applications*. Walter de Gruyter & Co., Berlin, 1996.

[Sol84] V.A. Solonnikov. Solvability of the problem of evolution of an isolated amount of a viscous incompressible capillary fluid. *Zap. Nauchn. Sem. Leningrad. Otdel. Mat. Inst. Steklov. (LOMI)*, 140:179–186, 1984. Mathematical questions in the theory of wave propagation, 14.

[Sol03a] V.A. Solonnikov. Lectures on evolution free boundary problems: classical solutions. In *Mathematical Aspects of Evolving Interfaces (Funchal, 2000)*, volume 1812 of *Lecture Notes in Math.*, pages 123–175. Springer, Berlin, 2003.

[Sol03b] V.A. Solonnikov. L_q-estimates for a solution to the problem about the evolution of an isolated amount of a fluid. *J. Math. Sci. (N.Y.)*, 117(3):4237–4259, 2003.

[SS01]   H.J. Schmeißer and W. Sickel. Traces, Gagliardo-Nirenberg inequalities and Sobolev type embeddings for vector-valued function spaces. *Jenaer Schriften zur Math. und Inf.*, January:pp. 58, 2001.

[SS05]   H.J. Schmeißer and W. Sickel. Vector-valued Sobolev spaces and Gagliardo-Nirenberg inequalities. In *Nonlinear Elliptic and Parabolic Problems*, volume 64 of *Progr. Nonlinear Differential Equations Appl.*, pages 463–472. Birkhäuser Verlag, Basel, 2005.

[SS07]   Y. Shibata and S. Shimizu. On a free boundary problem for the Navier-Stokes equations. *Differential Integral Equations*, 20(3):241–276, 2007.

[SS08]   Y. Shibata and S. Shimizu. On the L_p-L_q maximal regularity of the Neumann problem for the Stokes equations in a bounded domain. *J. Reine Angew. Math.*, 615:157–209, 2008.

[SS11a] Y. Shibata and S. Shimizu. Maximal L_p-L_q regularity for the two-phase Stokes equations; model problems. *J. Differential Equations*, 251(2):373–419, 2011.

[SS11b] Y. Shibata and S. Shimizu. Report on a local in time solvability of free surface problems for the Navier-Stokes equations with surface tension. *Appl. Anal.*, 90(1):201–214, 2011.

[Ste70] E.M. Stein. *Singular Integrals and Differentiability Properties of Functions*. Princeton Mathematical Series, No. 30. Princeton University Press, Princeton, N.J., 1970.

[Tri78] H. Triebel. *Interpolation Theory, Function Spaces, Differential Operators*, volume 18 of *North-Holland Mathematical Library*. North-Holland Publishing Co., Amsterdam, 1978.

[Tri82] H. Triebel. Characterizations of Besov-Hardy-Sobolev spaces via harmonic functions, temperatures, and related means. *J. Approx. Theory*, 35(3):275–297, 1982.

[Tri83] H. Triebel. *Theory of Function Spaces*, volume 78 of *Monographs in Mathematics*. Birkhäuser Verlag, Basel, 1983.

[Tri92] H. Triebel. *Theory of Function Spaces. II*, volume 84 of *Monographs in Mathematics*. Birkhäuser Verlag, Basel, 1992.

[Tri97] H. Triebel. *Fractals and Spectra*, volume 91 of *Monographs in Mathematics*. Birkhäuser Verlag, Basel, 1997.

[Vol63] L.R. Volevich. A problem in linear programming stemming from differential equations. *Uspehi Mat. Nauk*, 18(3 (111)):155–162, 1963.

[Vol65] L.R. Volevich. Solvability of boundary value problems for general elliptic systems. *Mat. Sb. (N.S.)*, 68(3):373–416, 1965. (Russian), Engl. transl. in Amer. Math. Soc. Transl. Ser. 2 67 (1968), 182–225.

[Vol01] L.R. Volevich. Newton polygon and general parameter-elliptic (parabolic) systems. *Russ. J. Math. Phys.*, 8(3):375–400, 2001.

[Wei02] P. Weidemaier. Maximal regularity for parabolic equations with inhomogeneous boundary conditions in Sobolev spaces with mixed L_p-norm. *Electron. Res. Announc. Amer. Math. Soc.*, 8:47–51, 2002.

[Wei05] P. Weidemaier. Vector-valued Lizorkin-Triebel spaces and sharp trace theory for functions in Sobolev spaces with mixed L_p-norm for parabolic problems. *Sbornik: Mathematics*, 196(6):777–790, 2005.

[Wil07] M. Wilke. *Analysis for Phase-Field Models of Cahn-Hilliard Type*. PhD thesis, Universität Halle, 2007.

List of symbols

Functions:

f_σ	Translated function $f_\sigma := f(\cdot + \sigma, \cdot)$, page 25
Λ_r	Symbol for the shift operator in space, page 37
Φ_N	Symbol of the order reduction operator corresponding to the Newton polygon N, page 117
ψ	Shift function for H^∞-calculus on H_P, page 19
$\psi_{n,N}$	Shift function for H^∞-calculus on H_P, page 19
Ψ_s	Symbol for the shift operator in time, page 51
$\lvert\xi\rvert_{\vec{a}}$	Anisotropic norm, page 152
$\lvert z\rvert_-$	Abbreviation for $(-\sum_{k=1}^n z_k^2)^{1/2}$, page 71

Function spaces:

$B^s_{pq}(\mathbb{R}^n, X)$	X-valued Besov space on $\mathbb{R}^n$, page 38
$B^s_{pq}(\mathbb{R}^n_+, X)$	X-valued Besov space on $\mathbb{R}^n_+$, page 41
$_0B^s_{pq}(\mathbb{R}^n_+, X)$	X-valued Besov space on $\mathbb{R}^n_+$ with vanishing traces, page 41
$_0B^s_{pq,\varrho}(\mathbb{R}_+, X)$	X-valued Besov space with exponential weight $\varrho \geq 0$, page 44
$\dot{B}^s_{pp}(\mathbb{R}^n)$	Homogeneous Besov space, page 189
$\mathscr{E}_K(S_\theta \times \Sigma^{n-1}_\delta, \mu)$	Class of extension symbols associated with μ, page 179
$E^{\mathcal{W}}[h]$	Operator defined by the extension symbol h, page 179
$\mathbb{F}_j$	Ground space for mixed order systems, page 141,172
$\mathbb{F}_j$	Ground space for mixed order systems, page 126
$\mathbb{F}_{ij}$	Space to describe the compatibility embeddings, page 126
$_0\mathcal{F}^s(\mathcal{K}^r)$	Space of mixed scales and mixed smoothness, page 44, 160

$_0\mathcal{F}^s_\varrho(\mathcal{K}^r)$	Space of mixed scales and mixed smoothness with exponential weight, page 44, 160
$F^{s,\vec{a}}_{\vec{p},q}(\mathbb{R}^n)$	Anisotropic Triebel-Lizorkin space with mixed norms in the sense of J. Johnsen and W. Sickel, page 152
$F^s_{pq}(\mathbb{R}^m, X)$	X-valued Triebel-Lizorkin space on $\mathbb{R}^m$, page 144
$F^s_{pq}(\mathbb{R}^m_+, X)$	X-valued Triebel-Lizorkin space on $\mathbb{R}^m_+$, page 145
$_0F^s_{p,q}(\mathbb{R}^m_+, X)$	X-valued Triebel-Lizorkin space on $\mathbb{R}^m_+$ with vanishing traces, page 145
$_0F^s_{pq,\varrho}(\mathbb{R}_+, X)$	X-valued Triebel-Lizorkin space with exponential weight $\varrho \geq 0$, page 145
$\mathbb{H}_i$	Ground space for mixed order systems, page 141, 172
$\mathbb{H}_i$	Ground space for mixed order systems, page 126
$\mathbb{H}_{ij}$	Space to describe the compatibility embeddings, page 126
$H^{-\infty}_p(\mathbb{R}^n, X)$	Union of all Bessel potential spaces, page 47
$H^{s,\vec{a}}_{\vec{p}}(\mathbb{R}^n)$	Anisotropic Bessel potential space with mixed norms, page 152
$H^r_p(\mathbb{R}^n, X)$	X-valued Bessel potential space on $\mathbb{R}^n$, page 37
$H^r_p(\mathbb{R}^n_+, X)$	X-valued Bessel potential space on $\mathbb{R}^n_+$, page 41
$_0H^s_p(\mathbb{R}^n_+, X)$	X-valued Bessel potential space on $\mathbb{R}^n_+$ with vanishing traces, page 41
$_0H^s_{p,\varrho}(\mathbb{R}_+, X)$	X-valued Bessel potential space with exponential weight $\varrho \geq 0$, page 44
$\dot{H}^s_p(\mathbb{R}^n)$	Homogeneous Bessel potential space, page 189
$H(\Omega, Y)$	Holomorphic functions on Ω, page 18
$H^\infty(\Omega, Y)$	Bounded holomorphic functions on Ω, page 18
$H^\infty_\mathcal{R}(\Omega, L(X))$	Bounded holomorphic functions on Ω with $\mathcal{R}(f(\Omega)) < \infty$, page 18
$H^\infty_0(\Omega, Y)$	Holomorphic functions on Ω with asymptotical condition, page 18
$H_P(\Omega, Y)$	Polynomially bounded holomorphic functions on Ω, page 18
$\mathscr{K}_K(S_\theta \times \Sigma^{n-1}_\delta, \mu)$	Class of kernel functions associated with μ, page 174
$\ell^s_q(X)$	Weighted ℓ_q-space, page 146
$L^{\vec{r}}_{\vec{p},q}(\mathbb{R}^n)$	Anisotropic Triebel-Lizorkin space with mixed norms in the sense of M.Z. Berkolaiko, page 152

List of Symbols 241

$L_{\vec{p}}(\mathbb{R}^n, X)$	Lebesgue space with mixed norms, page 151
$L_p(\mathbb{R}, X)$	Bochner space of X-valued L_p-functions, page 14
$\widetilde{S}(L_t \times L_x)$	Symbol class polynomially-shaped functions, page 74
$S(L_t \times L_x)$	Symbol class of symbols with regular representation, page 77
$S_N(L_t \times L_x)$	Symbol class of N-parameter-elliptic symbols, page 92
$\mathscr{S}'(\mathbb{R}^n, X)$	X-valued tempered distributions, page 36
$\mathscr{S}(\mathbb{R}^n, X)$	X-valued Schwartz functions, page 36
$\mathscr{S}(\mathbb{R}^n_+, X)$	X-valued Schwartz functions on the half space $\mathbb{R}^n_+$, page 39
$_0\mathscr{S}(\mathbb{R}^n_+, X)$	X-valued Schwartz functions with vanishing traces, page 39
$\mathscr{S}'(\mathbb{R}^n_+, X)$	X-valued tempered distributions supported in $\mathbb{R}^n_+$, page 40
$S^{(N)}(L)$	Set of all non-vanishing functions in $C(L, \mathbb{C})$ which are homogeneous of degree N, page 71
$S^{(\rho, N)}(L_t \times L_x)$	Set of all non-vanishing functions which are ρ-homogeneous of degree N, page 71
$_0\mathscr{S}'(\mathbb{R}^n_+, X)$	X-valued tempered distributions on the half space, page 40
$_0\mathcal{F}^s_\varrho(\mathcal{K}^r)$	Space of mixed scales and mixed smoothness, page 160
$W^{\vec{m}}_{\vec{p}}(\mathbb{R}^n)$	Anisotropic Sobolev space with mixed norms, page 153
$W^k_p(\Omega, X)$	X-valued Sobolev space of order $k \in \mathbb{N}_0$, page 37
$[X_0, X_1]_\theta$	Complex interpolation space of X_0 and X_1, page 28
$(X_0, X_1)_{\theta, p}$	Real interpolation space of X_0 and X_1, page 28
$\mathscr{Z}(\mathbb{R}^n)$	Ground space for homogeneous function spaces, page 188
$\mathscr{Z}'(\mathbb{R}^n)$	Ground space for homogeneous function spaces, page 188

Newton polygons and order functions:

α_μ	Value associated with μ, page 83
β_μ	Value associated with μ, page 83
$b_\ell(\mu)$	Value associated with the order function μ, page 81
γ_j	Value associated with the exterior normal q_j, page 80
$I(\mu)$	Index of the order function μ, page 89
$\kappa_1(N)$	Indicator of the regularity in time of the Newton polygon N, page 98
$\kappa_2(N)$	Indicator of the regularity in space of the Newton polygon N, page 98

$m_\ell(\mu)$	Value associated with the order function μ, page 81
μ_+	Strictly positive order function associated to μ, page 83
$N(\nu)$	Newton polygon of the finite set $\nu \subseteq [0,\infty)^2$, page 82
$N(\mu)$	Newton polygon corresponding to the positive order function μ, page 82
$N(P)$	Newton polygon corresponding to the order-representative symbol P, page 78
$\nu(\mu)$	Set of tuples corresponding to the positive order function μ, page 82
$\nu(P)$	Set of tuples corresponding to the symbol P, page 78
N_V	Set of vertices of the Newton polgon N, page 78
$\mu_{\mathbb{F}_j}$	Order function corresponding to the regularity of $\mathbb{F}_j$, page 127, 140, 170, 172
$\mu_{\mathbb{H}_i}$	Order function corresponding to the regularity of $\mathbb{H}_i$, page 127, 140, 170, 172
μ_N	Order function corresponding to the Newton polygon N, page 82
$\mu(P)$	Order function such that P admits a two-sided estimate by $W_{\mu(P)}$, page 93
μ	Order function, page 81
q_j	Exterior normal to an edge of a Newton polygon, page 80
$q_j^\perp$	Vector orthogonal to q_j, page 80
s_j	Order function which represents the row part of the upper order structure of a Douglis-Nirenberg system, page 124
supp μ	Support of the order function μ, page 89
t_j	Order function which represents the column part of the upper order structure of a Douglis-Nirenberg system, page 124
v_j	Vertices of a Newton polygon, page 78
W_N	Weight function corresponding to the Newton polygon N, page 78
W_ν	Weight function corresponding to the set $\nu \subseteq [0,\infty)^2$, page 78
W_μ	Weight function corresponding to the order function μ, page 83, 84

Operators:

A^+	Natural extension of the operator A, page 59, 161
e^+	Smooth extension operator, page 39, 40
e_0^+	Trivial extension operator, page 39, 40
$\mathscr{F}$	Fourier transform, page 36
$f(\mathcal{D}_+)$	Maximal realization of the formal term $f(\nabla_+)$, page 64
$f(\mathbf{T})$	H^∞-calculus of the tuple $\mathbf{T}$ for a holomorphic function f, page 19
$f(\mathbf{T})$	H^∞-calculus of the tuple $\mathbf{T}$ for a holomorphic function f, page 18
$G^{\mathcal{N}}[\lambda, k]g$	Parameter-dependent singular integral operator, page 174
$G^{\mathcal{W}}[k]$	Singular integral operator on L_p-L_q induced by the kernel k, page 176
H	Hilbert transform, page 16
H_+	One-sided Hilbert transform, page 174
$\dot{J}_\sigma$	Shift operator on homogeneous spaces, page 189
$\mathscr{L}(\mathcal{D}_+)$	Maximal realization of the formal system $(\mathscr{L}_{j,k}(\nabla_+))_{j,k=1,\ldots,m}$, page 125
$\Lambda_r(\mathcal{D}_+)$	Shift operator in space, page 65
$\Lambda_{r,\vec{a}}$	Anisotropic shift operator, page 153
$\Lambda_r(\mathcal{D}_+^{\mathcal{W}})$	Shift operator in space, page 164
$\mathscr{M}_\varrho$	Multiplication operator with exponential weight function, page 44
$\mathcal{D}_+$	Abbreviation for the operator tuple $(\mathcal{D}_t, \mathcal{D})$, page 63
$\mathcal{D}_{+,\sigma}^{\mathcal{W}}$	Abbreviation for the operator tuple $(\sigma + \mathcal{D}_t^{\mathcal{W}}, (\mathcal{D}^{\mathcal{N}})^+)$ with $\sigma \geq 0$, page 162
$\operatorname{op}[\Lambda_r]$	Shift operator in space, page 37
$\operatorname{op}_+[\Psi_s]$	Shift operator in time, page 51
$\mathcal{D}$	Abbreviation for the tuple $(\mathcal{D}_1, \ldots, \mathcal{D}_n)$ of realizations of the partial derivatives, page 52
$\mathcal{D}_j$	Realization of the partial derivative ∂_j, page 52
$\mathcal{D}^{\mathcal{N}}$	Abbreviation for the tuple $(\mathcal{D}_1, \ldots, \mathcal{D}_n)$ of $\mathcal{N}$-realizations of the partial derivatives, page 52

$\mathcal{D}_t$	Realization of the time derivative, page 55
$\mathcal{D}_t^{\mathcal{M}}$	Time derivative realized on the ground space $\mathcal{M}$, page 161
$\mathcal{D}_t^{\mathcal{N}}$	$\mathcal{N}$-realization of the time derivative on $\mathbb{R}$, page 55
$\widetilde{\mathcal{D}}_t$	Realization of the time derivative on $\mathbb{R}$, page 55
$\Phi_N(\mathcal{D}_{+,\sigma})$	Order reduction operator corresponding to the Newton polygon N, page 117
$\Phi_N(\nabla_{+,\mu}^{W})$	Shift operator corresponding to the Newton polygon N, page 165
op$[m]$	L_p-Fourier multiplier, page 45
$\Psi_s(\mathcal{D}_+)$	Shift operator in time, page 65
$\Psi_s(\mathcal{D}_+^{W})$	Shift operator in time, page 164
R	Riesz transform, page 224
r^+	Point-wise restriction operator, page 39, 41
r_0^+	Restriction operator, page 39, 41
U	Rearranging operator, page 175

Others:

$C_j(\varepsilon_0)$	A part of a logarithmic partition, page 94
$\#I$	Cardinality of the set I, page 110
χ_A	Characteristic function of the set A, page 82
$d_\gamma(\psi)$	γ-order of the symbol ψ, page 72
$\widetilde{G}_j(\varepsilon_0)$	A part of a logarithmic partition, page 94
$G_\ell(\varepsilon_0, \varepsilon_1)$	A part of a logarithmic partition, page 94
Γ_φ	Admissible curve, page 12
$d_\gamma(P)$	γ-order of the symbol P, page 75
$\pi_\gamma P$	γ-principal part of the symbol P, page 75
I_γ	Index set of the γ-principal part, page 75
$\ker T$	Kernel of the operator T, page 12
L_t, L_x	Closed cones, page 70
$L(X)$	Space of bounded linear operators from X to X, page 12
$L(X,Y)$	Space of all bounded linear operators from X to Y, page 14
$\{P(y)\}_{y \geq 0}$	Cauchy-Poisson semigroup, page 213

List of Symbols

$[\![\varphi]\!]$	Jump of the function φ between two phases, page 215
φ_T	Spectral angle of the sectorial operator T, page 13
$\varphi_T^{(bi)}$	Spectral angle of the bisectorial operator T, page 13
φ_T^∞	H^∞-angle of the sectorial operator T, page 20
$\varphi_T^{\infty,(bi)}$	H^∞-angle of the bisectorial operator T, page 20
$\varphi_T^{\mathcal{R},\infty}$	$\mathcal{R}$-H^∞-angle of the sectorial operator T, page 20
$\varphi_T^{\mathcal{R},\infty,(bi)}$	$\mathcal{R}$-H^∞-angle of the bisectorial operator T, page 20
$\pi_\gamma \psi$	γ-principal part of the symbol ψ, page 72
$\dot{\mathbb{R}}^{n+1}$	Abbreviation for $\mathbb{R}^n \times (\mathbb{R} \setminus \{0\})$, page 215
$\rho(T)$	Resolvent set of the operator T, page 12
$\mathcal{R}_p(\mathcal{T})$	Constant in the definition of $\mathcal{R}$-boundedness, page 14
$\mathbb{R}_+$	The set of positive real numbers, page 12
$R(T)$	Range of the operator T, page 12
$S_\ell(\varepsilon_0, \varepsilon_1)$	A part of a logarithmic partition, page 94
S_θ	Sector concentrated on $\mathbb{R}_+$, page 12
Σ_δ	Bisector concentrated on $i\mathbb{R}$, page 12
$\sigma_c(T)$	Continuous spectrum of the operator T, page 55
$\sigma_p(T)$	Point spectrum of the operator T, page 55
$\sigma_r(T)$	Residual spectrum of the operator T, page 55
supp	Support of a function or distribution, page 48
T'	Banach space adjoint operator of T, page 40
$[v_j v_{j+1}]$	The line segment connecting v_j and v_{j-1}, page 79
$\{X_0, X_1\}$	Interpolation couple, page 27
$Y \stackrel{d}{\hookrightarrow} X$	Dense embedding of Y into X, page 29
$x \otimes y$	Abbreviation for $x \cdot y^T$, page 200
$Y \hookrightarrow X$	Continuous embedding of the space Y into the space X, page 25

Index

Admissible curve, 12
Admissible operator tuple, 17
Admissible scale, 138, 140, 171, 172

Banach space
    of class $\mathcal{HT}$, 16
    quasi-linearizable, 134
    UMD, 16
    with property (α), 17
Besov space
    homogeneous, 189, 213
    on $\mathbb{R}_+^n$, 41
    vanishing traces, 42
    with exponential weight, 44
Bessel potential space
    anisotropic with mixed norm, 152
    homogeneous, 189
    on $\mathbb{R}^n$, 37
    on $\mathbb{R}_+^n$, 41
    vanishing traces, 42
    with exponential weight, 44
Bisector, 12
Bisectorial, 13
Bounded H^∞-calculus, 19
Boussinesq-Scriven surface, 214

Cauchy-Poisson semigroup, 213
Characterization by vanishing traces, 42, 149
Commutator, 18
Compatibility conditions, 133, 138, 171
Compatibility of H^∞-calculus, 25
Compatible tuple of spaces, 126, 170
Complex interpolation space, 28

Compressible fluid model of Korteweg type, 202
Coretraction, 34

Douglis-Nirenberg system, 124
Dunford calculus, 224
Dyadic decomposition, 37, 144

Embedding, 25
Embedding condition, 126, 132, 170
Embedding result, 118, 135, 137, 138, 166, 168
Exact interpolation functor, 28
Exponential weight, 131, 145
Extension operator, 39
Extension symbol, 179

Finite time interval, 213
Fourier multiplier
    on L_p, 45
    on $L_{\vec{p}}$, 155
    with holomorphic symbol, 46–48
Fourier transform, 36
Free boundary, 2, 187, 208, 214

γ-order, 72, 75
γ-principal part, 72, 75
Generalized Stokes problem, 188
Generalized thermo-elastic plate equations, 196

H^∞-calculus
    for polynomially bounded functions, 19
    operator-valued, 18
H^∞-calculus and interpolation, 30

H^∞-calculus and isomorphisms, 23
H^∞-calculus of a shifted operator, 24, 25
Hanzawa transform, 2, 208
Heat equation in L_p-L_q, 185
Helmholtz projection, 67
Hilbert transform, 16, 66
    one-sided, 174
H^∞-angle, 20
Homogeneous of degree N, 71

Image space, 34
Index of an order function, 89
Integral operator, 16
Interpolation
    and isomorphism, 35
    and $\mathcal{R}$-boundedness, 30
    and retraction, 34
    of Bessel potential spaces, 38
Interpolation couple, 27
Interpolation functor, 27
    complex interpolation method, 28
    exact, 28
    real interpolation method, 28
Interpolation of ℓ_p^s, 29
Interpolation of L_p, 29
Interpolation of Bessel potential spaces, 42
Interpolation of Triebel-Lizorkin spaces, 146–148, 158
Intersection problem, 134
Iterative calculus, 21

Joint time-space H^∞-calculus, 63, 162

Kahane's contraction principle, 15
Kalton-Weis theorem, 21

Lebesgue space with mixed norms, 151
Linear Cahn-Hilliard-Gurtin problem, 199
Linear three-phase problem, 205
Liouville space, 153

Locally convex space, 34, 39, 188
Logarithmic partitions, 94
Lopatinskii matrix, 3, 187
Lower order function, 84

Maximal realization of $f(\mathcal{D}_+)$, 64, 125
Mixed-order system, 3, 124, 128, 171

N-parabolic, 92
N-parabolic mixed-order system, 125, 128, 140, 171, 172
N-parameter-elliptic, 92
Natural extension, 59, 161
    and H^∞-calculus, 61
    of bounded operator, 162
    of bounded operators, 60
Newton polygon, 77, 78
    associated order function, 82
    associated weight function, 78
    regular, 79
    regular in space, 79
    regular in time, 79

Operator
    bisectorial, 13
    sectorial, 12
    strongly elliptic, 190
Operator tuple
    admissible, 17
    bounded H^∞-calculus, 19
    $\mathcal{R}$-bounded H^∞-calculus, 19
Order function, 81
    associated Newton polygon, 82
    associated strictly positive order function, 83
    associated weight function, 83
    concave, 81, 165
    convex, 81, 165
    decreasing, 81, 121
    increasing, 81, 120
    index, 89
    lower, 84
    strictly negative, 81

Index

strictly positive, 81
support, 89
upper, 84, 120, 121, 124, 165
Order reduction operator associated with Newton polygon, 117

Partition of the co-variable space, 94
Point-wise restriction operator, 39
Property (α), 17

Quasi-homogeneous, 71
Quasi-linearizable, 134
Quotient norm, 34

$\mathcal{R}$-bounded H^∞-calculus, 19
Rademacher functions, 14
$\mathcal{R}$-bound, 14
$\mathcal{R}$-bounded H^∞-calculus for $\mathcal{D}$, 52
$\mathcal{R}$-bounded H^∞-calculus for $\mathcal{D}_t$, 58, 63
$\mathcal{R}$-bounded H^∞-calculus for $\mathcal{D}$, 63
$\mathcal{R}$-boundedness, 14
Real interpolation space, 28
Rearranging operator, 175
Regular representation of a symbol, 76
Reiteration theorem, 29
Representation by intersections, 154
Resolvent set, 12
Restriction operator, 39
Retraction, 34
$\mathcal{R}$-H^∞-angle, 20
ρ-homogeneous of degree N, 71
Riesz transform, 67, 224

Schwartz function
on $\mathbb{R}^n$, 36
on $\mathbb{R}^n_+$, 39
Sector, 12
Sectorial, 12
Shift operator associated with Newton polygon, 165
Singular integral operator, 174, 176
Sobolev embedding, 42

Sobolev space, 37
anisotropic with mixed norm, 153
Space-derivative operator, 52
Spaces of mixed scales, 44, 160
Spectral angle, 13
Spin-coating process, 207
Square function estimate in L_p, 15
Strongly elliptic operator, 190
Support, 48
Support of an order function, 89
Symbol
N-parabolic, 92, 123, 125, 166
N-parameter-elliptic, 92, 114
regular, 79
regular in space, 79
regular in time, 79
regular representation, 76
ρ-homogeneous, 71
Symbol class $\widetilde{S}(L_t \times L_x)$, 74

Tempered distribution
on $\mathbb{R}^n$, 36
on $\mathbb{R}^n_+$, 40
Theorem
of Kalton-Weis, 21
of Lions-Magenes, 149
of Weis, 46
Sobolev's embedding, 42
Time-derivative operator, 55, 161
Trace result in anisotropic space with mixed norm, 153, 154
Triebel-Lizorkin space, 143
anisotropic with mixed norm, 152
on $\mathbb{R}^m_+$, 145
on $\mathbb{R}^m$, 144
vanishing traces, 149
with exponential weight, 145
Trivial extension operator, 39
Two-phase Navier-Stokes equations, 214
Two-phase Stefan problem, 225

UMD space, 16
Upper order function, 84

Volevich trick, 173

Weight function, 78, 83, 141

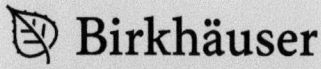 www.birkhauser-science.com

Operator Theory: Advances and Applications (OT)

This series is devoted to the publication of current research in operator theory, with particular emphasis on applications to classical analysis and the theory of integral equations, as well as to numerical analysis, mathematical physics and mathematical methods in electrical engineering.

Edited by
Joseph A. Ball (Blacksburg, VA, USA), Harry Dym (Rehovot, Israel),
Marinus A. Kaashoek (Amsterdam, The Netherlands), Heinz Langer (Vienna, Austria),
Christiane Tretter (Bern, Switzerland)

■ **OT 238: Edmunds, D.E. / Evans, W.D.**, Representations of Linear Operators Between Banach Spaces (2013).
ISBN 978-3-0348-0641-1

■ **OT 237: Kaashoek, M.A. / Rodman, L. / Woerdeman, H.J.** (Eds.), Advances in Structured Operator Theory and Related Areas. The Leonid Lerer Anniversary Volume (2013).
ISBN 978-3-0348-0638-1

■ **OT 236: Cepedello Boiso, M. / Hedenmalm, H. / Kaashoek, M.A. / Montes Rodríguez, A. / Treil, S.** (Eds.), Concrete Operators, Spectral Theory, Operators in Harmonic Analysis and Approximation. 22nd International Workshop in Operator Theory and its Applications, Sevilla, July 2011 (IWOTA11) (2013).
ISBN 978-3-0348-0647-3

■ **OT 234/OT235: Eidelman, Y. / Gohberg, I. / Haimovici, I.** (Eds.), Separable Type Representations of Matrices and Fast Algorithms.
Vol. 1. Basics. Completion problems. Multiplication and inversion algorithms (2013).
ISBN 978-3-0348-0605-3
Vol 2. Eigenvalue method (2013).
ISBN 978-3-0348-0611-4

■ **OT 233: Todorov, I.G. / Turowska, L.** (Eds.), Algebraic Methods in Functional Analysis. The Victor Shulman Anniversary Volume (2013).
ISBN 978-3-0348-0501-8

■ **OT 232: Demuth, M. / Kirsch, W.** (Eds.), Mathematical Physics, Spectral Theory and Stochastic Analysis (2013).
ISBN 978-3-0348-0590-2

■ **OT 231: Molahajloo, S. / Pilipović, S. / Toft, J. / Wong, M.W.** (Eds.), Pseudo-Differential Operators, Generalized Functions and Asymptotics (2013).
ISBN 978-3-0348-0584-1

■ **OT 230: Brown, B.M. / Eastham, M.S.P. / Schmidt, K.M.**, Periodic Differential Operators (2013).
ISBN 978-3-0348-0527-8

■ **OT 229: Almeida, A. / Castro, L. / Speck, F.-O.** (Eds.), Advances in Harmonic Analysis and Operator Theory. The Stefan Samko Anniversary Volume (2013).
ISBN 978-3-0348-0515-5

■ **OT 228: Karlovich, Y.I. / Rodino, L. / Silbermann, B. / Spitkovsky, I.M.** (Eds.), Operator Theory, Pseudo-Differential Equations, and Mathematical Physics (2013).
ISBN 978-3-0348-0536-0

■ **OT 227: Janas, J. / Kurasov, P. / Laptev, A. / Naboko, S.** (Eds.), Operator Methods in Mathematical Physics. Conference on Operator Theory, Analysis and Mathematical Physics (OTAMP) 2010, Bedlewo, Poland (2013).
ISBN 978-3-0348-0530-8

■ **OT 226: Alpay, D. / Kirstein, B.**, Interpolation, Schur Functions and Moment Problems II (2012).
ISBN 978-3-0348-0427-1

■ **OT 225: Sakhnovich, L. A.**, Levy Processes, Integral Equations, Statistical Physics: Connections and Interactions (2012).
ISBN 978-3-0348-0355-7